CREATIVITY IN ORGANIC SYNTHESIS

Volume 1

ACADEMIC PRESS RAPID MANUSCRIPT REPRODUCTION

CREATIVITY IN ORGANIC SYNTHESIS
Volume 1

Jasjit S. Bindra
Ranjna Bindra
Pfizer Inc.
Groton, Connecticut

foreword by

E.J.Corey
Harvard University

ACADEMIC PRESS, INC.
New York San Francisco London 1975
A Subsidiary of Harcourt Brace Jovanovich, Publishers

For Ratna Junita

ACADEMIC PRESS, INC.
111 Fifth Avenue, New York, New York 10003

United Kingdom Edition published by
ACADEMIC PRESS, INC. (LONDON) LTD.
24/28 Oval Road, London NW1

Library of Congress Cataloging in Publication Data

Bindra, Jasjit S
Creativity in organic synthesis.

Bibliography: p.
Includes index.
1. Chemistry, Organic–Synthesis. I. Bindra, Ranjna, joint author. II. Title.
QD262.B497 547'.2 75-1147
ISBN 0–12–099450–X

PRINTED IN THE UNITED STATES OF AMERICA

CONTENTS

FOREWORD

The advances in the science of chemical synthesis of complex organic molecules now pour forth at a pace greatly exceeding any seen in the past decades. An almost breathtaking affirmation is set forth graphically in this beautifully done volume. There has been no "golden age" of synthesis, no quantum leap in scope, no revolutionary doctrine, but simply a relentless and accelerating growth. Unusually gifted and capable young people are attracted to synthesis in considerable numbers by the beauty of its past accomplishments, the challenge of new horizons, and the historical suggestion that their own success will eclipse past achievement.

Nature continues to be exceedingly generous to the synthetic chemist in providing ample opportunity for discovery and creative endeavor of highest magnitude and in surrounding him with an incredible variety of fascinating and complicated molecular structures. These structures are being revealed to us at an ever increasing rate, especially since the advent of computers and techniques of X-ray crystallographic analysis.

The creative possibilities of synthesis are manifold and embrace not only conceptual and strategic information but also methodological discovery in the form of new chemistry and new techniques for experimental execution.

All these considerations give force to the conviction that the systematization of synthetic chemistry and its literature are of enormous importance. The unparalleled effectiveness of the organic chemist's graphical language looms especially large in this connection, enabling what is seen to be a remarkable ease of communication as compared to other intellectual disciplines. And it is no exaggeration to say that the intellectual potential of the human mind depends on the power of the language engaged in communication. The pages which follow abound with important information for the synthetic chemist, presented in crystal-clear, concise, flow-chart form, taking full advantage of the superb attributes of chemical graphics. Although the interspersed comments are less extensive than the charts, they have been written so skillfully as to be of no less value to the user of this book.

One fervently hopes that the Drs. Bindra will extend this enterprise to a series of additional volumes, well assured of their merit and value but tantalized by a curiosity to know what new marvels of chemical synthesis will be contained therein.

E. J. Corey

PREFACE

The ability to synthesize complex organic molecules continues to occupy an increasingly important place in the repertoire of the organic chemist. New synthetic methods and reactions, characterized by exquisite selectivity and stereochemical control, are being continually developed and quickly find application in natural products synthesis. In *Creativity in Organic Synthesis* we have sought to present some of the outstanding accomplishments of natural products synthesis during the last five years, hoping that it would provide a useful commentary on the state of the art. To the practicing chemist it should provide a wealth of information on selective transformations, the efficacy of which has been proven under the most exacting and often highly complex situations, serving also as a guide to the selection of proper reagents and reaction conditions and as a valuable source of model transformations. To the student of organic chemistry, since synthesis involves the application of knowledge and techniques of the entire science, it provides an excellent opportunity to study the subject as it actually is.

The presentation of each synthesis, using structural formulae and easily readable flow charts, follows the format employed successfully in *Art in Organic Synthesis* (Anand, Bindra and Ranganathan, Holden-Day). Each synthesis is preceded by a brief introductory paragraph, highlighting the salient features of the synthesis, its novel aspects and the strategy employed. Since emphasis is on economy of words, liberal use has been made of three-dimensional formulae and perspective drawings in order to illustrate the force of arguments predicting the selectivity or stereochemical outcome of key reactions. Wherever necessary, obscure or unusual reactions have been discussed at greater length in footnotes and pertinent literature and review references have been given. Literature coverage is from the middle of 1969 to mid-1974.

Jasjit S. Bindra
Ranjna Bindra

PREFACE

ACKNOWLEDGEMENT

We wish to acknowlege with sincere thanks the generous help of several people in the preparation of this book. We are deeply indebted to Dr. Charles J. V. Scanio for critically reading the manuscript and making several useful comments. Thanks are due to Drs. G. R. Evanega, E. Hamanaka, M. R. Johnson, J. J. Plattner and T. K. Schaaf who read parts of the manuscript and made helpful comments; to Dr. G. R. Frysinger who brought together the facilities we used for preparation of the manuscript; to Messrs John P. Stratton and Charles J. Kenney for technical advice on preparation of the camera-ready copy; and to Dr. J. Buckley and his helpful staff for library facilities. We are most grateful to Mrs. Anita L. Parker who cheerfully and with great skill typed the unending stream of structures and text that finally emerged as the camera-ready copy.

It is a pleasure to thank Prof. E. J. Corey for helpful discussions and advice, and Dr. H.-J. Hess for continued encouragement and guidance, without which the writing of this book would have been impossible.

Jasjit S. Bindra
Ranjna Bindra

GLOSSARY OF ABBREVIATIONS

Ac	acetyl
Am	amyl
Aq	aqueous
Ar	aryl
B^-	base
Bu	butyl
Bz	benzyl
Chf	chloroform
conc	concentrated
DBN	1,5-diazabicyclo-[4.3.0]nonene-5
DBU	1,5-diazabicyclo-[5.4.0]undecene-5
DCC	dicyclohexyl-carbodiimide
DDQ	2,3-dichloro-5,6-dicyano-1,4-benzoquinone
DHP	dihydropyran
diglyme	Diethyleneglycol dimethyl ether
Dibal	diisobutylaluminum hydride
dil	dilute
Diox	dioxane
DMA	dimethylacetamide
DME	1,2-dimethoxyethane
DMF	dimethylformamide
DMSO	dimethylsulfoxide
Et	ethyl
EVK	ethyl vinyl ketone
Glyme	1,2-dimethoxyethane
Hex	hexane
HMPA	hexamethylphosphor-amide
H_3O^+	aqueous acid
hr	hour
i	iso
LAH	lithium aluminum hydride
Liq	liquid
m	meta
Me	methyl
MEK	methyl ethyl ketone
min	minutes
Ms	methanesulfonyl
MVK	methyl vinyl ketone
n	normal
NBA	N-bromoacetamide
NBS	N-bromosuccinimide
NCS	N-chlorosuccinimide
Oxdn	oxidation
Ph	phenyl
PPA	polyphosphoric acid
Pr	propyl
Py	pyridine
rt	room temperature
Sia_2BH	disiamylborane
t	tertiary
TEA	triethylamine
TEG	triethylene glycol
TFA	trifluoroacetic acid
THF	tetrahydrofuran
THP	tetrahydropyranyl
TMS	trimethylsilyl
Tol	toluene
Ts	p-toluenesulfonyl

Xy	xylene	Δ	reflux(or heat)

Although only one enantiomer is depicted in structural formulae all compounds are racemates, unless otherwise stated. In polycyclic compounds the configuration is shown as follows:

1) A solid thick line indicates β-configuration
2) A broken line indicates α-configuration
3) A wavy or straight line indicates either unknown or unspecified configuration

The numbering of carbon atoms followed for intermediates has no connection with conventional numbering applicable to that particular ring system but refers to the numbering of the final product.

Yields are mentioned in italics under the arrows. When a reaction results in a mixture of epimers, only the predominant epimer is depicted. Formation of both epimers and their separation is noted only when critical to the synthetic plan.

The book by one of us, *Art in Organic Synthesis,* Holden-Day, Inc., San Francisco (1970), is referred to as *Art Org. Synth.*

β - ACORENOL

β-Acorenol is a spirocyclic sesquiterpene, isolated from the wood of *Juniperus rigida*. The synthesis of racemic β-acorenol by Oppolzer[1] is based on a stereocontrolled construction of five-membered ring systems using an intramolecular ene-reaction.[2,3]

CO_2Et Br + → [280°, Tol (sealed tube)][3] → 65%

[CrO_3] → 64%

[MeLi, Et_2O, -70° → $ClCH_2OMe$, HMPA-Et_2O] →

---

1. W. Oppolzer, Helv., **56**, 1812 (1973).

2. W. Oppolzer, E. Pfenninger and K. Keller, Helv., **56**, 1807 (1973).

3. For a review of the ene-reaction, see H. M. R. Hoffmann, Angew. Chem. Int. Ed., **8**, 556 (1969).

CO_2Et OCH_2OCH_3 — [280°][4] → CO_2Et H O OCH_3

→ CO_2Et — [MeLi, Et_2O, 25°] →

OH

β-Acorenol

4. Retro-ene reaction, cf. E. Mutterer, J. M. Morgan, J. M. Biedermann, J. P. Fleury and F. Weiss, Tetrahedron, **26**, 477 (1970).

AFLATOXIN M_1

H

MeO

O O

OH

O

O O

The aflatoxins are a group of notoriously toxic and highly carcinogenic mold metabolites produced by some members of *Aspergillus* and *Penicillium* species.[1] The aflatoxins, originally discovered in peanut meal exported from Africa and South America, were later found in cottonseed meal and other cereal products. The toxins are readily produced in agricultural products under appropriate conditions of temperature and moisture and also make their way into edible animal products when the animals are fed rations containing sublethal levels of aflatoxins. The first chemical synthesis of an aflatoxin was reported in 1966, but was accomplished in rather low overall yield.[2] Subsequently, Büchi and Weinreb[3] developed an improved route to aflatoxins based on a mild new coumarin synthesis, specially applicable to acid-sensitive phenols such as A. This approach is portrayed in the synthesis of aflatoxin M[1], the milk toxin, outlined below.[3]

HO O OH O

[Me_2SO_4, K_2CO_3, DME, Δ] →

MeO O O OMe

[$AlCl_3$, CH_2Cl_2, Δ] →

64%

MeO O O OH

[BzBr, K_2CO_3 DMF-DME, Δ] →

---

1. Review: C. P. Mathew, Chem. Ind., 913 (1970); "Aflatoxin", ed., L. A. Goldblatt, Academic Press, New York (1969).

2. G. Büchi, D. M. Foulkes, M. Kurono, G. F. Mitchell and R. S. Schneider, J. Amer. Chem. Soc., **88**, 4534 (1966); **89**, 6745 (1967); Art. Org. Synth., 6 (1970).

3. G. Büchi and S. M. Weinreb, J. Amer. Chem. Soc., **91**, 5408 (1969); **93**, 746 (1971).

MeO, OBz benzofuranone — [$PhN^+Me_3Br_3^-$, THF] → 88% → MeO, OBz, Br bromoketone

— [BzOH, $CaCO_3$] → MeO, OBz, OBz ketone — [allylMgBr, THF-Et_2O][4] →

MeO, H, OBz, OBz, OH allyl carbinol — [OsO_4 → $NaIO_4$] → MeO, H, OBz, CHO, OBz, OH

— [1. H_2, Pd-C, Ac_2O-C_6H_6, NaOAc; 2. H_2, Pd-C, EtOAc]* → [MeO, H, OH, CHO, OAc, OH]

4. The Grignard reagent attacks from the side opposite to that of the bulky benzyloxy group resulting in predominant formation of the *trans* isomer. For a discussion of the stereochemistry of addition of organometallics to α-alkoxy ketones, see D. J. Cram and D. R. Wilson, J. Amer. Chem. Soc., **85**, 1245 (1963).

**The first hydrogenation cleaves the benzyl group on the aromatic ring to give a monoacetate, which is further hydrogenated in a different solvent for removal of the second benzyl group.*

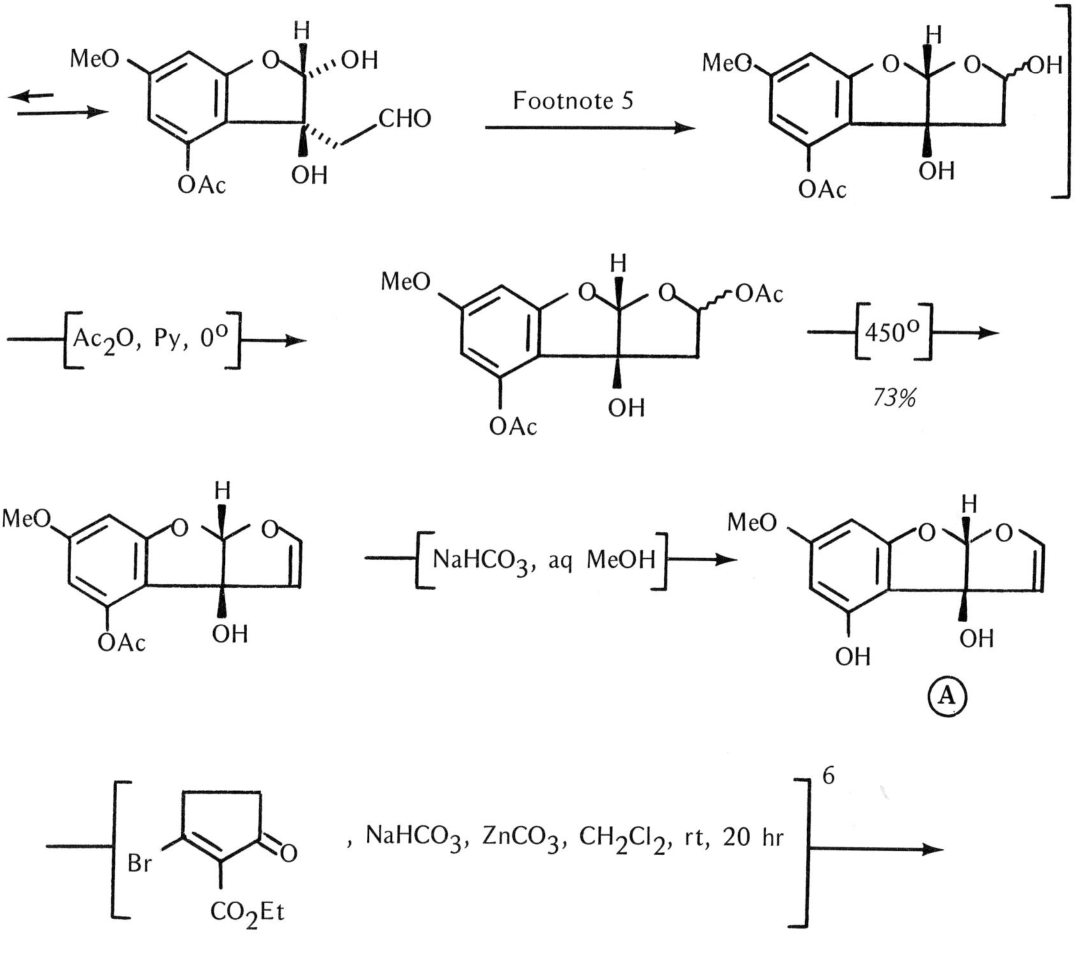
MeO
H
OH
CHO
OAc
OH
Footnote 5
O
Ac2O, Py, 0°
OAc
450°
73%
NaHCO3, aq MeOH
A
Br
CO2Et
, NaHCO3, ZnCO3, CH2Cl2, rt, 20 hr
6

5. The more stable *cis*-ring fused furobenzofuran is favored over the *trans*-fused system.

6. Owing to exceptional sensitivity of the tricyclic phenol (A) to acidic reagents, direct condensation with 2-carbethoxycyclopentane-l,3-dione (i) under von Pechmann coumarin synthesis conditions proved impossible. However, the bromide (ii) readily underwent condensation to give the desired coumarin ring D when zinc carbonate was used both as catalyst and acid scavenger.

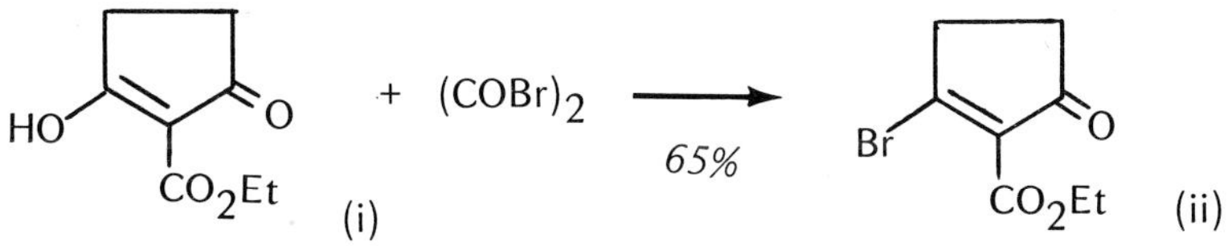
HO
O
CO2Et
(i)
+ (COBr)2
65%
Br
(ii)

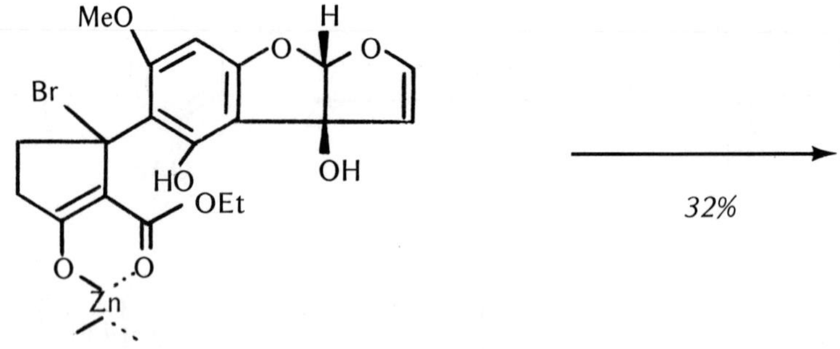

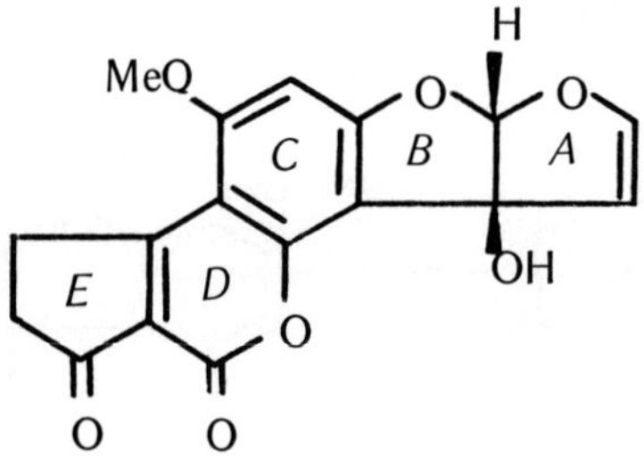

MeO Br HO OEt O O Zn H O O OH

32%

MeO H O O C B A E D OH O O O

Aflatoxin M_1

AGAROSPIROL

OH

The synthesis of agarospirol (epihinesol)[1] portrays an efficient stereoselective approach to spiro[4·5] decanes based on the intramolecular keto-carbene insertion reaction[2] (A→B) and cleavage of the derived cyclopropyl alcohol (C) to give a spiroketone.

O

CO_2Et

[1. $(CH_2OH)_2$, H^+ 2. LAH]*

**Note double bond migration during ketalization of α,β-unsaturated ketone.*

O O

CH_2OH

[2N HCl, THF]

O

[1. $NaCH(CO_2Et)_2$ 2. 2N HCl 3. CH_2N_2]

65%

O

CO_2Me

[1. $(CH_2OH)_2$, H^+ 2. B^- 3. $(COCl)_2$, Py 4. CH_2N_2]

O O

$COCHN_2$

(A)

---

1. M. Mongrain, J. Lafontaine, A. Belanger and P. Deslongchamps, Canad. J. Chem., **48**, 3273 (1970).

2. G. Stork and J. Ficini, J. Amer. Chem. Soc., **83**, 4678 (1961).

[Cu, C_6H_6][2]

40%
(based on ester)

+

[$CO(OMe)_2$, NaH]

1:9

(B)[3]

[$NaBH_4$]

O

CO_2Me

OH

CO_2Me

[MeMgI]

H

OH

H^+

OH

(C)

[2N HCl, DME]

3. The synthesis of (B) by an essentially identical scheme was reported independently by P. M. McCurry, Tetrahedron Lett., 1845 (1971).

O

1. LAH
2. Ac_2O, Py

OAc

HO

HO

Li, $EtNH_2$

HO

Agarospirol

AJMALICINE

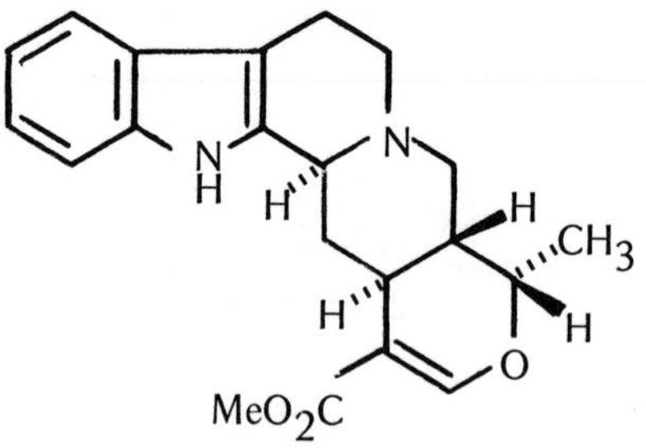

Synthesis of the heteroyohimbine alkaloid, ajmalicine,[1] outlined below is a classic example of van Tamelen's biogenetic-type synthesis of natural products.[2] The non-aromatic part of the molecule is condensed with formaldehyde and tryptamine in a biogenetically patterned key step, serving to assemble the essential skeletal features of the alkaloid at the very outset.

Preparation of Keto-triester:

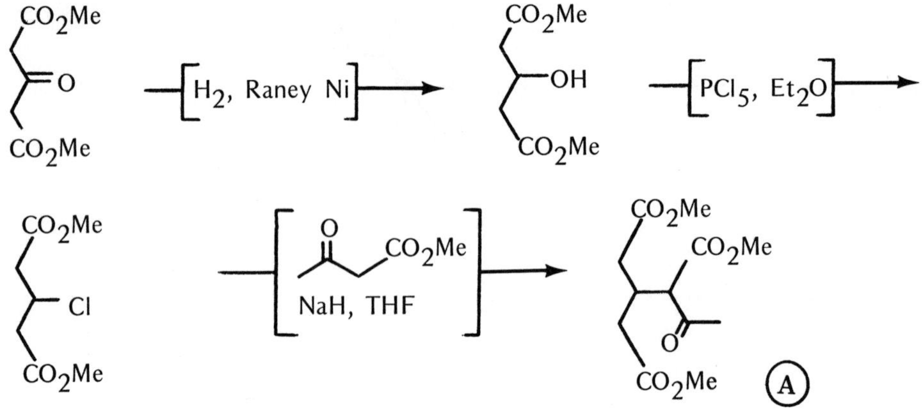

H_2, Raney Ni

PCl_5, Et_2O

NaH, THF

1. E. E. van Tamelen, C. Placeway, G. P. Schiemenz and I. G. Wright, J. Amer. Chem. Soc., **83**, 2594 (1961); **91**, 7359 (1969).

2. E. E. van Tamelen, Fortschr. Chem. Org. Naturst., **19**, 242 (1961).

Preparation of Ajmalicine:

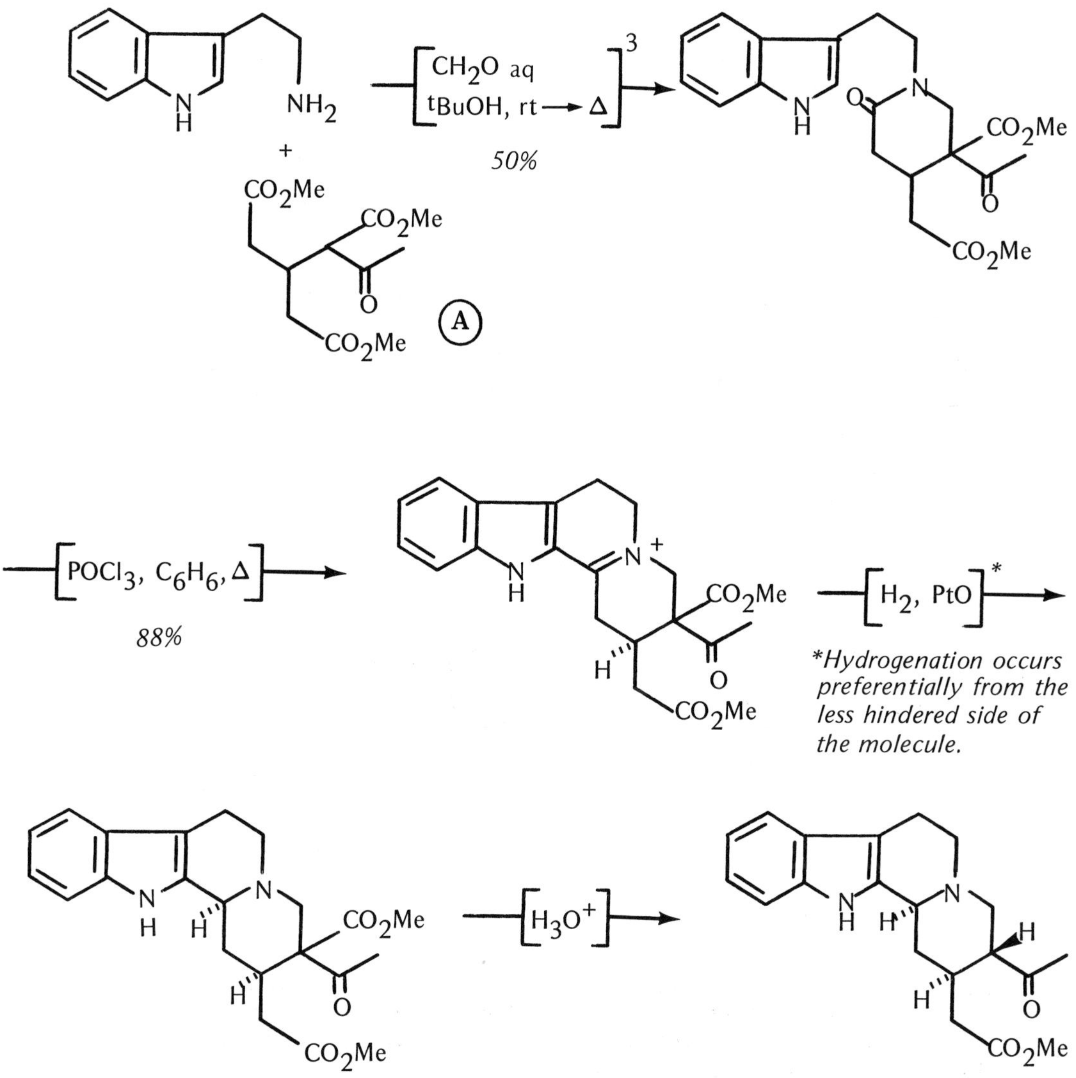

NH$_2$ + (A) —[CH_2O aq, tBuOH, rt → Δ][3]→ 50%

—[$POCl_3$, C_6H_6, Δ]→ 88%

—[H_2, PtO]*→

**Hydrogenation occurs preferentially from the less hindered side of the molecule.*

—[H_3O^+]→

3. Presence of the customary strong acid in this reaction suppresses removal of the acidic proton from the β-keto triester, thus effectively preventing its participation in the desired Mannich condensation. The reaction was therefore carried out in the absence of acid, and one of the basic components (e.g. tryptamine) served to abstract the acidic proton from the β-keto ester, allowing the desired Mannich reaction to proceed.

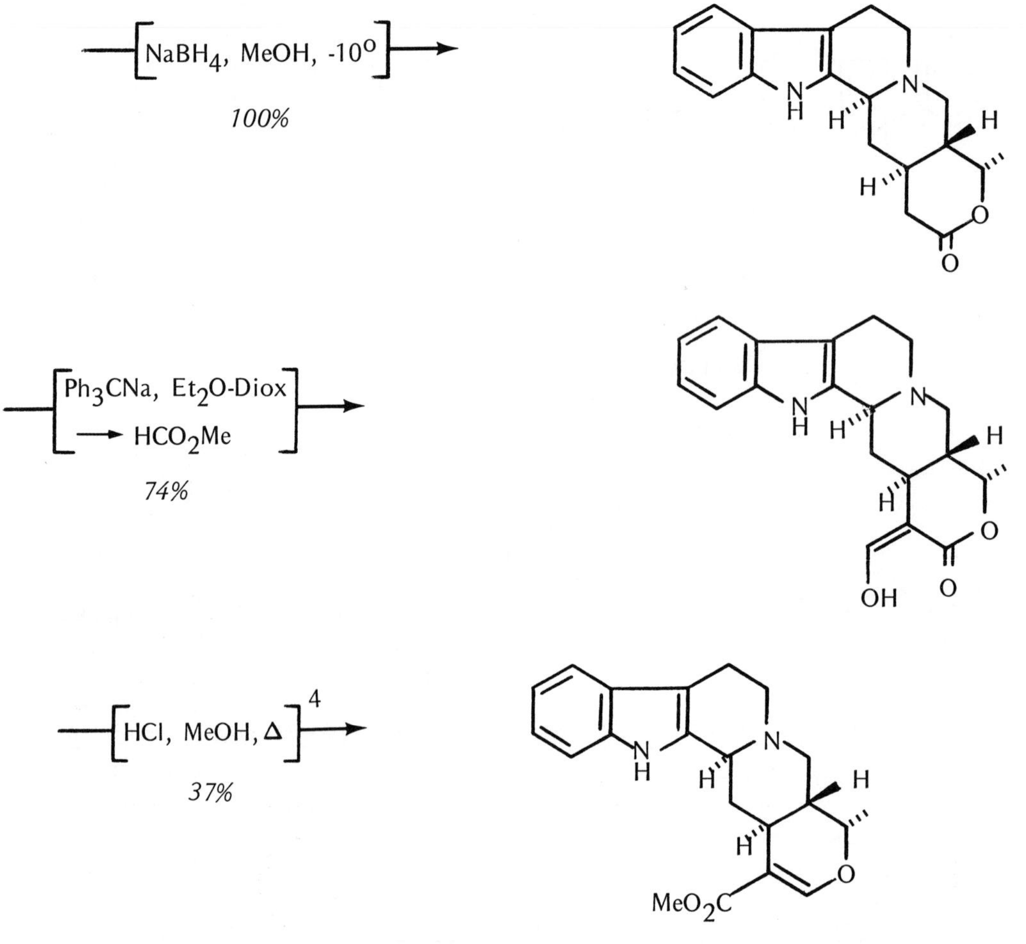

[$NaBH_4$, MeOH, -10°] →

100%

[Ph_3CNa, Et_2O-Diox → HCO_2Me] →

74%

[HCl, MeOH, Δ][4] →

37%

Ajmalicine

---

4. The "acyl lactone rearrangement", namely the conversion of an α-acyl-δ-lactone (i) to the corresponding dihydropyran carboxylic ester (ii) is a facile and convenient process, see F. Korte and K. H. Buchel, Angew. Chem., **71**, 709 (1959).

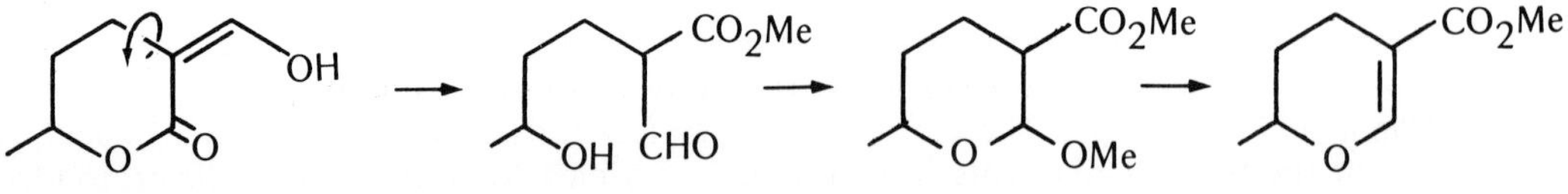

(i) → (ii)

Recently, the Hoffman-La Roche group[5] reported an efficient convergent synthesis of ajmalicine, in which the D/E-rings of the molecule have been constructed from *trans*-3-vinyl-4-piperidine acetic acid.[6]

Preparation of D/E Component:

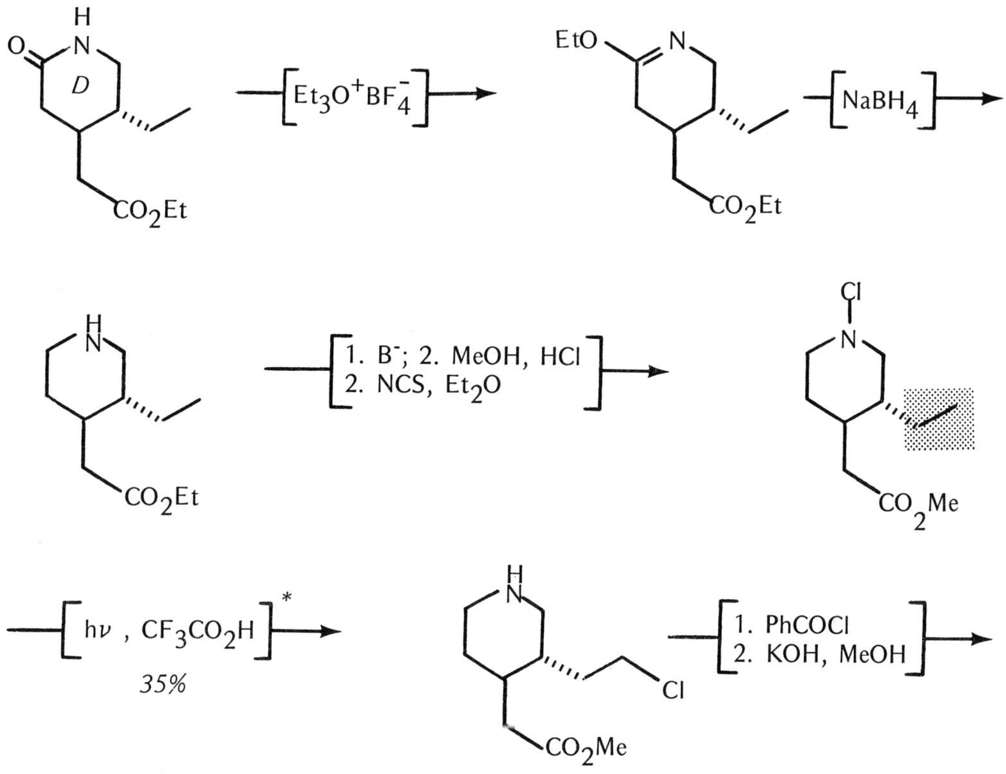
H
O
N
D
CO2Et
Et3O+BF4-
EtO
N
CO2Et
NaBH4
H
N
CO2Et
1. B-; 2. MeOH, HCl
2. NCS, Et2O
Cl
N
CO2Me
hν , CF3CO2H
*
35%
H
N
Cl
CO2Me
1. PhCOCl
2. KOH, MeOH

---

5. J. Gutzwiller, G. Pizzolato and M. Uskoković, J. Amer. Chem. Soc., **93**, 5907 (1971).

6. M. Uskoković, C. Reese, H. L. Lee, G. Grethe and J. Gutzwiller, J. Amer. Chem. Soc., **93**, 5902 (1971).

**Note conversion of the ethyl group to vinyl side chain by means of the photolytic Löffler-Freytag reaction.*

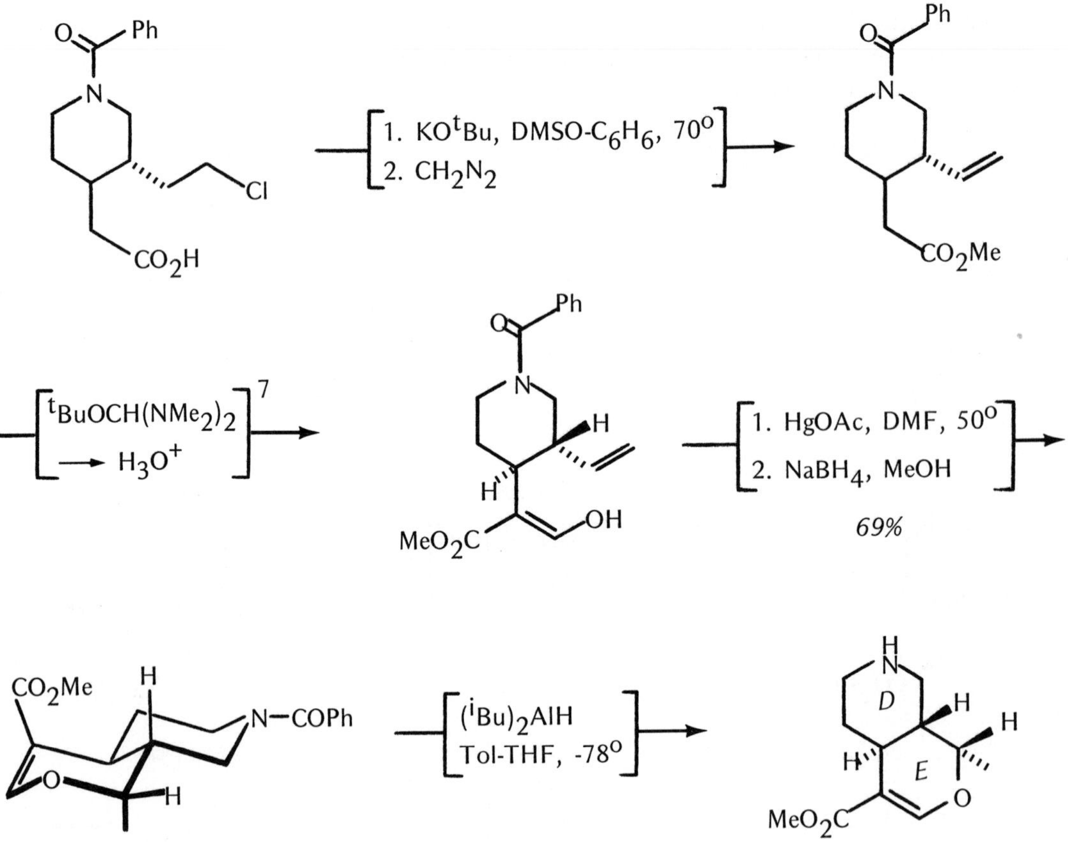

Ph, O, N, Cl, CO_2H

[1. KO^tBu, DMSO-C_6H_6, 70°; 2. CH_2N_2] →

Ph, O, N, CO_2Me

[$^tBuOCH(NMe_2)_2$ → H_3O^+][7] →

Ph, O, N, H, H, MeO_2C, OH

[1. HgOAc, DMF, 50°; 2. $NaBH_4$, MeOH] →

69%

CO_2Me, H, O, N—COPh, H

[$(^iBu)_2AlH$; Tol-THF, -78°] →

H, N, D, H, H, H, E, O, MeO_2C

Preparation of Ajmalicine:

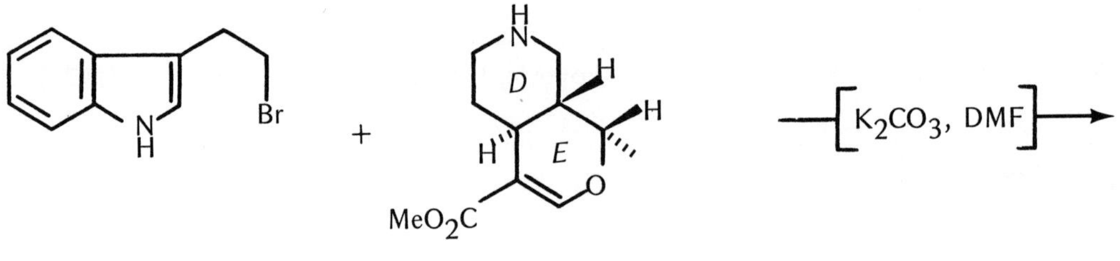

N, H, Br + H, N, D, H, H, H, E, O, MeO_2C

[K_2CO_3, DMF] →

---

7. Bis(dimethylamino)-*tert*-butoxymethane, see H. Bredereck, G. Simchen, S. Reesdat, W. Kautlehner, P. Horn, R. Wahl, H. Hoffmann and P. Grieshaber, Chem. Ber., **101**, 41 (1968).

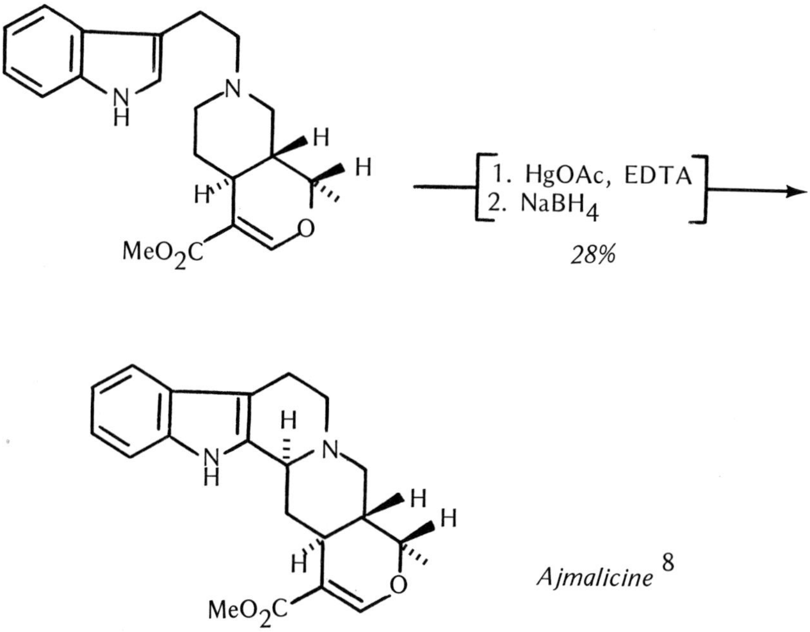

1. HgOAc, EDTA
2. $NaBH_4$

28%

Ajmalicine[8]

8. For a synthesis of ajmalicine from elenolic acid and tryptamine, see F. A. Mackellar, R. C. Kelly, E. E. van Tamelen and C. Dorschel, J. Amer. Chem. Soc., **95**, 7155 (1973).

AJMALINE

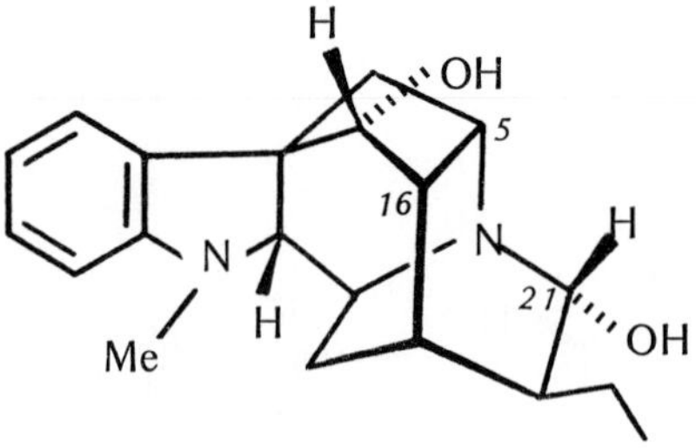

The biogenetic-type total synthesis of ajmaline[1,2] commences from N-methyl-tryptophan and a suitable "C_9" precursor, incorporating a cyclopentane nucleus which serves as a latent dialdehyde functionality for the Pictet-Spengler cyclization step (A→B). The iminium salt (C) required for the critical C-5, C-16 bond formation has been generated by an ingenious decarbonylation reaction[3] of the β-carboline carboxylic acid (B). Final stages of the synthesis utilize the known deoxyajmalal-A→21-deoxyajmaline transformation[4], and functionalization of the latter at C-21 by a phenyl chloroformate ring opening-oxidative ring closure sequence.[5]

Preparation of cyclopentyl "C_9" precursor:

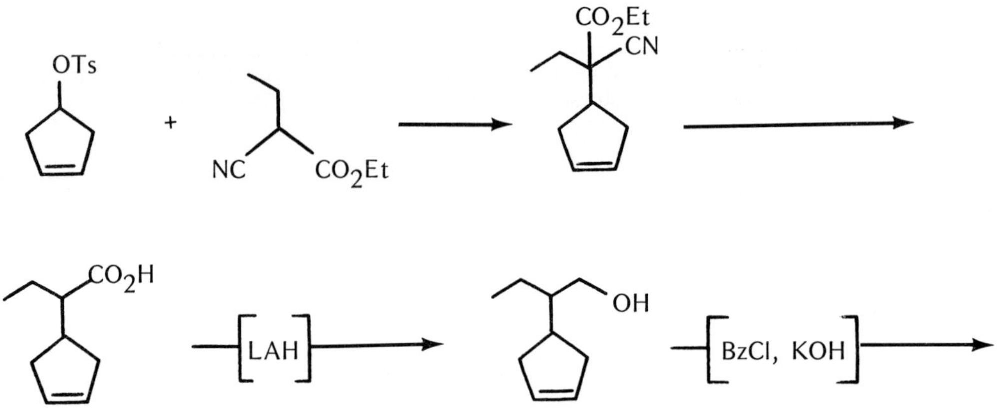

LAH

BzCl, KOH

1. The first total synthesis of ajmaline was reported by S. Masamune, S. K. Ang, C. Egli, N. Nakatsuka, S. K. Sarkar and Y. Yasunari, J. Amer. Chem. Soc., **89**, 2506 (1967); Art Org. Synth., 8 (1970).

2. E. E. van Tamelen and L. K. Oliver, J. Amer. Chem. Soc., **92**, 2136 (1970).

OBz — [OsO_4] → OBz, OH OH — [$(MeO)_2CO$, NaOMe] →

OBz — [H_2, Pd-C, Et_2OH] → OH — [CrO_3, Py, CH_2Cl_2] →

CHO

Preparation of Ajmaline:

CO_2H, NH_2, N, Me + CHO — [H_2, Pd-C, EtOH] →

CO_2H HN N Me

[KOH, MeOH]

CO_2H HN N Me HO OH

[$NaIO_4$, aq NaOAc]

CO_2H HN N Me CHO CHO

(A)

CO_2H N N Me H H H CHO

(B)

[DCC, TsOH, Diox, 80°]

85%

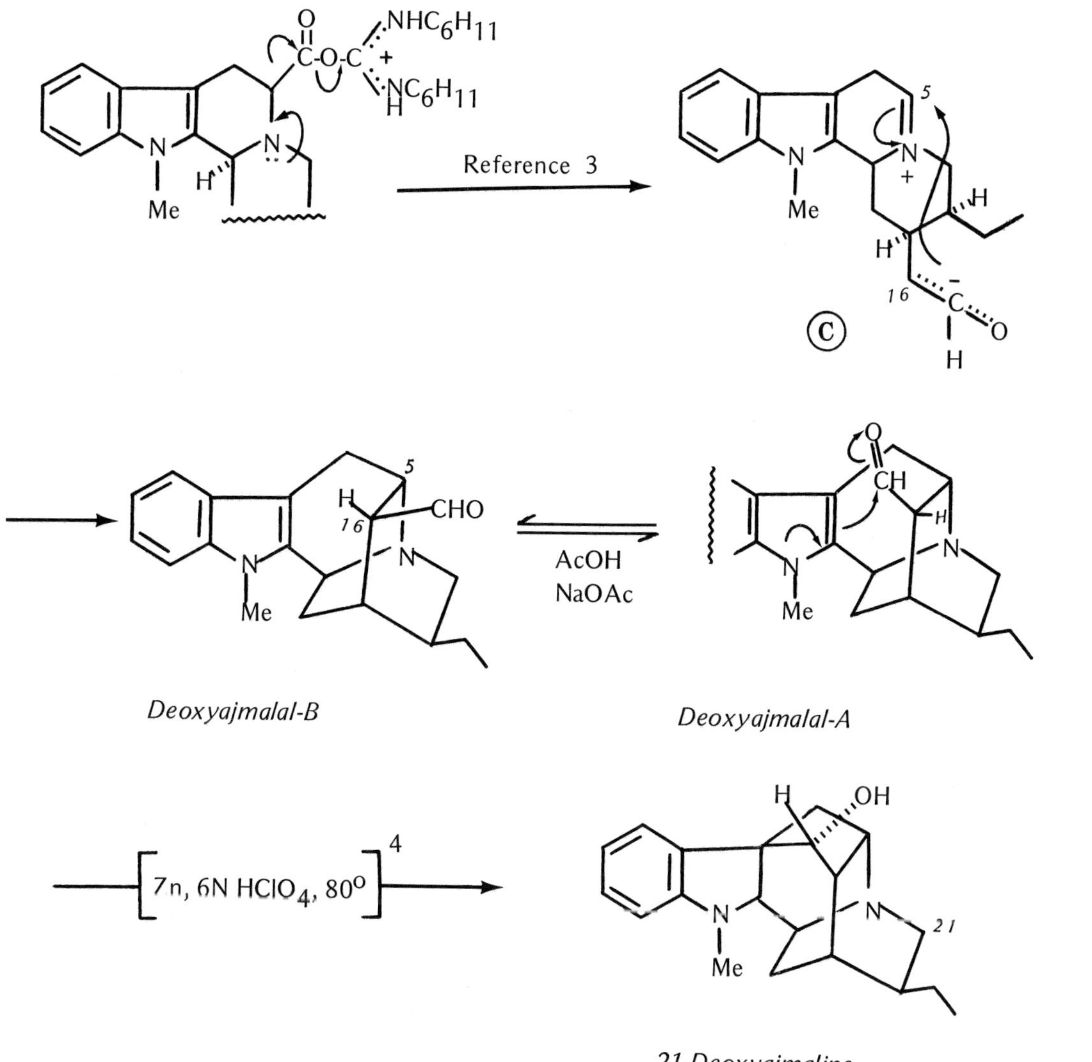

Reference 3

Ⓒ

AcOH
NaOAc

Deoxyajmalal-B

Deoxyajmalal-A

[Zn, 6N $HClO_4$, 80°][4]

21-Deoxyajmaline

---

3. Cf. V. I. Maksimov, Tetrahedron, **21**, 687 (1965).

4. M. F. Bartlett, B. F. Lambert, H. M. Werblood and W. I. Taylor, J. Amer. Chem. Soc., **85**, 475 (1963).

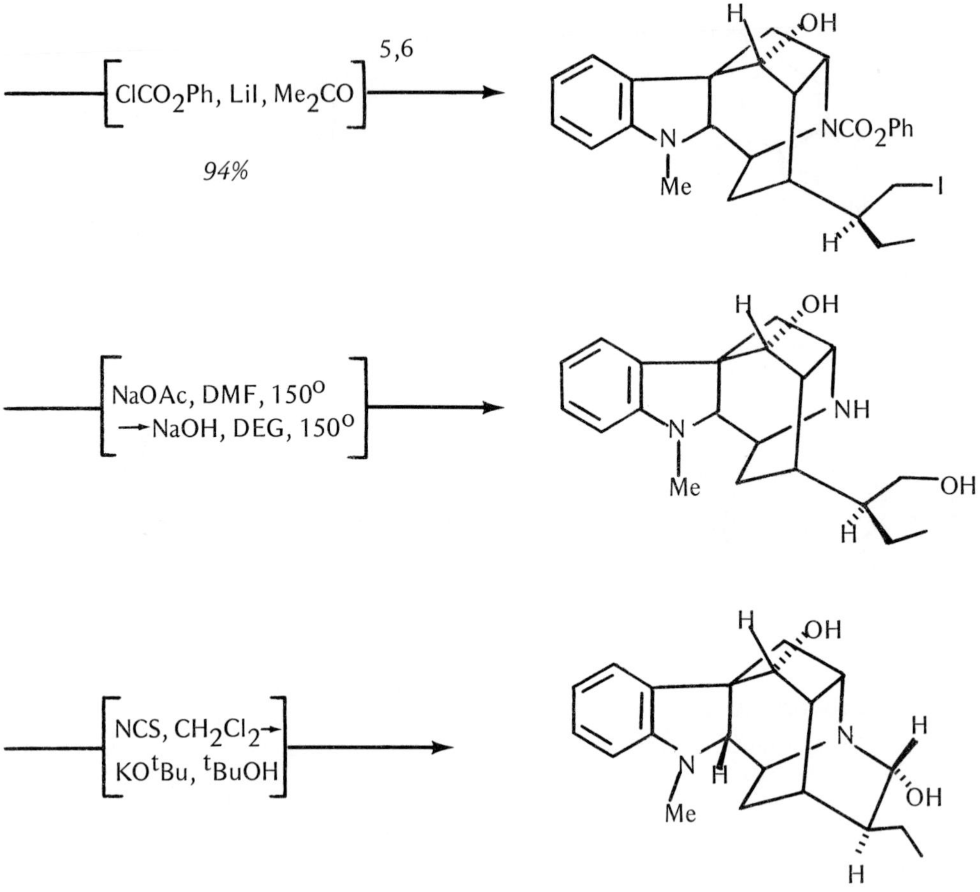

5,6

[$ClCO_2Ph$, LiI, Me_2CO]

94%

[NaOAc, DMF, 150°
→ NaOH, DEG, 150°]

[NCS, CH_2Cl_2 →
KO^tBu, ^tBuOH]

Ajmaline

5. J. D. Hobson and J. G. McCluskey, J. Chem. Soc., 2015 (1967).

6. The reaction of phenyl chloroformate with tertiary aliphatic and alicyclic amines to give the corresponding N-carboxylates (pathway a) is an efficient general procedure for cleavage of tertiary amines, and provides a convenient alternative to the well-known von Braun cyanogen bromide cleavage. The choice of phenyl chloroformate over alkyl esters in this reaction probably results in suppression of the undesired competitive reaction pathway b.

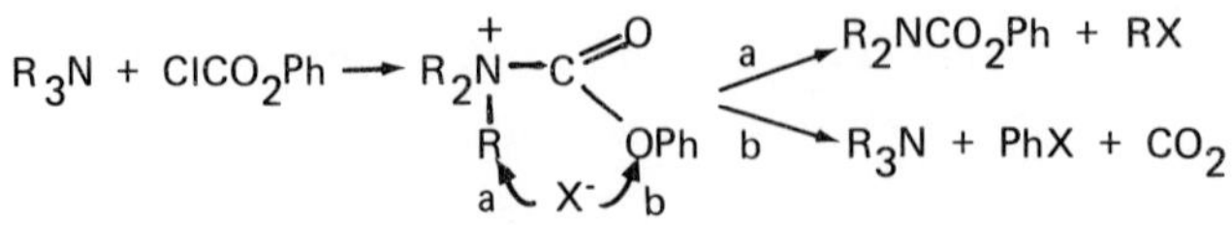

$$R_3N + ClCO_2Ph \rightarrow R_2\overset{+}{N}(R)-C(=O)OPh \xrightarrow{X^-} \begin{cases} \text{a: } R_2NCO_2Ph + RX \\ \text{b: } R_3N + PhX + CO_2 \end{cases}$$

ALNUSENONE

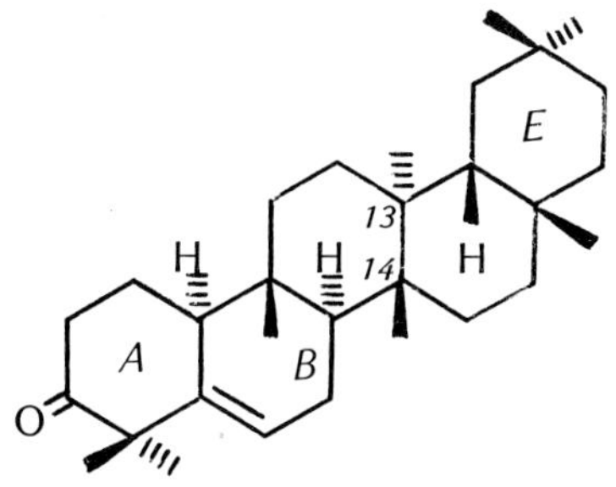

Even though several of the problems of total synthesis of pentacyclic triterpenes have already been solved during the development of steroid total synthesis, the intricately disposed angular methyl groups and difficult stereochemical features of the triterpenes still present a formidable synthetic challenge. One solution to the problem, which emerged from a series of elegant experiments in the laboratories of R. E. Ireland, culminated in the synthesis of alnusenone and is outlined below.[1] The synthetic plan involved establishing the desired *trans*-orientation of the C-13 and C-14 angular methyl groups, potentially the most difficult structural feature of the pentacyclic nucleus, in an appropriate tricyclic intermediate (A) early in the synthesis.[2] Closure of ring B in (A) by an acid catalyzed cyclization affords the B/C *trans*-fused pentacyclic diether (B) which furnishes alnusenone after sequential modification of the terminal aromatic rings.[3,4]

---

1. R. E. Ireland and S. C. Welch, J. Amer. Chem. Soc., **92** 7232 (1970); R. E. Ireland, M. I. Dawson, S. C. Welch, A. Hagenbach, J. Bordner and B. Trus, J. Amer. Chem. Soc., **95**, 7829 (1973).

2. R. E. Ireland, D. A. Evans, D. Glover, G. M. Rubottom and H. Young, J. Org. Chem., **34**, 3717 (1969).

3. Model studies on tetracyclic analogs predicted that the Friedel-Crafts cyclialkylation reaction would result predominantly in generation of the desired *trans*-fused system (Ref. 4).

4. The aromatic rings serve not only to facilitate construction of the key decahydropicene intermediates through the use of a cyclialkylation reaction [R. E. Ireland, S. W. Baldwin and S. C. Welch, J. Amer. Chem. Soc., **94**, 2056 (1972)], but also serve as suitable substrates from which to build the terpenoid substitution pattern in the terminal rings. Since the synthesis entails sequential incorporation of the two aromatic rings, this provides an opportunity to differentiate the two terminal oxygen functions (as methoxy and ethoxy) for preparation of the desired 3-alkoxy-10-hydroxy-decahydropicene nucleus (cf. B).

NEt$_2$ Ref. 5 MeO MeO O + OEt O [KOH, MeOH] 72%

OEt 13 14 O MeO (A) [Et_3Al, HCN THF] 6,7 85% OEt O MeO CN

[1. MeMgI 2. $SOCl_2$, Py, 0° 3. $(^iBu)_2AlH$, C_6H_6 4. N_2H_4, KOH, TEG, Δ] 73% OEt 13 14 MeO [TsOH, Tol, Δ] 70%

5. G. H. Douglas, J. M. H. Graves, D. Hartley, G. A. Huges, B. J. McLoughlin, J. Siddall and H. Smith, J. Chem. Soc., 5072 (1963).

6. The 1,4-conjugate hydrocyanation of α,β-unsaturated ketones using alkylaluminium compounds is an important method for preparation of angular cyano ketones, which may be converted to polycyclic compounds bearing angular substituents such as carboxyl, formyl, methyl, vinyl, aminomethyl and hydroxymethyl or to a variety of bridged ring systems, W. Nagata and M. Yoshioka, Tetrahedron Lett., 1913 (1966). For a series of definitive papers, see W. Nagata, M. Yoshioka and S. Hirai, J. Amer. Chem. Soc., **94**,4635 (1972); W. Nagata, M. Yoshioka and M. Murakami, J. Amer. Chem. Soc., **94**, 4644 (1972); J. Amer. Chem. Soc., **94**, 4654 (1972); W. Nagata, M. Yoshioka and T. Terasawa, J. Amer. Chem. Soc., **94**, 4672 (1972).

7. While the thermodynamically more stable *cis*-cyano ketone was obtained under aprotic conditions (Et_2AlCN-C_6H_6), the desired *trans* isomer was virtually the sole product of kinetically controlled cyanide attack in the protic medium (Et_3Al-HCN-THF).

OEt

[$LiP(C_6H_5)_2$, THF][9]

87%

OEt

E

H

A

HO

MeO

H

(B)[8]

[Li, NH_3, glyme → MeI → H_3O^+][10]

42%

O

H

H

MeO

[1. $LiAl(^tBuO)_3H$, C_6H_6 2. CH_2I_2, Zn(Cu) Et_2O-DME][11]

OH

H

H

MeO

[1. CrO_3, Py 2. K^+BuO, THF ▸ MeI][12]

O

H

H

MeO

8. Stereochemistry of the pentacyclic structure (B) was confirmed by single X-ray structure analysis.

9. Selective methoxy group cleavage in presence of the ethoxy aromatic system was accomplished using the procedure described by F. G. Mann and M. J. Pragnell, Chem. Ind., 1386 (1964).

10. While ring E is readily reduced, the phenolic ring A is protected against Birch reduction by virtue of higher reduction potential of the derived phenoxide [Cf. J. F. Fried, N. A. Abraham and T. S. Santhanakrishnan, J. Amer. Chem. Soc., **89**, 1044 (1967)]. Treatment of the reaction mixture with methyl iodide regenerates the methoxy ether.

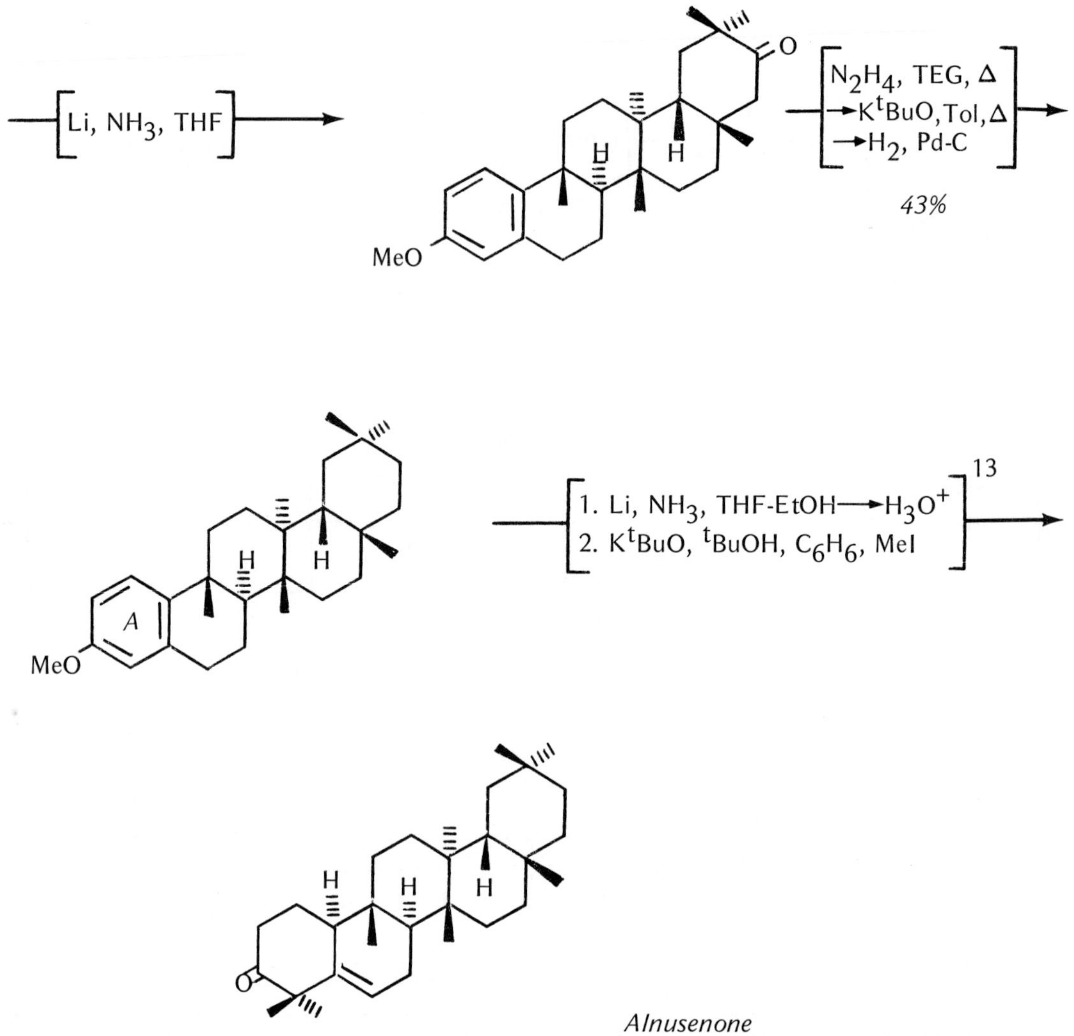

[Li, NH_3, THF] →

MeO

[N_2H_4, TEG, Δ → K^tBuO, Tol, Δ → H_2, Pd-C] →

43%

MeO

A

[1. Li, NH_3, THF-EtOH → H_3O^+ 2. K^tBuO, ^tBuOH, C_6H_6, MeI] 13 →

Alnusenone

11. The equatorial hydroxyl function serves to direct the Simmons–Smith methylenation in the desired β-stereochemical sense, H. E. Simmons and R. D. Smith, J. Amer. Chem. Soc., **80**, 5323 (1958); **81**, 4256 (1959).

12. The cyclopropyl group serves not only as the latent angular methyl group but also allows direct methylation of the ketone for introduction of the gem dimethyl function.

13. The sequence of reactions used for modification of the methoxylated aromatic A ring follows procedures developed earlier in the synthesis of *dl*-rimuene by R. E. Ireland and L. N. Mander, J. Org. Chem., **32**, 689 (1967).

δ-AMYRIN

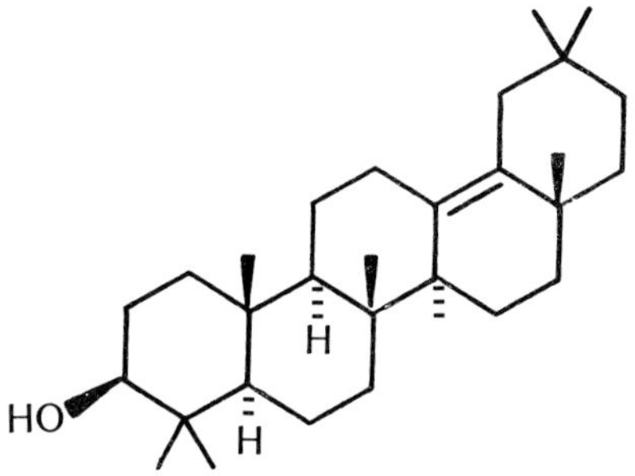

Squalene 2,3-oxide has been proposed to account for a vast array of polycyclic triterpenes in nature.[1] The synthesis of δ-amyrin by van Tamelen and his associates mimics the biological cyclization process, featuring the stereoselective generation of five asymmetric centres during a polyolefinic cyclization of the terminal epoxide (D).[2,3]

Preparation of D/E Component:

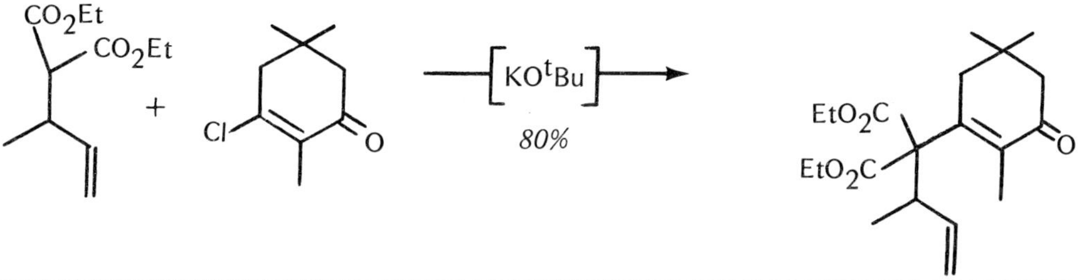

CO_2Et CO_2Et + Cl O —[KO^{t}Bu]→ *80%* EtO_2C EtO_2C O

---

1. The function of squalene 2,3-oxide in nature has been reviewed by E. E. van Tamelen, Accts. Chem. Res., **1**, 111 (1968) See also E. J. Corey and P. R. Ortiz de Montellano, J. Amer. Chem. Soc., **89**, 3362 (1967).

2. E. E. van Tamelen, M. P. Seiler and W. Wierenga, J. Amer. Chem. Soc., **94**, 8229 (1972).

3. For related biogenetic-type total syntheses, see E. E. van Tamelen, G. M. Milne, M. I. Suffness, M. C. Rudler Chauvin, R.J. Anderson and R. S. Achini, J. Amer. Chem. Soc., **92**, 7202 (1970); E. E. van Tamelen and R. J. Anderson, J. Amer. Chem. Soc., **94**, 8225 (1972); E. E. van Tamelen, R. A. Holton, R. E. Hopla and W. E. Konz, J. Amer. Chem. Soc., **94**, 8229 (1972).

[NaBH$_4$] → [distil] →

[$BF_3 \cdot EtO$, C_6H_6] → 60%

[NaCN, DMSO 160°] → 70% (A) [NBS, CCl_4, Δ → DBN, C_6H_6][4] →

[H_2, Pd-C, EtOH] → (B)

4. Since the β,γ-unsaturated ester (A) could not be converted directly by base or acid to the α,β isomer (B), the desired transformation was effected by indirect means.

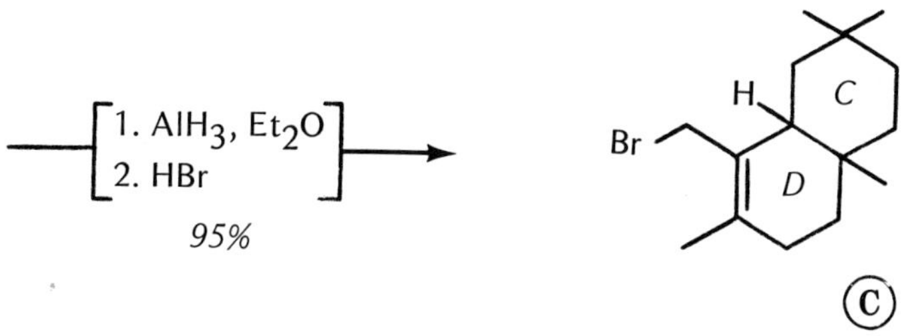

[1. AlH_3, Et_2O; 2. HBr] → 95%

Br, H, C, D

(C)

Preparation of δ-amyrin:

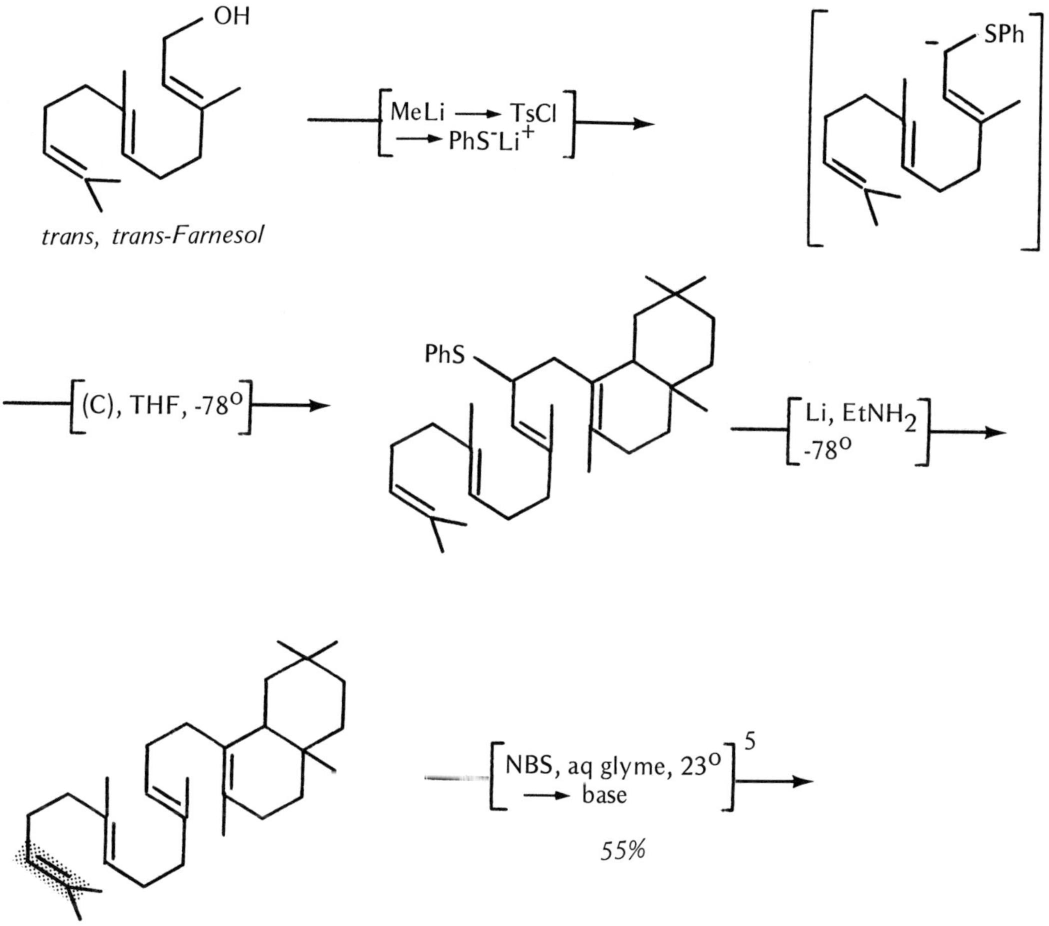

OH

trans, trans-Farnesol

[MeLi → TsCl → PhS^-Li^+] →

SPh

[(C), THF, -78°] →

PhS

[Li, $EtNH_2$, -78°] →

[NBS, aq glyme, 23° → base][5] →

55%

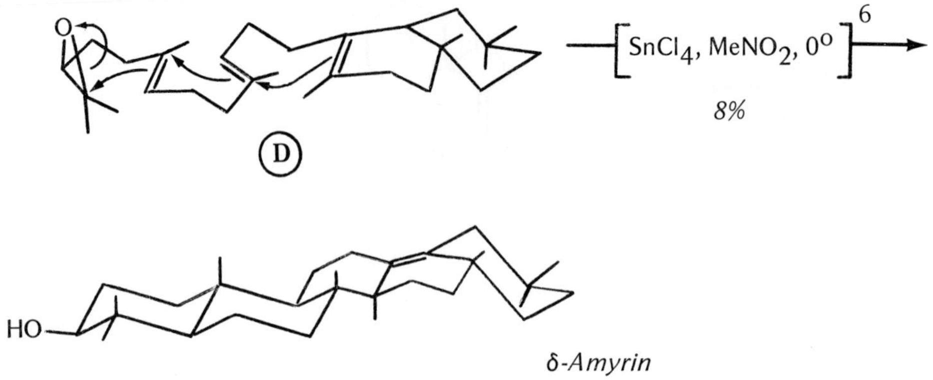

[$SnCl_4$, $MeNO_2$, 0°][6]

8%

D

HO

δ-Amyrin

---

5. Selective oxidation of the terminal double bond in the starting hydrocarbon should not be possible in a solvent of low polarity because the molecule would exist in an uncoiled, fully extended state, with all of the double bonds equally vulnerable to attack by the oxidizing agent. However, in a more polar medium the molecule is expected to assume a highly coiled, compact conformation such that the system of internal hydrogen bonds would be disrupted as little as possible. Under these circumstances the internal double bonds would be sterically shielded and therefore chemically less reactive, while the terminal olefinic links would remain exposed and available for reaction.

6. For definitive papers on stereochemistry of polyene cyclization, see G. Stork and A. W. Burgstahler, J. Amer. Chem. Soc., **77**, 5068 (1955); A. Eschenmoser, L. Ruzicka, O. Jeger and D. Arigoni, Helv., **38**, 1890 (1955).

ANDROCYMBINE

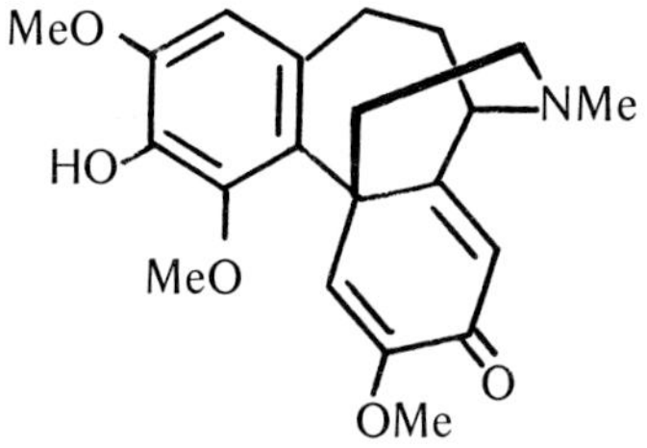

Androcymbine is a homomorphinandienone alkaloid which occupies a key position along the biosynthetic pathway to colchicine.[1] Biogenetically, androcymbine itself is derived from autumnaline, a bisphenolic 1-phenethylisoquinoline, by phenolic oxidative coupling.[2,3] While there is ample laboratory analogy for the application of phenol oxidation for preparation of homomorphinandienone-type compounds, the method has not yet been applied successfully to the synthesis of androcymbine.[4,5] One approach to androcymbine, in which coupling of the two aromatic rings is accomplished using a photo-Pschorr reaction, is illustrated below.[6]

---

1. A. R. Battersby, R. Ramage, A. F. Cameron, C. Hannaway and F. Santávy, J. Chem. Soc. C, 3514 (1971).

2. D. H. R. Barton and T. Cohen in "Festschrift Arthur Stoll," Birkhauser, Basel (1957) p. 117.

3. A. C. Baker, A. R. Battersby, E. McDonald, R. Ramage and J. H. Clements, Chem. Commun., 390 (1967).

4. T. Kametani, Y. Satoh and K. Fukumoto, J. Chem. Soc. Perkin I, 2160 (1970).

5. Three synthetic methods have been developed for androcymbine-type compounds: phenol oxidation [T. Kametani, K. Fukumoto, M. Koizumi and A. Kozuka, J. Chem. Soc. C., 1295 (1969)], Pschorr reaction [T. Kametani, K. Fukumoto, F. Satoh and H. Yagi, J. Chem. Soc. C., 3084 (1968)], and photolysis of diazonium salts [T. Kametani, M. Koizumi and K. Fukumoto, J. Chem. Soc. C., 1792 (1971)].

6. T. Kametani, M. Koizumi and K. Fukumoto, J. Org. Chem., **36**, 3729 (1971).

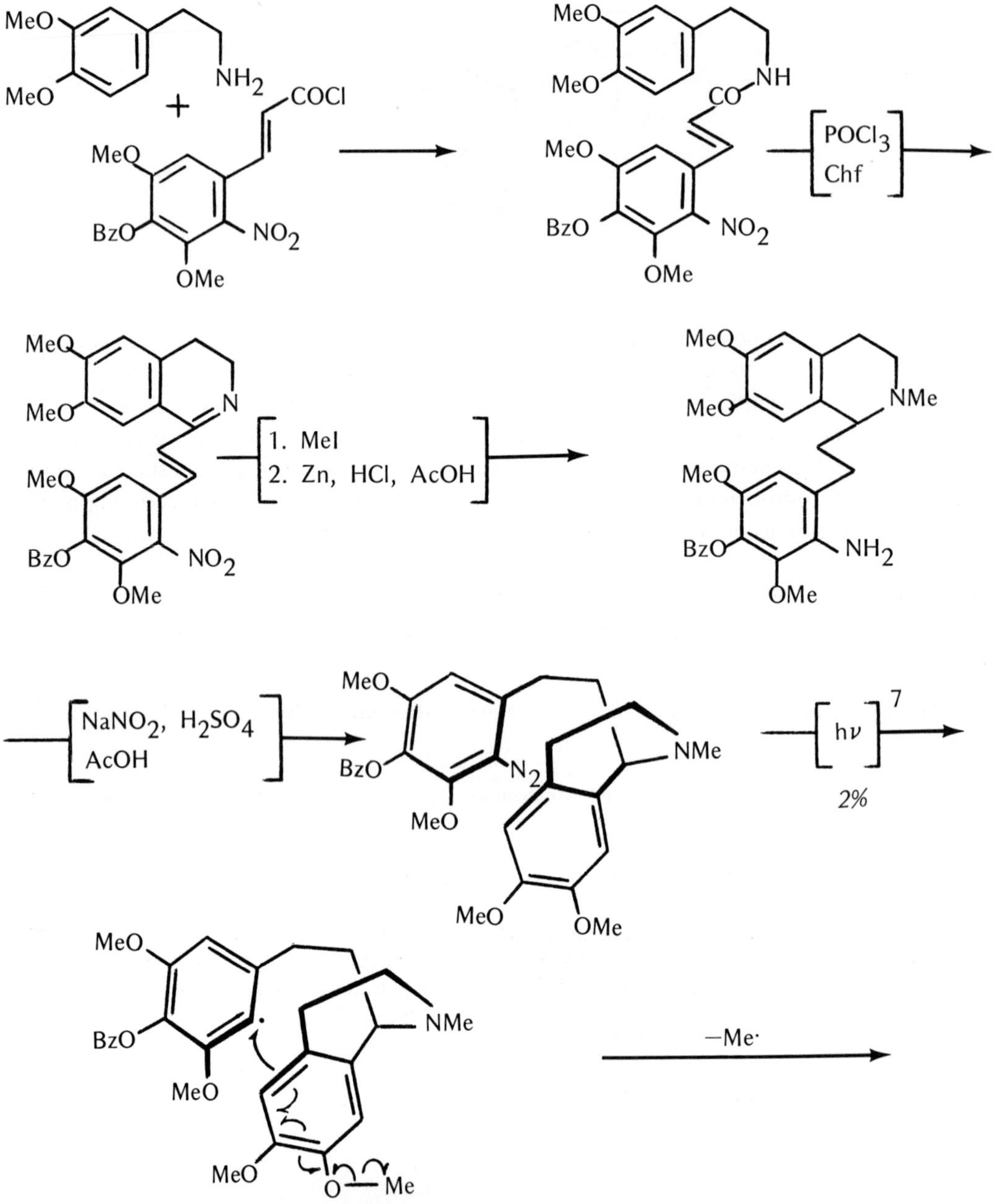

MeO MeO NH_2 + COCl MeO BzO OMe NO_2 → MeO MeO NH CO MeO BzO OMe NO_2 [$POCl_3$ Chf] →

MeO MeO N MeO BzO OMe NO_2 [1. MeI 2. Zn, HCl, AcOH] → MeO MeO NMe MeO BzO OMe NH_2

[$NaNO_2$, H_2SO_4 AcOH] → MeO BzO MeO N_2 NMe MeO OMe [hν][7] 2%

MeO BzO MeO NMe MeO O Me —Me· →

7. The aromatic radical formed by decomposition of the diazonium group participates in the coupling of both aromatic rings.

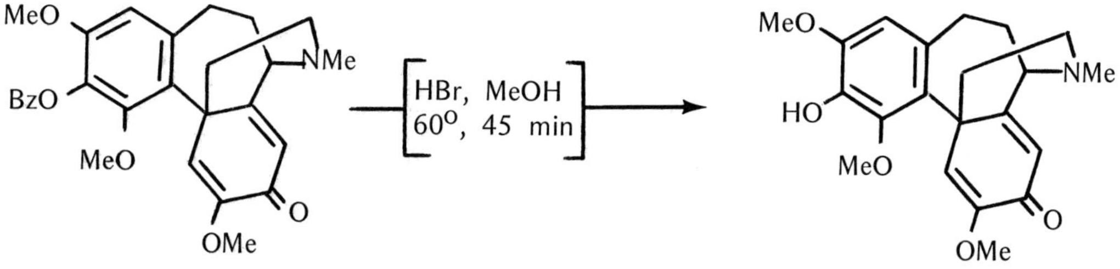

MeO
BzO
MeO
NMe
OMe
O

HBr, MeOH
60°, 45 min

MeO
HO
MeO
NMe
OMe
O

Androcymbine

AVENACIOLIDE

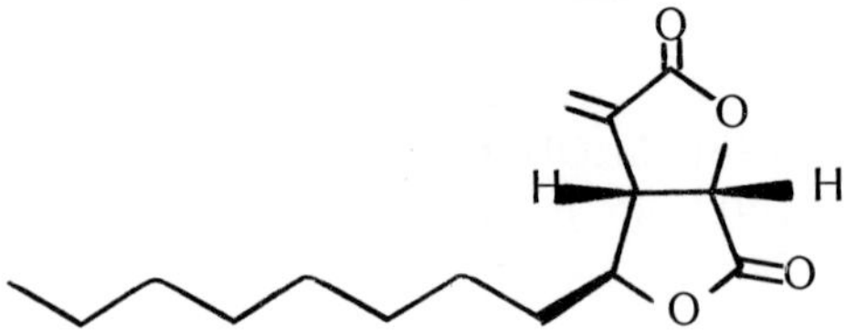

Avenaciolide is a novel fungicidal bislactone first isolated from cultures of *Aspergillus avenaceus* G. Smith. The synthesis[1] of avenaciolide, outlined below, is based on a new method for synthesis of α-methylene-γ-butyrolactones developed at the Dow Chemical Company.[2]

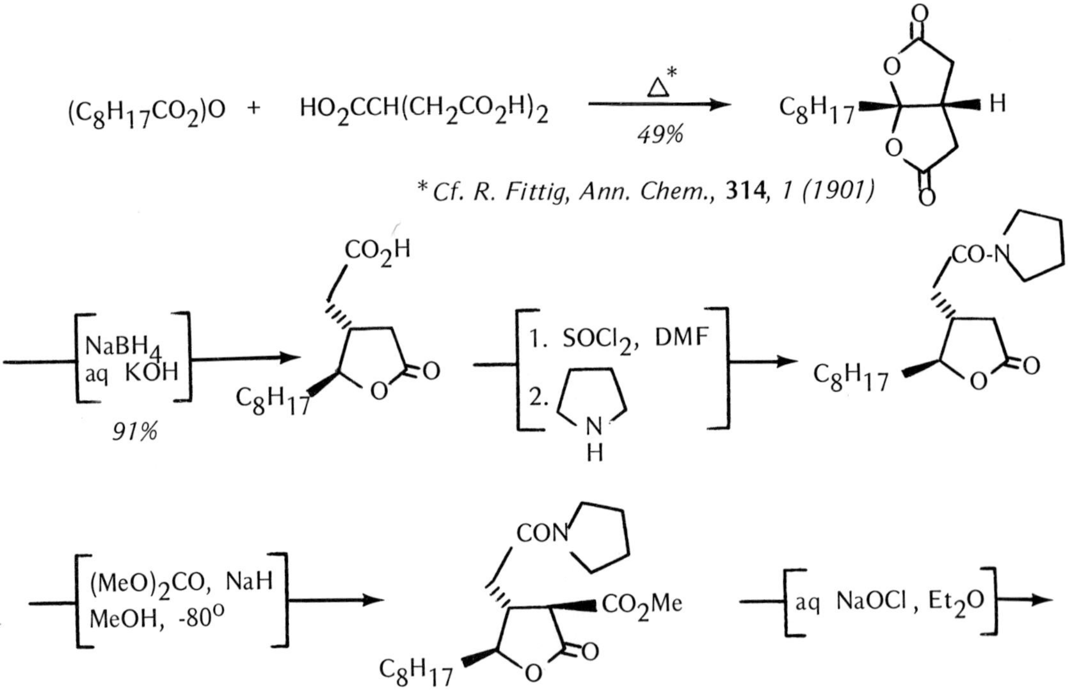

$(C_8H_{17}CO_2)O$ + $HO_2CCH(CH_2CO_2H)_2$ $\xrightarrow{\Delta^*}$ 49%

* *Cf. R. Fittig, Ann. Chem.,* **314**, *1 (1901)*

$NaBH_4$, aq KOH; 91%

1. $SOCl_2$, DMF
2. pyrrolidine

$(MeO)_2CO$, NaH, MeOH, -80°

aq NaOCl, Et_2O

---

1. W. L. Parker and F. Johnson, J. Amer. Chem. Soc., **91**, 7208 (1969); J. Org. Chem., **38**, 2489 (1973).

2. J. Martin, P. C. Watts and F. Johnson, Chem. Commun., 27 (1970).

aq HBr, Diox, Δ
57%

Stiles Reagent*
DMF, 120°
75%

1. CH_2O, Et_2NH
2. NaOAc, AcOH
66%

Avenaciolide

**Methyl methoxymagnesium carbonate*

Recently, Schlessinger *et al.*[3] reported a notably efficient synthesis of avenaciolide based on a novel conjugate addition-halogenation sequence (A→B). The resulting α-methyl-α-thiomethylbutyrolactone (C) readily oxidizes to the sulfoxide and undergoes thermal elimination to give the desired α-methylene lactone.

$C_8H_{17}CHO + LiC{\equiv}C\text{-}CO_2Et \rightarrow C_8H_{17}CHOHC{\equiv}C\text{-}CO_2Et \longrightarrow$

$C_8H_{17}CHOHC{\equiv}C\text{-}CO_2H$ — [H_2, Pd-$BaSO_4$] →

3. J. L. Herrmann, M. H. Berger and R. H. Schlessinger, J. Amer. Chem. Soc., **95**, 7923 (1973).

4. A. Nobuhara, Agr. Biol. Chem., **34**, 1745 (1970).

THF, -78° → I_2 [5]

93%

(A) → (B)

1. TsOH, C_6H_6, Δ
2. $NaHCO_3$

95%

(C)

1. m-$ClC_6H_4CO_3H$
2. Succinic anhydride, 140° → 10% HCl [6]

80%

Avenaciolide [7]

5. See R. J. Cregge, J. L. Hermann, C. S. Lee, J. E. Richman and R. H. Schlessinger, Tetrahedron Lett., 2425 (1973), for a description of the preparation and handling of lithium enolates of simple ester systems.

6. For related sulfoxide to olefin transformations, see B. M. Trost and T. N. Salzmann, J. Amer. Chem. Soc., **95**, 5321 (1973). Recently, P. A. Grieco and J. J. Reap [Tetrahedron Lett., 1097 (1974)] have reported a novel α-methylenation sequence, based on the thermal elimination of α-phenylsulfinyl lactones to α-methylene-γ-butyrolactones (i→ii).

(i) → (ii)

7. The wide variety and biological activity of natural products containing the α-methylene-γ-butyrolactone feature has created considerable interest in developing syntheses of this functional group. For an extensive compilation of existing syntheses of α-methylene lactones, see P. F. Hurdlik, L. R. Rudnick and S. H. Korzeniowski, J. Amer. Chem. Soc., **95**, 6848 (1973).

BAKKENOLIDE-A

The stereoselective synthesis[1] of bakkenolide-A illustrates a novel lactone spiroannelation reaction based on [2,3]sigmatropic rearrangement of nucleophilic carbenes.[2,3] The desired sulfur-stabilized carbene (A) has been generated by a Bamford-Stevens reaction[4], and readily undergoes rearrangement across the less sterically biased convex face of the molecule, establishing the asymmetric center at C-7 and the necessary functionality for further elaboration of the β-methylene-γ-butyrolactone ring.

[HCO_2Et, NaOMe / C_6H_6] 90%; CH_2OH; [^nBuSH, TsOH / C_6H_6] 87%

1. D. A. Evans and C. L. Sims, Tetrahedron Lett., 4691 (1973).

2. Sulfur stabilized carbenes readily undergo [2,3] sigmatropic rearrangement to dithioesters (i→ii) and provide the basis of a new carbon-carbon bond synthesis, J. E. Baldwin and J. A. Walker, Chem. Commun., 354 (1972).

MeS; S+; MeS; :S; (i); MeS; S; (ii)

3. For stereoselectivity of such bond reorganization processes, see G. Andrews and D. A. Evans, Tetrahedron Lett., 5121 (1972); J. E. Baldwin and J. E. Patrick, Chem. Commun., 354 (1972).

4. Cf. V. Schollkopf and E. Wiskott, Annalen, **694**, 44 (1966); W. R. Bamford and T. S. Stevens, J. Chem. Soc., 4735 (1952).

KO^{t}Bu, ^{t}BuOH

Cl

CHS^{n}Bu

87%

*5:1 epimeric mixture

CHS^{n}Bu

aq KOH
DEG, Δ

96%

Ref. 5

OsO_4, $NaIO_4$ **
aq ^{t}BuOH

93%

**Separation of epimeric diketones was carried out chromatographically at this stage.

KO^{t}Bu
^{t}BuOH

65%

H_2, Pd-C [6]
EtOH

Li, Et_2O

70%

PBr_3, Et_2O, 0°

Ts-NH-N=C(S⁻ Na⁺)(SMe)

75%

TsNHN=C(SMe)–S

NaH, THF

60%

5. The methallylcyclohexanone has previously been synthesized by E. Piers, R. W. Britton and W. de Waal, Canad. J. Chem., 47, 831 (1969).

6. Catalytic hydrogenation of hydrindenones invariably gives the *cis*-ring fusion, R. L. Augustine, "Catalytic Hydrogenation," Marcel Dekker, New York (1965) p. 61.

SMe

S C:

H

(A)

Δ

S

MeS

H

$HgO, HgCl_2$
Me_2CO, Δ
75%

O

MeS

H

SeO_2, C_6H_6, Δ

O

O

H

Bakkenolide-A[7]

7. For a different approach to bakkenolide-A, see K. Naya, M. Hayashi, I. Takagi, S. Nakamura and M. Kobayashi, Bull. Soc. Chim. Japan, **45**, 3673 (1972); K. Hayashi, H. Nakamura and H. Mitsuhashi, Chem. Pharm. Bull., **21**, 2807 (1973).

BYSSOCHLAMIC ACID

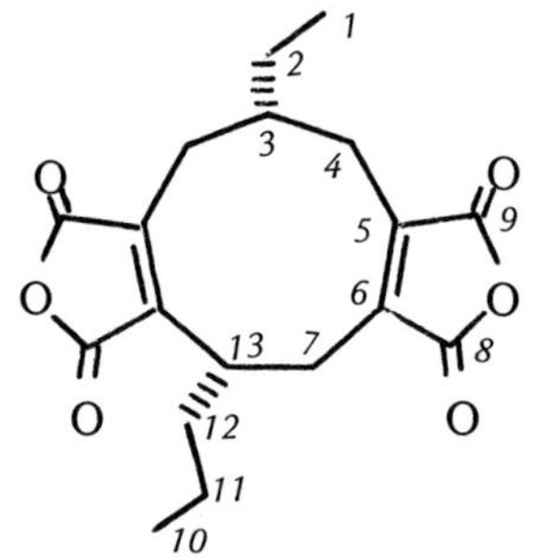

Byssochlamic acid belongs to a small but unusual group of mold metabolites called the nonadrides, characterized by the presence of a nine-membered alicyclic ring fused to two five-membered anhydride systems.[1,2] In the synthesis of byssochlamic acid by Stork *et al.*,[3] the problem of constructing the key nine-membered ring bishydroquinone dimethyl ether (D) has been neatly solved by fragmentation of a bicyclo[4·3·1]decanone derivative (C). The aromatic rings in (D) serve not only as the latent anhydride functions, but also as a means to direct establishment of *cis*-stereochemistry of the C-3, C-13 alkyl substituents.

Preparation of dichloromethyl dimethoxy benzene:

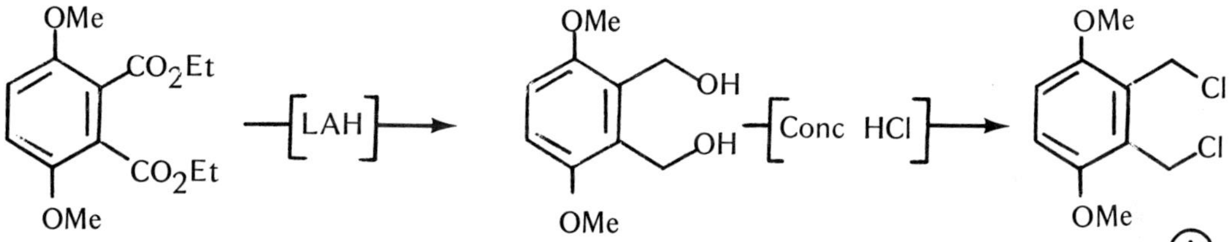

1. D. H. R. Barton and J. K. Sutherland, J. Chem. Soc., 1769 (1965).

2. The work that led to the constitution and stereochemistry of byssochlamic acid has been reviewed by J. K. Sutherland, Fortschr. Chem. Org. Naturst., **25**, 131 (1967).

3. G. Stork, J M. Tabak and J. F. Blount, J. Amer. Chem. Soc., **94**, 4735 (1972).

Preparation of tetralone (B):

OMe

OMe

OMe

N

OMe

[I ; MeOH, Δ] 80%

OMe

O

OMe

(B)

Preparation of byssochlamic acid:

OMe OMe

Cl + Cl

O

OMe OMe

[NaH, glyme, Δ] 80%

OMe OMe

O

OMe OMe

(A) (B) (C)

[1. NH_2OH ; 2. $POCl_3$, Py, 0°]*

OMe CN OMe

3

OMe OMe

(D)

[1. MeLi, Et_2O* ; 2. NH_2NH_2, KOH] 80%

*Beckmann fragmentation.
Desired isomer formed in 46% yield was separated by crystallization from material with exocyclic double bond.

*(–CN → –$COCH_3$)

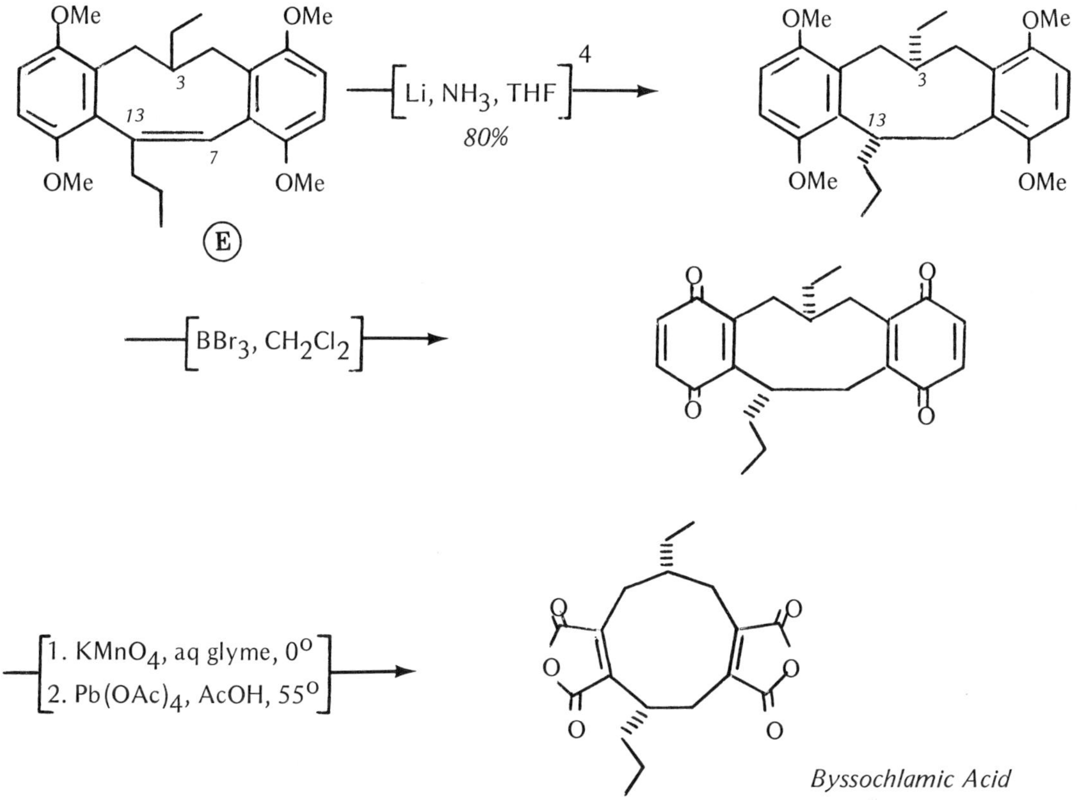
OMe
OMe
3
13
7
OMe
OMe
E
Li, NH3, THF
4
80%
OMe
OMe
3
13
OMe
OMe
BBr3, CH2Cl2
O
O
O
O
1. KMnO4, aq glyme, 0°
2. Pb(OAc)4, AcOH, 55°
O
O
O
O
O
O

Byssochlamic Acid

4. Catalytic hydrogenation of the stilbene derivative (E) was precluded due to tub-shape of the molecule and projecting ring substituents which prevented close approach of the catalyst to the olefinic linkage. The reduction, however, could be carried out using metal and ammonia and resulted in addition of an electron to the double bond with some rotation around the original C-7, C-13 linkage, thus allowing partial stabilization of the radical ion by the aromatic rings. Establishment of the necessary *cis* stereochemistry of the two alkyl chains during reduction was a consequence of the fact that formation of the unwanted *trans* isomer would have resulted in a serious eclipsing interaction between the propyl and methoxyl groups, an interaction which was avoided in the *cis* transition state (i).

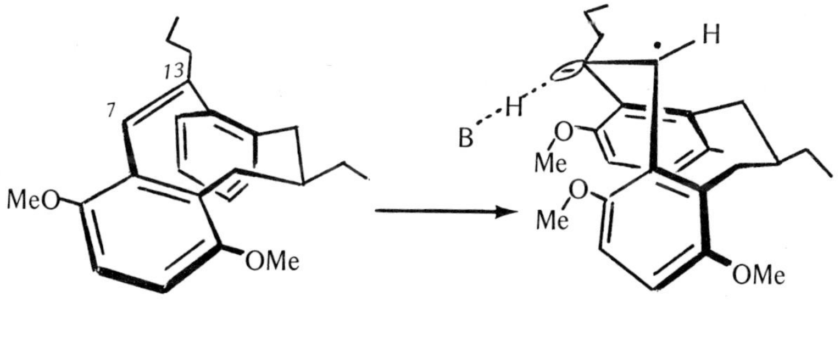
13
7
MeO
OMe
H
B
H
O
Me
O
Me
OMe

(i)

CAMPHERENONE

Recently there has been much interest in devising synthetic routes to sesquiterpenoids along plausible biogenetic pathways. Money's route to bicyclo[2·2·1]heptane systems using monocyclic enol acetate intermediates[1] is illustrated by the efficient conversion of dihydrocarvone to *dl*-campherenone.[2]

Dihydrocarvone

[Ph_3P OTHP]

[$(CO_2H)_2$, $(CH_2OH)_2$, C_6H_6]

OTHP

OH

[CCl_4, $(^nOctyl)_3P$][3]

Cl

1. J. C. Farlie, G. L. Hodgson and T. Money, Chem. Commun., 1196 (1969); J. Chem. Soc. Perkin I, 2109 (1973).

2. G. L. Hodgson, D. F. MacSweeney and T. Money, Chem. Commun., 766 (1971).

3. This useful procedure for converting alcohols to halides was developed by J. Hooz and S. S. H. Gilani, Canad. J. Chem., **46**, 86 (1968).

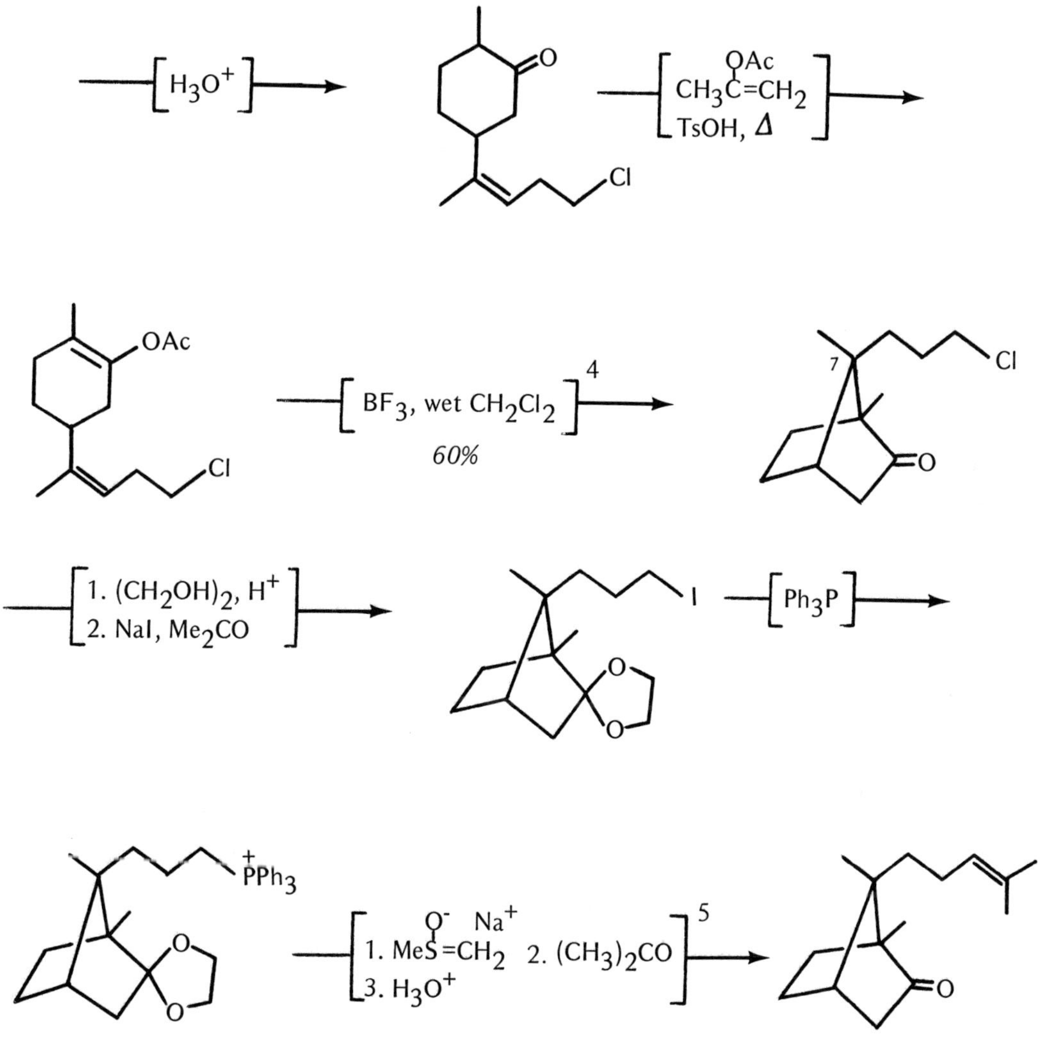

[H_3O^+] →

[$CH_3C(OAc)=CH_2$, TsOH, Δ] →

[BF_3, wet CH_2Cl_2][4] → 60%

7

[1. $(CH_2OH)_2$, H^+ 2. NaI, Me_2CO] →

[Ph_3P] →

$\overset{+}{P}Ph_3$

[1. $MeS(O^-Na^+)=CH_2$ 2. $(CH_3)_2CO$ 3. H_3O^+][5] →

Campherenone

---

4. A mixture of bicyclic ketones, epimeric at C-7, was obtained in this reaction and carried through to the final stage of the synthesis. Only one isomer is depicted in the sequence above.

5. The mixture of campherenone and 7-epicampherenone ethylene ketals was separated by preparative g.l.c. prior to hydrolysis.

CAMPTOTHECIN

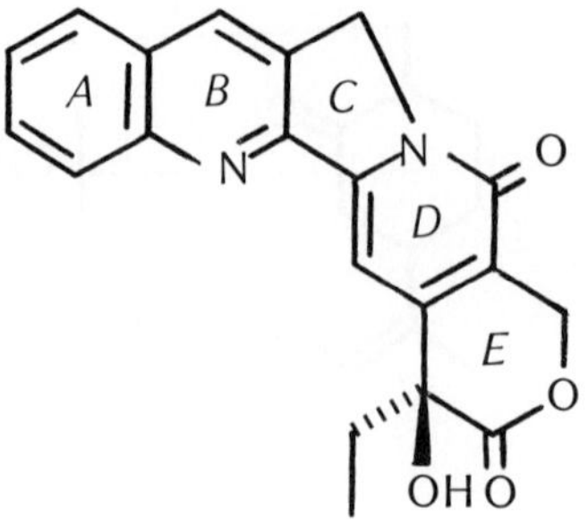

Camptothecin, the potent antitumor and antileukemic agent isolated from the stem wood of the Chinese tree *Camptotheca acuminata* Nyssaceae, has been the target of numerous chemical syntheses.[1] The first successful synthesis of this unusual alkaloid, characterized by an α-hydroxy-lactone and a pyridone system fused to a five-membered ring, was reported by Stork and Schultz in 1971.[2] A key feature of this synthesis is construction of ring E by a novel lactone annelation involving addition-cyclization of the carbonate of an α-hydroxy ester to the tetracyclic α,β-unsaturated lactam (B).

Preparation of pyrrolidone(A):

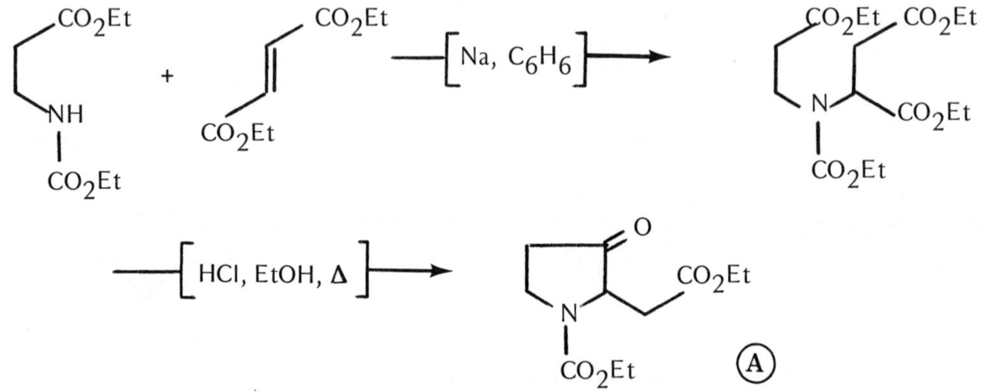

[Na, C_6H_6]

[HCl, EtOH, Δ]

(A)

---

1. For excellent and comprehensive reviews of the chemistry and biology of camptothecin, see A. G. Schultz, Chem. Rev., **73**, 385 (1973), and M. Shamma and V. St. Georgiev, J. Pharm. Sci., **63**, 163 (1974).

2. G. Stork and A. G. Schultz, **93**, 4074 (1971).

Preparation of Camptothecin:

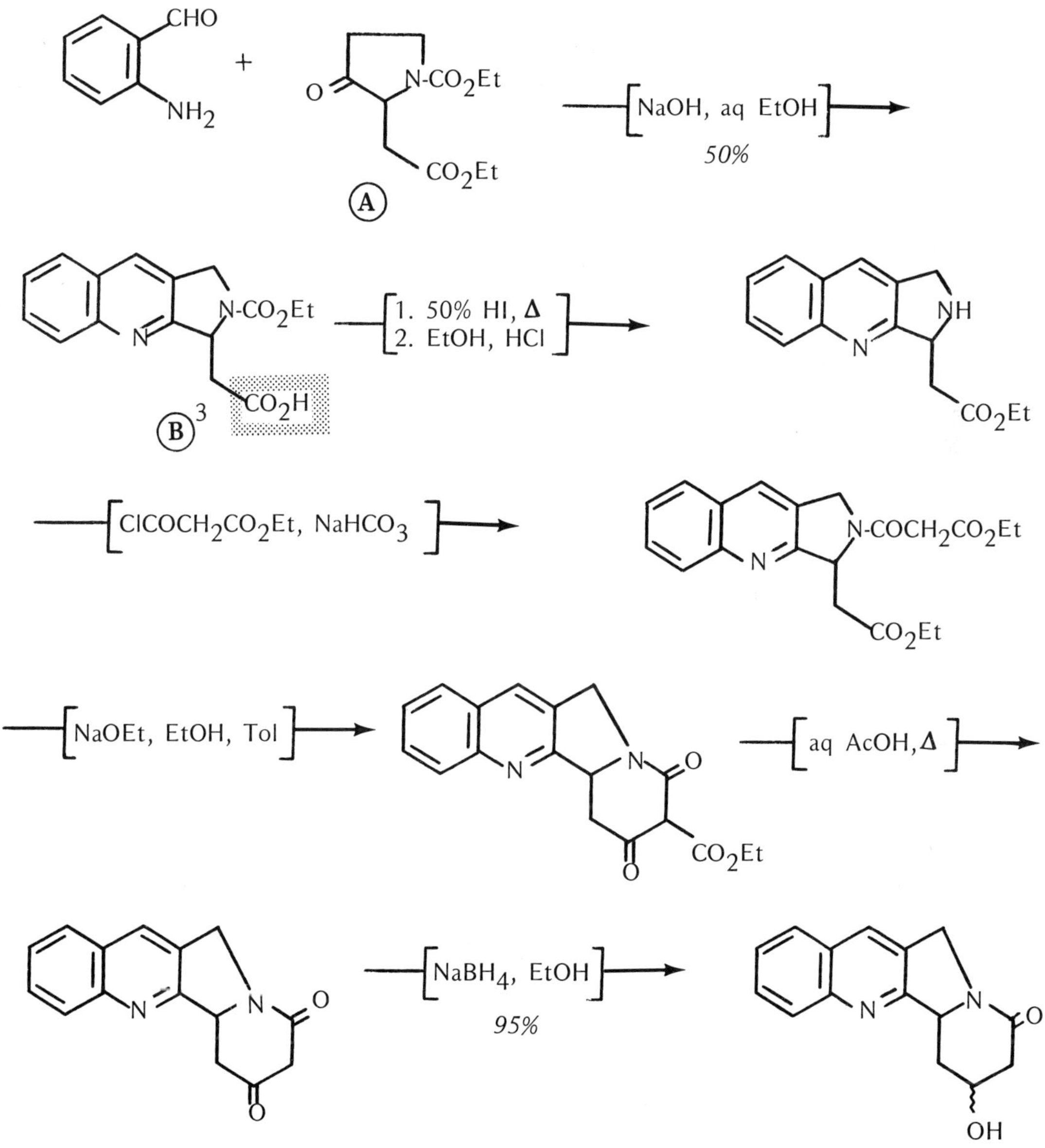

CHO

NH_2

+

O

$N\text{-}CO_2Et$

CO_2Et

(A)

[NaOH, aq EtOH] →

50%

$N\text{-}CO_2Et$

N

(B)[3] CO_2H

[1. 50% HI, Δ
2. EtOH, HCl] →

NH

N

CO_2Et

[$ClCOCH_2CO_2Et$, $NaHCO_3$] →

$N\text{-}COCH_2CO_2Et$

N

CO_2Et

[NaOEt, EtOH, Tol] →

N

N

O

O

CO_2Et

[aq AcOH, Δ] →

N

N

O

O

[$NaBH_4$, EtOH] →

95%

N

N

O

OH

---

3. Previously prepared by J. A. Kepler, M. C. Wani, J. M. McNaull, M. E. Wall and S. G. Levine, J. Org. Chem., **34**, 3853 (1969).

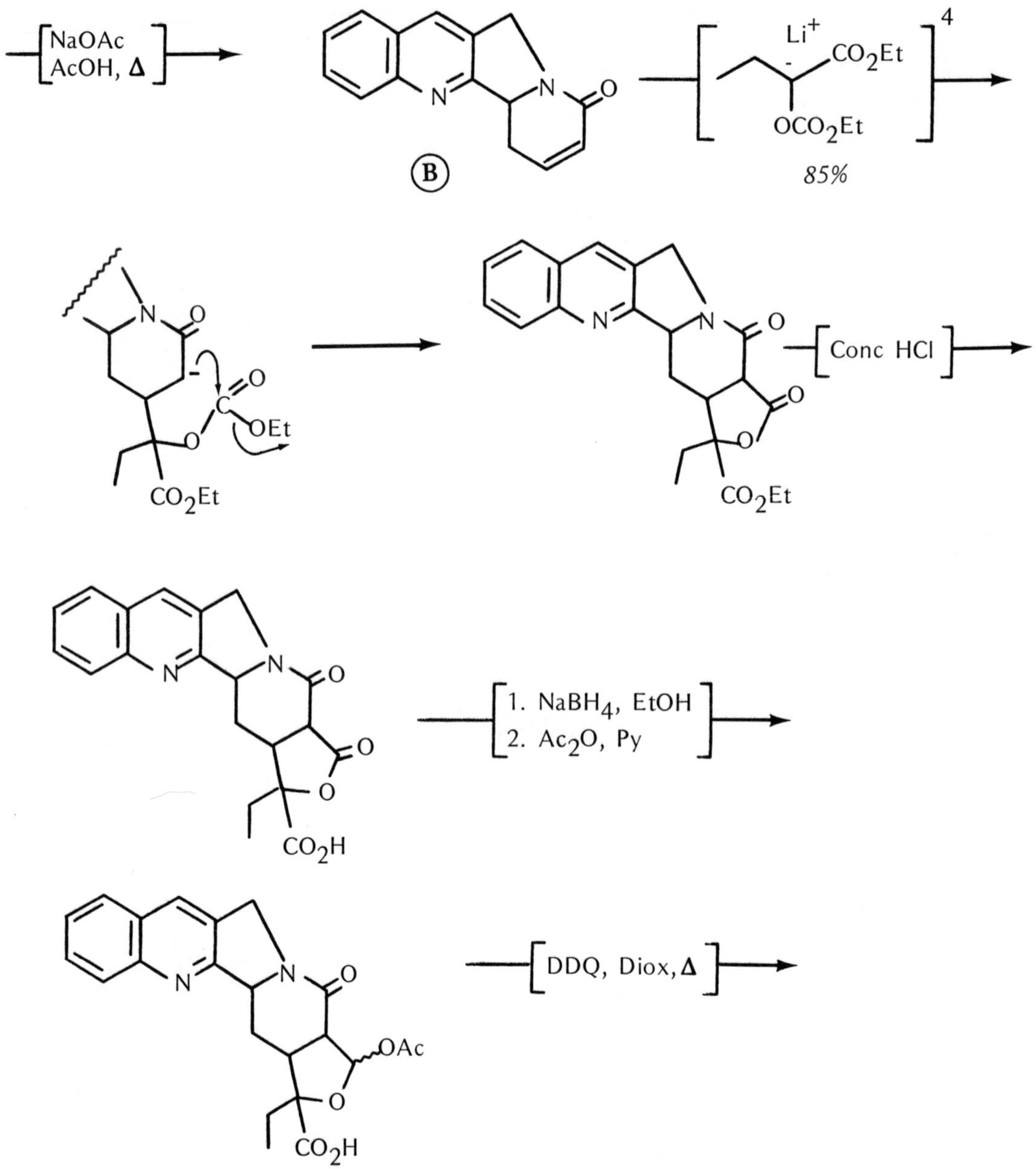

NaOAc
AcOH, Δ

(B)

Li^+
CO_2Et
OCO_2Et
4

85%

O
OEt
CO_2Et

Conc HCl

CO_2Et

1. $NaBH_4$, EtOH
2. Ac_2O, Py

CO_2H

DDQ, Diox, Δ

OAc
CO_2H

---

4. Anion formation from the ester carbonate was carried out using lithium diisopropylamide in THF at -70°.

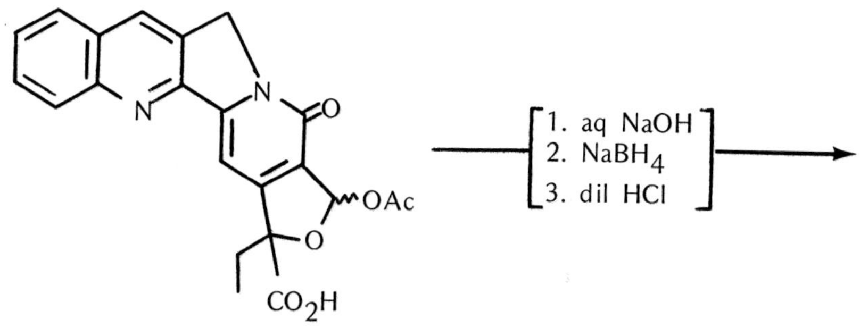

1. aq NaOH
2. $NaBH_4$
3. dil HCl

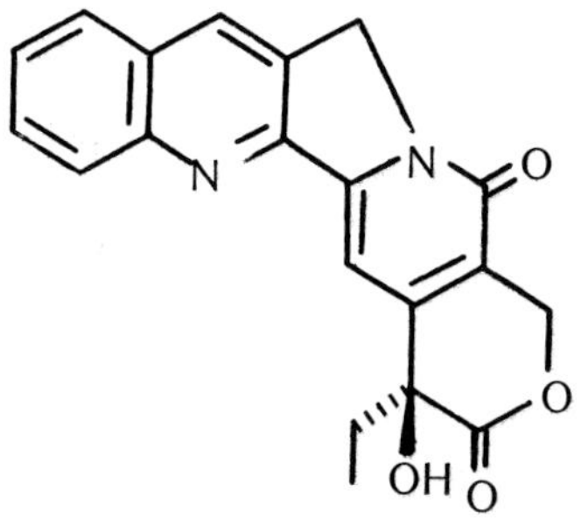

Camptothecin

The approach to camptothecin by Danishefsky *et al.*[5] revolves around an interesting synthesis of pyridones *via* nucleophilic addition to 1,3-dicarbethoxy allene.[6] The C ring of camptothecin has been established by a Dieckmann closure followed by attachment of the quinoline ring using the Friedlander synthesis. Construction of ring E gives desoxycamptothecin, which readily undergoes oxidation to camptothecin.

Preparation of C/D Rings:

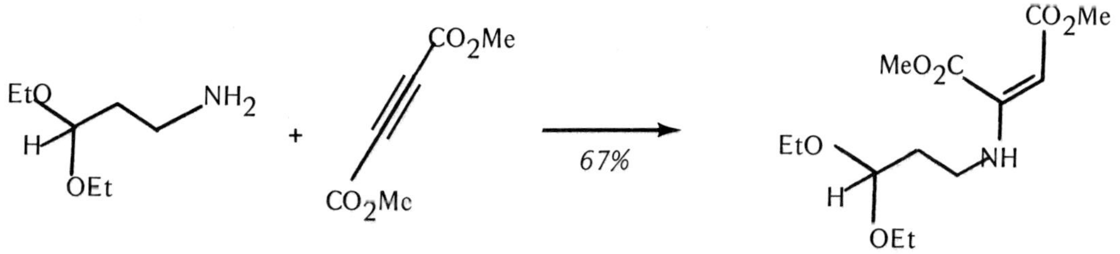

67%

5. R. Volkmann, S. Danishefsky, J. Eggler and D. M. Solomon, J. Amer. Chem. Soc., **93**, 5576 (1971).

6. S. Danishefsky, S. J. Etherdge, R. Volkmann, J. Eggler and J. Quick, J. Amer. Chem. Soc., **93**, 5575 (1971).

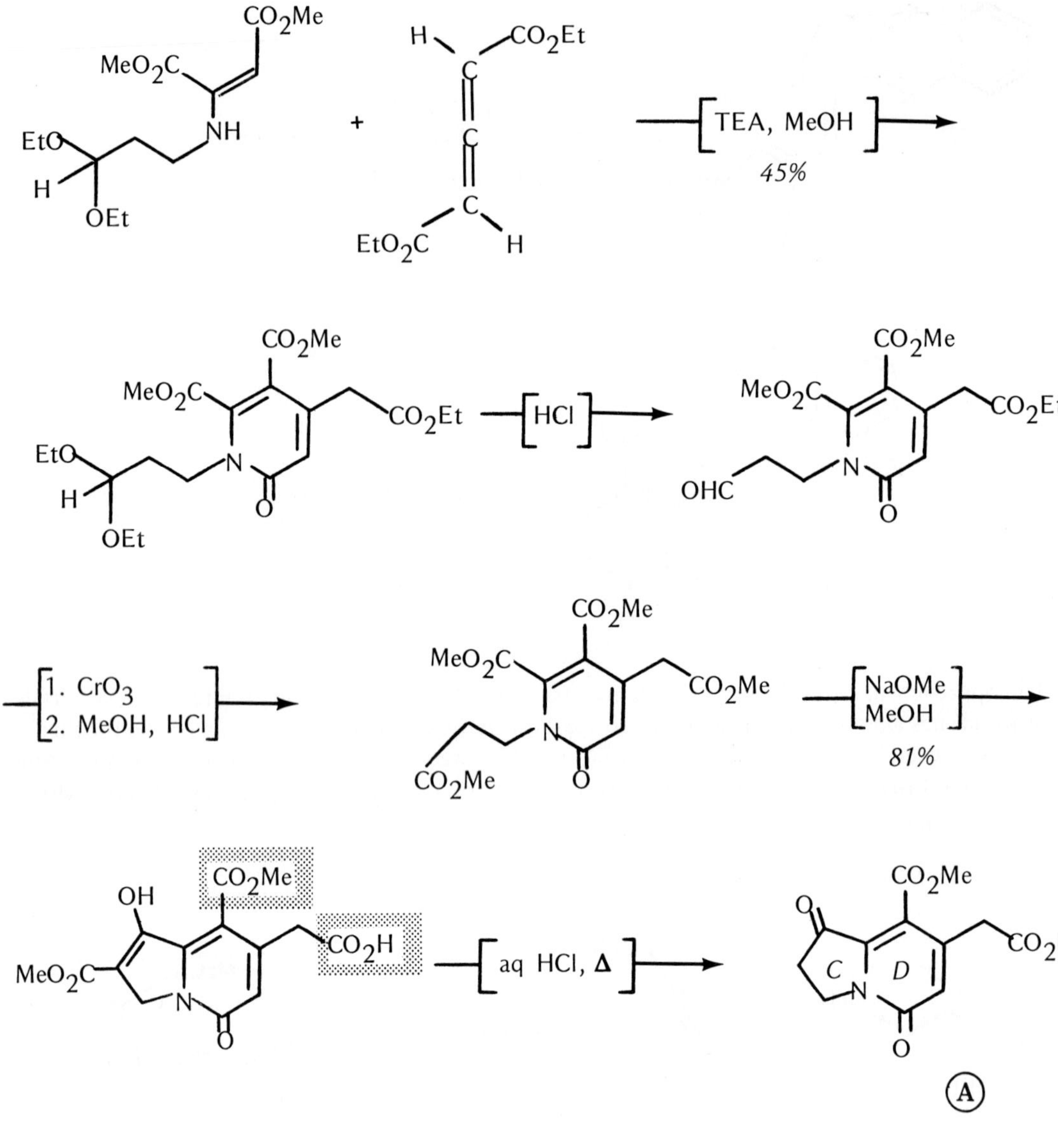

CO_2Me MeO$_2$C NH EtO H OEt + H CO_2Et C C C EtO_2C H —[TEA, MeOH]→ *45%*

CO_2Me MeO_2C CO_2Et EtO H N O OEt —[HCl]→ CO_2Me MeO_2C CO_2Et OHC N O

—[1. CrO_3 2. MeOH, HCl]→ CO_2Me MeO_2C CO_2Me N CO_2Me O —[NaOMe MeOH]→ *81%*

OH CO_2Me CO_2H MeO_2C N O —[aq HCl, Δ]→ O CO_2Me CO_2H C D N O (A)

Preparation of camptothecin:

CHO
NH_2
+
N O
O
CO_2Me CO_2H
(A)
[aq NaOH, Δ]

N N O
CO_2H
CO_2H
[MeOH, HCl]
N N O
CO_2H
CO_2Me

[CuO, 240°]
N N O
CO_2Me
[NaH, DME → EtI]
20%

N N O
CO_2Me
[HCHO, Diox, H_2SO_4, Δ]

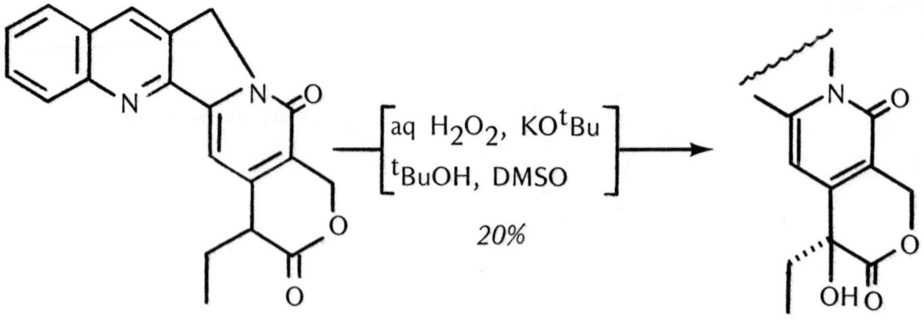

[aq H_2O_2, KO^tBu / ^tBuOH, DMSO] 20%

Desoxycamptothecin *Camptothecin*

The synthesis of camptothecin developed at Research Triangle Institute by Wall, Wani and associates[7] utilizes a Michael addition[8] for delivering elements of rings D and E to the tricyclic precursor (B).

Preparation of α, β-unsaturated ketone:

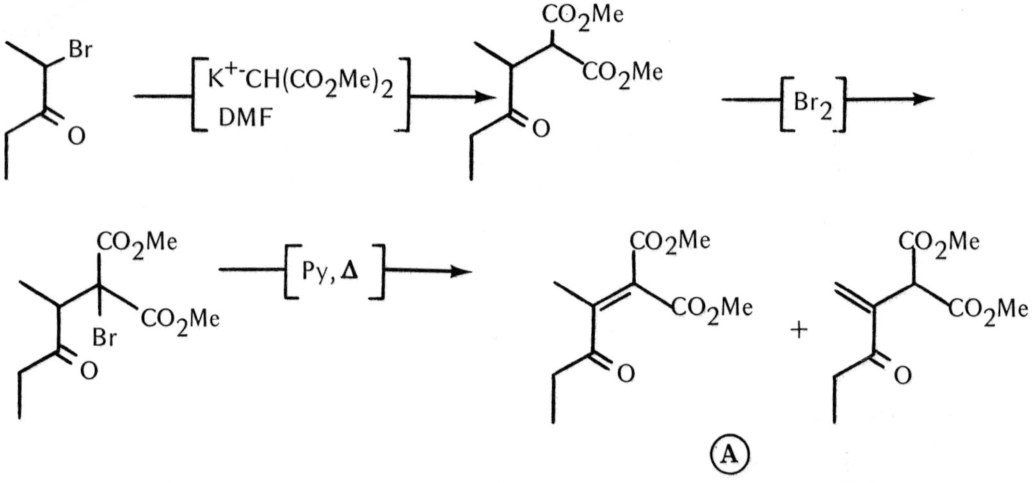

[$K^{+-}CH(CO_2Me)_2$ / DMF] → [Br_2] → [Py, Δ] → A +

---

7. M. C. Wani, H. F. Campbell, G. A. Brine, J. A. Kepler and M. E. Wall, J. Amer. Chem. Soc., **94**, 3631 (1971).

8. M. C. Wani, J. A. Kepler, J. B. Thompson, M. E. Wall and S. G. Levine, Chem. Commun., 404 (1970); M. E. Wall, H. F. Campbell, M. C. Wani and S. G. Levine, J. Amer. Chem. Soc., **94**, 3632 (1972).

Preparation of Camptothecin:

B + A — [110°] → 81%

— [HCN, KCN] → 82%

— [HCl, MeOH] → 76%

— [HBr, AcOH] →

— [NaOMe, MeOH → C_6H_6, Δ] →

CO_2Me

OH

$CONH_2$

+

$CONH_2$

TsOH, C_6H_6, Δ

[DDQ, Diox, Δ]

$CONH_2$

[1. $LiBH_4$, THF, Δ
2. dil HCl, 100°]

OH

Camptothecin

The versatile approach to camptothecin by Rapoport *et al.*[9,10] is based on a series of alternate rearrangement-oxidation reactions. Noteworthy features of this synthesis are the α-methylene lactam rearrangement[11] of a nipecotic acid (A) to a methylene piperidone (B), conversion to the pyridone allylic alcohol (C), and introduction of the lactone E ring carbons by a Claisen rearrangement (C→D).

Preparation of C/D-Rings:

$[Br(CH_2)_2CO_2Et]$

1. NaOEt
2. H_3O^+

$NaBH_4$, MeOH, 0°
86%

A

Ac_2O, Δ [11]
84%

B

SeO_2, AcOH, Δ
58%

9. C. Tang and H. Rapoport, J. Amer. Chem. Soc., **94**, 8615 (1972).

10. J. J. Plattner, R. D. Gless and H. Rapoport, J. Amer. Chem. Soc., **94**, 8613 (1972).

11. M. Ferles, Coll. Czech. Chem. Commun., **29**, 2323 (1964).

Johnson's modification of Claisen thermal rearrangement. For a detailed discussion see page 259.

[K_2CO_3, MeOH] → (C)

(C) [$PrC(OMe)_3$, $EtCO_2H$ →, K_2CO_3, MeOH] 12* →

(D) [DCC, DMSO, PPA] 76% → (E)

Preparation of camptothecin:

(E) + 75% →

[SeO_2, AcOH, Δ] →

12. W. S. Johnson, L. Werthemann, W. R. Bartlett, T. J. Brocksom, T. Li, D. J. Faulkner and M. R. Peterson, J. Amer. Chem. Soc., **92**, 741 (1970).

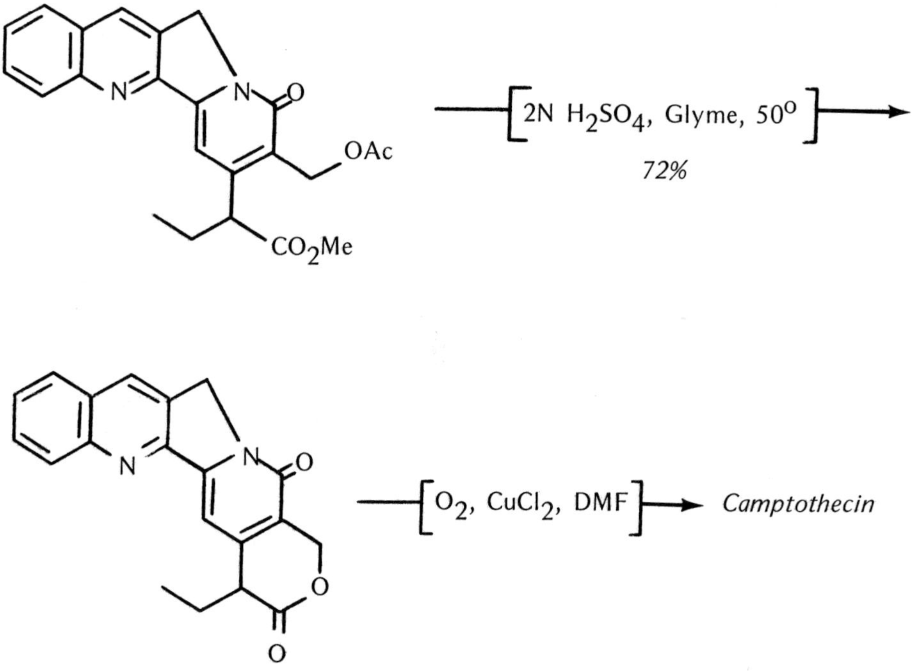

OAc

CO_2Me

[2N H_2SO_4, Glyme, 50°]

72%

[O_2, $CuCl_2$, DMF] *Camptothecin*

A biogenetically[13] oriented synthesis of camptothecin by Winterfeldt *et al.*[14] is based on the efficient indole→quinolone conversion (A→B).[15]

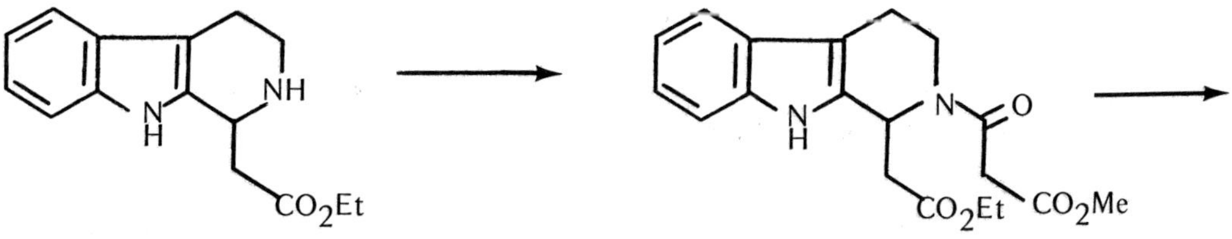

CO_2Et

CO_2Et CO_2Me

---

13. E. Wenkert, K. G. Dave, R. G. Lewis and P. W. Sprague, J. Amer. Chem. Soc., **89**, 6741 (1967).

14. E. Winterfeldt, T. Korth, D. Pike and M. Boch, Angew. Chem. Int. Ed., **11**, 289 (1972).

15. E. Winterfeldt and H. Randuz, Chem. Commun., 374 (1971); J. Warneke and E. Winterfeldt, Chem. Ber., **105**, 2120 (1972).

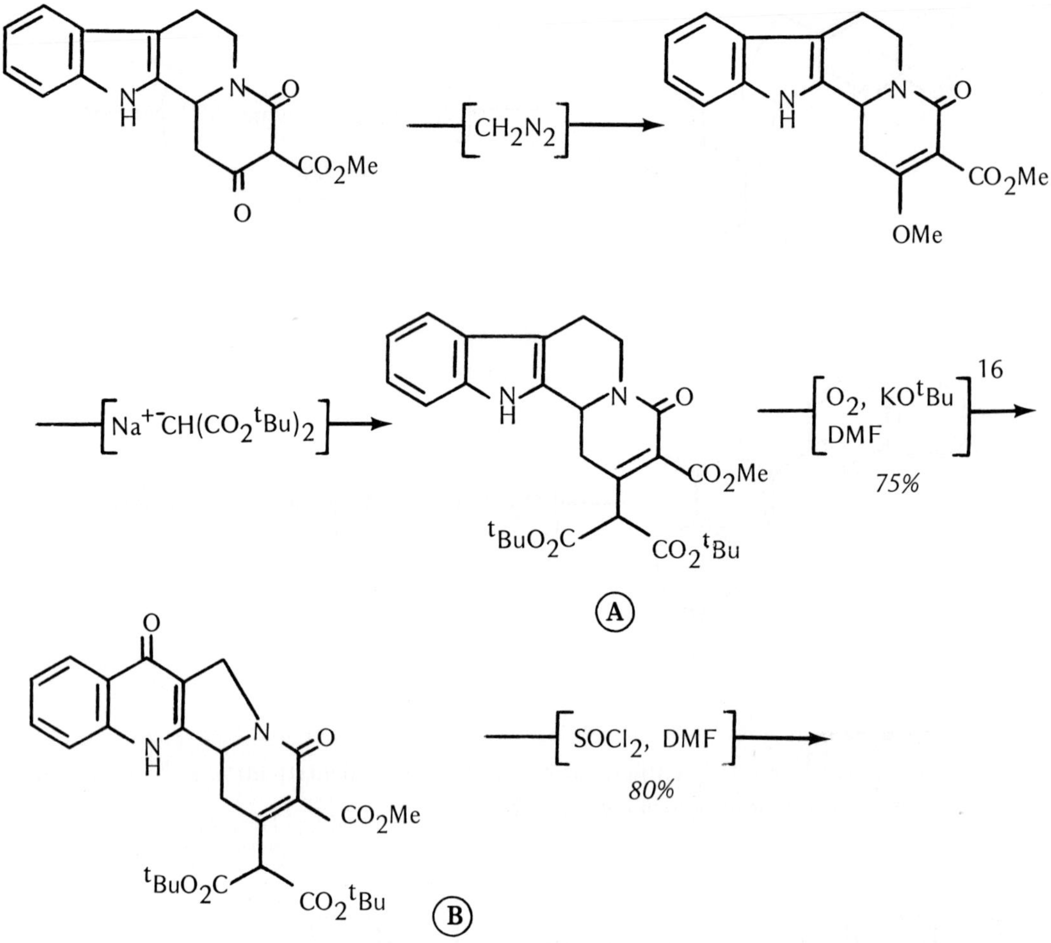

[CH_2N_2]

[$Na^{+-}CH(CO_2{}^tBu)_2$]

(A)

[O_2, KO^tBu DMF][16]

75%

(B)

[$SOCl_2$, DMF]

80%

---

16. E. Winterfeldt, Ann., **745**, 23 (1971).

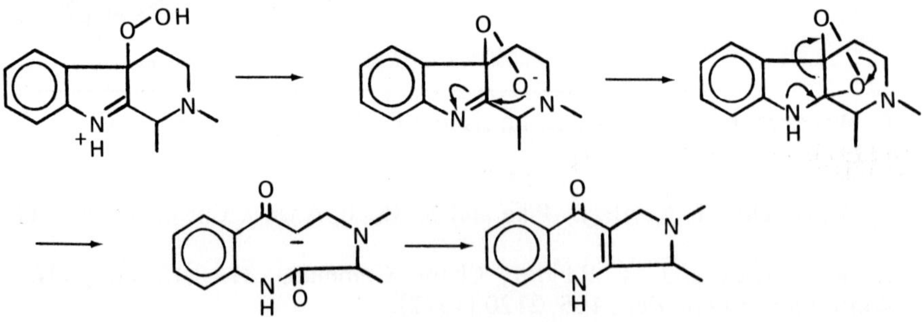

Cl

CO_2Me —[H_2, Pd-$BaSO_4$]→ CO_2Me

80%

${}^tBuO_2C$ $CO_2{}^tBu$ ${}^tBuO_2C$ $CO_2{}^tBu$

—[$({}^iBu)_2AlH$]→ OH —[TFA]→

${}^tBuO_2C$ $CO_2{}^tBu$

—[NaH, DMF —→EtI]→

—[O_2, $CuCl_2$, DMF]→ *Camptothecin*

---

17. Several other excellent syntheses of camptothecin have been reported recently: A. S. Kende, T. J. Bentley, R. W. Draper, J. K. Jenkins, M. Joyeux and I. Kubo, Tetrahedron Lett., 1307 (1973); M. Shamma, D. A. Smithers and V. St. Georgiev, Tetrahedron, **29**, 1949 (1973); T. Sugasawa, T. Toyoda and K. Sasakura, Tetrahedron Lett., 5109 (1972); A. I. Meyers, R. L. Nolen, E. W. Collington, T. A. Narwid and R. C. Strickland, J. Org. Chem., **38**, 1974 (1973).

CARPANONE

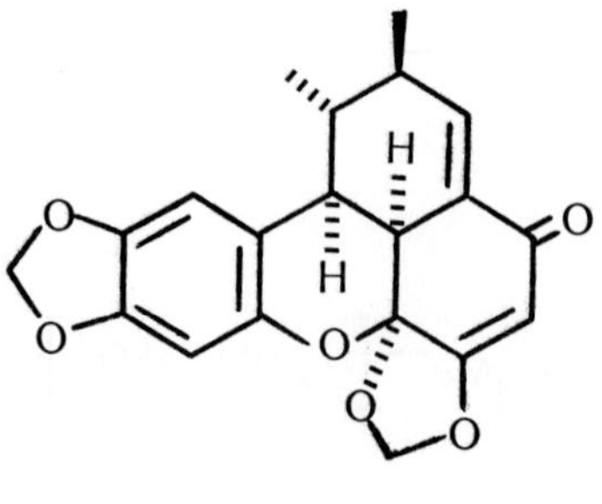

The remarkable synthesis of carpanone,[1] a lignan obtained from the bark of the carpano tree, is based on an intramolecular version of the reaction of o-quinonemethides with nucleophilic olefins.[2] The requisite o-quinonemethide has been generated *in situ* by phenolic coupling of two molecules of 2-(*trans*-1-propenyl)-4,5-methylenedioxyphenol. The synthesis features the introduction of five contiguous asymmetric centres in a single step.

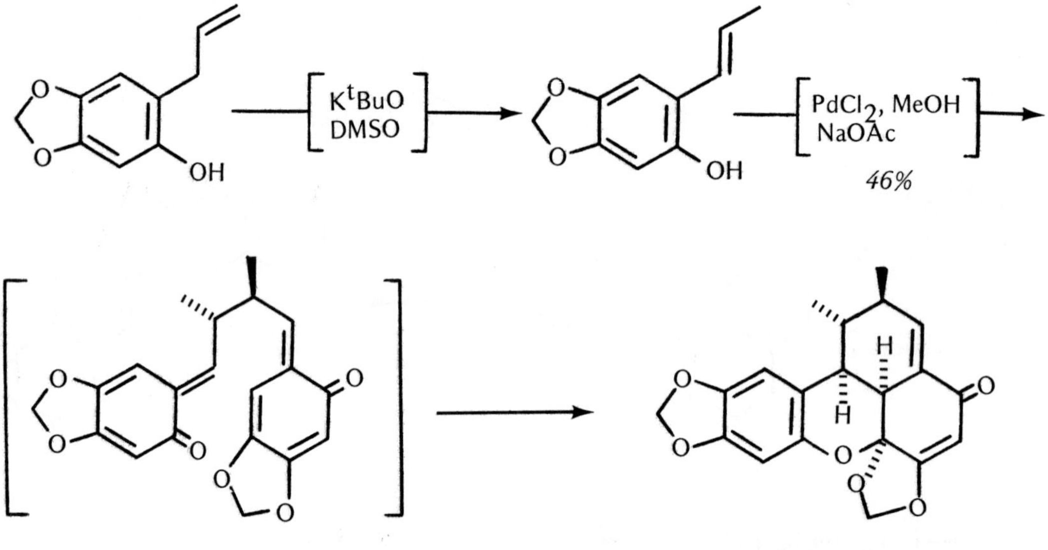

K^tBuO, DMSO

$PdCl_2$, MeOH, NaOAc

46%

Carpanone

1. O. L. Chapman, M. R. Engel, J. P. Springer and J. C. Clardy, J. Amer. Chem. Soc., **93** 6696 (1971).

2. O. L. Chapman and C. L. McIntosh, Chem. Commun., 383 (1971).

CEDRENE

Based on the fascinating biogenetic relationships that exist among monocyclic, spiro and tricyclic sesquiterpenes,[1] several unusually simples routes have been devised to the cedrane skeleton.[2-3,8,9] One such synthesis of α-cedrene is outlined below, and proceeds from nerolidol, presumably along lines of the proposed biogenesis.[3]

OH

[HCO_2H, n-C_5H_{12}, 25°]

H β α

Nerolidol

[CF_3CO_2H, n-C_5H_{12}, 25°]

H^+

+

γ-Bisabolene

1. W. Parker, J. S. Roberts and R. Ramage, Quart. Rev., **21**, 331 (1967).

2. For earlier syntheses of cedrene and cedrol, see G. Stork and F. H. Clarke, J. Amer. Chem. Soc., **83**, 3114 (1961); E. J. Corey, N. N. Girotra and C. T. Mathew, J. Amer. Chem. Soc., **91**, 1557 (1969); T. G. Crandall and R. G. Lawton, J. Amer. Chem. Soc., **91**, 2127 (1969). For a summary, see Art Org. Synth., 83 (1970).

3. N. H. Anderson and D. D. Syrdal, Tetrahedron Lett., 2455 (1972).

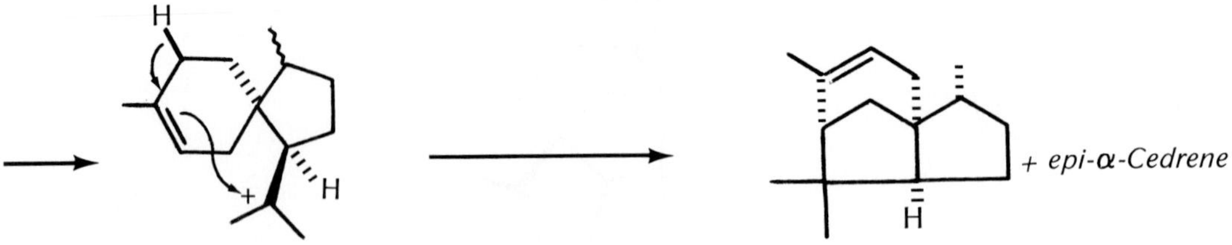

H

H

+ *epi-α-Cedrene*

H

α-Cedrene
(20% overall)

A somewhat lengthier but nonetheless unique stepwise transformation of nerolidol to cedrene has been described by Demole *et al.*[4] The reaction of nerolidol with N-bromosuccinimide and dehydrobromination furnishes a cycloheptenone (A) which has been converted to the spirotriene (B). The latter may be cyclized to 2,8-cedradiene (C) and thence to a mixture of α-cedrene and 2-epi-α-cedrene.

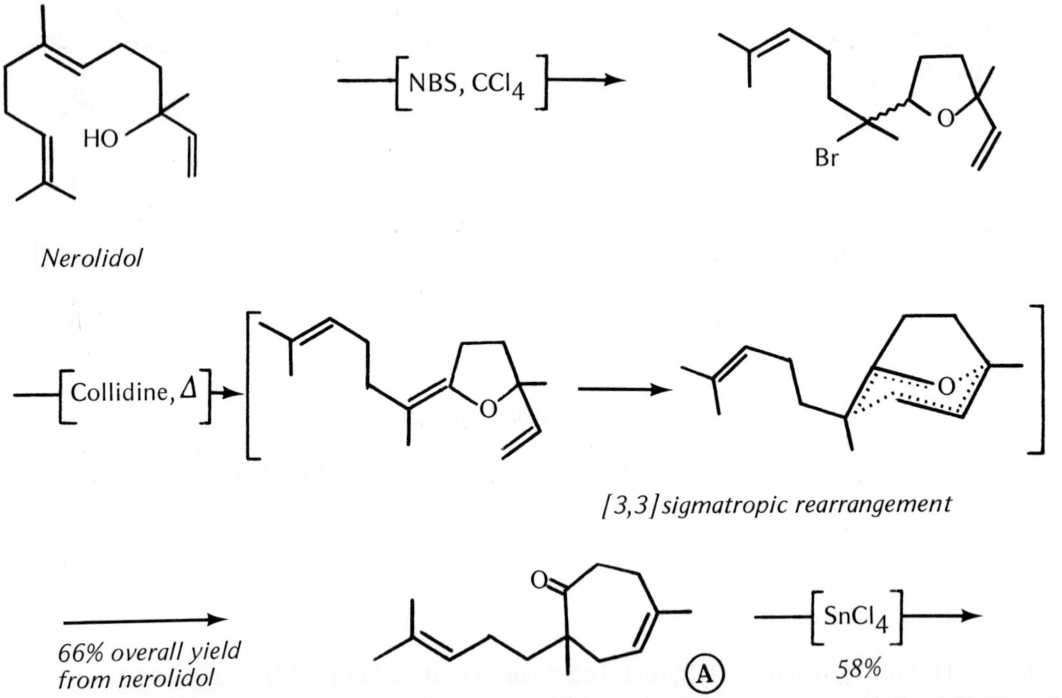

HO

[NBS, CCl_4]

Br

O

Nerolidol

[Collidine, Δ]

O

O

[3,3] sigmatropic rearrangement

66% overall yield from nerolidol

O

(A)

[$SnCl_4$]

58%

4. E. Demole, P. Enggist and C. Borer, Helv., **54**, 1845 (1971).

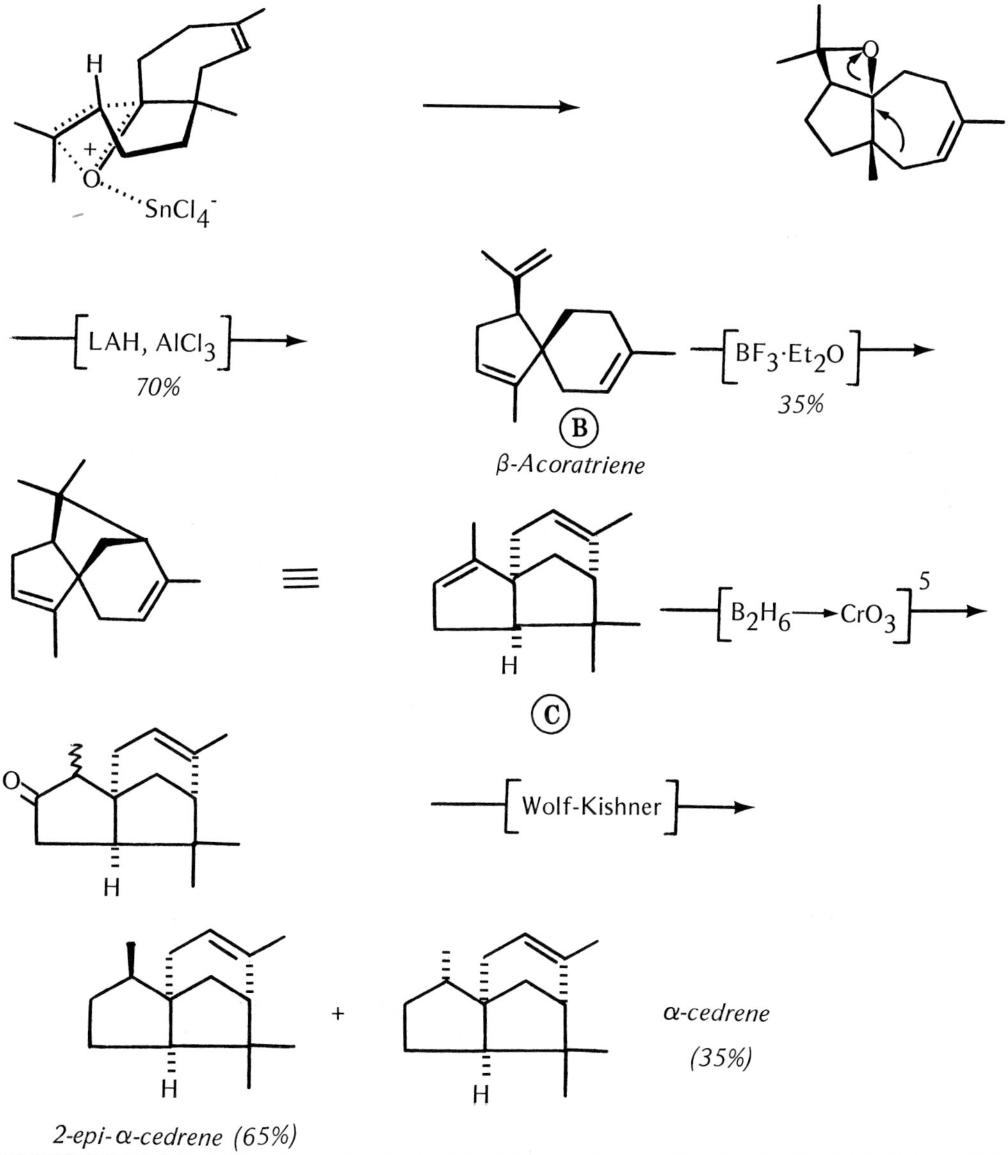

$SnCl_4^-$

[LAH, $AlCl_3$]
70%

(B)
β-Acoratriene

[$BF_3 \cdot Et_2O$]
35%

≡

[$B_2H_6 \rightarrow CrO_3$][5]

(C)

[Wolf-Kishner]

H

+

α-cedrene
(35%)

2-epi-α-cedrene (65%)

---

5. Selective catalytic hydrogenation of the 2,3-double bond in (C) afforded only 2-epi-α-cedrene, and may be explained by preferential addition of hydrogen to the double bond from the less hindered side of the molecule (i).

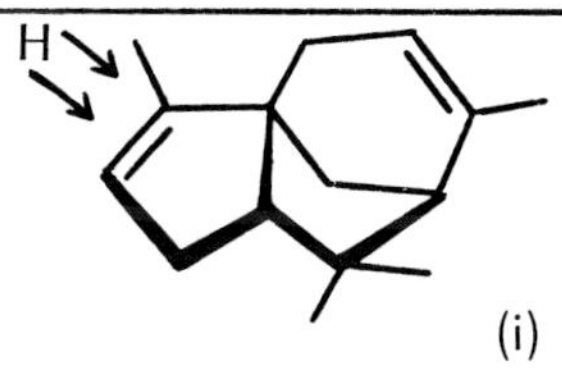

(i)

An alternative approach to the spirotriene (B), reported by Naegeli and Kaiser[6] is outlined below.

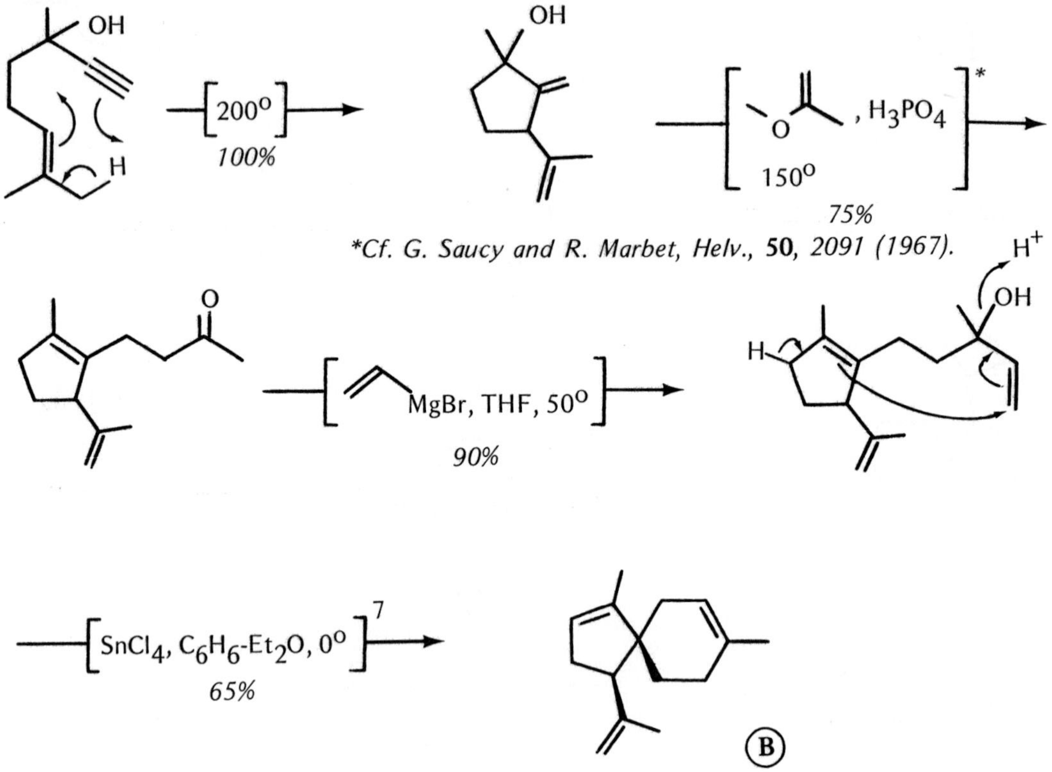

[200°] 100%

[$CH_3OC(=CH_2)CH_3$, H_3PO_4, 150°]* 75%

*Cf. G. Saucy and R. Marbet, *Helv.*, **50**, 2091 (1967).

[$CH_2{=}CHMgBr$, THF, 50°] 90%

[$SnCl_4$, C_6H_6-Et_2O, 0°][7] 65%

(B)

---

6. P. Naegli and R. Kaiser, Tetrahedron Lett., 2013 (1972).

7. This cyclization proceeds with remarkable stereospecificity. The isopropenyl group seems to direct the approach of the end of the side chain to the unhindered side of the cyclopentene ring.

Recently, Corey and Balanson[8] reported an ingeniously simple approach to cedrene and cedrol, utilizing a synchronous double annulation process. The key step involves direct conversion of the cationic enol ester (B), generated from the readily accessible cyclopropyl ketone (A), to give cedrone.

OMe

Li

+

O

-78°

O

$[(^iBu)_2AlH$ / $C_6H_6, 5^o]$

OH

$[CH_2I_2\text{-}Zn(Cu)]$

OH

$[CrO_3, Py, CH_2Cl_2]$

O

CH=C(CH$_3$)$_2$

(A)

$[AcOMs^*, CH_2Cl_2, -40^o$ → aqueous workup$]$

Acetyl methanesulfonate

OAc

+

CH=C(CH$_3$)$_2$

(B)

O

H

Cedrone

$[MeLi]$

OH

H

Cedrol

H

Cedrene

8. E. J. Corey and R. D. Balanson, Tetrahedron Lett., 3153 (1973).

In a remarkable and efficient total synthesis of α-cedrene by Lansbury *et al.*,[9] the chloroalkene moiety serves as a latent ketone[10, 11] during the cationic spiroannelation reaction. To avoid the potentially disastrous homoallyl to allyl cation (ii→iii) rearrangement during generation of the acorane skeleton (ii→ i), introduction of unsaturation into the cation B has been deferred until after the spiroannelation reaction.

a
b
Nu (i)
Nu (ii)
Nu (iii)

Synthesis of α-cedrene:

+ I Cl Ⓐ[12] — [^tBuLi, Et_2O, -78°] →

HO Cl — [1. HCO_2H, Ac_2O, 100° 2. NaOMe, MeOH, Δ] → + Cl

9. P. T. Lansbury, V. R. Haddon and R. C. Stewart, J. Amer. Chem. Soc., **96**, 896 (1974).

10. For a review of the application of latent functionality in organic synthesis, see D. Lednicer, Adv. Org. Chem., **8**, 179 (1972).

11. P. T. Lansbury, Accts. Chem. Res., **5**, 311 (1972).

12. Obtained in two steps from methyl cyclopropyl ketone: PCl_5 fragmentation followed by iodide displacement.

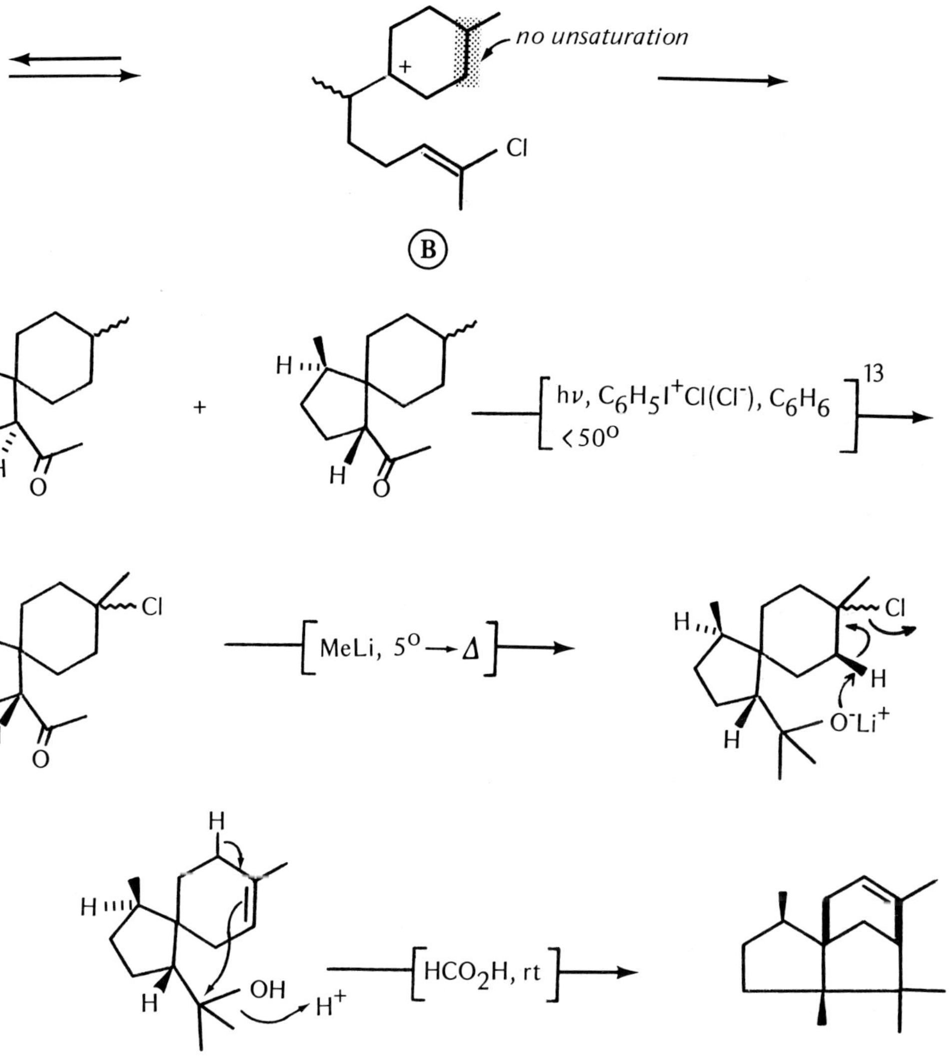

no unsaturation

Cl

(B)

H

H

H

H

O

O

+

[hν, $C_6H_5I^+Cl(Cl^-)$, C_6H_6 <50°][13]

H

H

Cl

O

[MeLi, 5° → Δ]

H

H

Cl

H

O^-Li^+

H

H

H

OH

H^+

[HCO_2H, rt]

α-Cedrene

---

13. Photochemical chlorination with iodobenzene dichloride occurs at the less hindered cyclohexyl tertiary C-H bond, cf. R. Breslow, J. A. Dale, P. Kalicky, S. Y. Liu and W. N. Washburn, J. Amer. Chem. Soc., **94**, 3276 (1972).

CEPHALOTAXINE

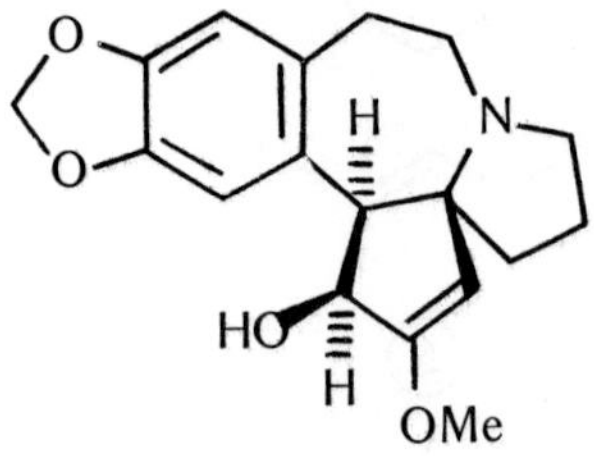

Cephalotaxine, characterized by an unusual 1-azaspiro[4.4] nonene structural feature, is the parent member of a group of minor alkaloids isolated from the Japanese plum-yew (*Cephalotaxus harringtonia*).[1] The finding of promising antileukemic activity in several of the naturally occurring esters of cephalotaxine, the harringtonines, and relative scarcity of the natural material kindled interest in developing synthetic approaches to these alkaloids. The first successful synthesis of cephalotaxine, accomplished by Auerbach and Weinreb[2] is based on construction of the key tricyclic enamine (A), followed by introduction of a 3-carbon pyruvyl fragment and an intramolecular Michael reaction for annelation of the 5-membered ring to afford the pentacyclic desmethylcephalotaxinone.

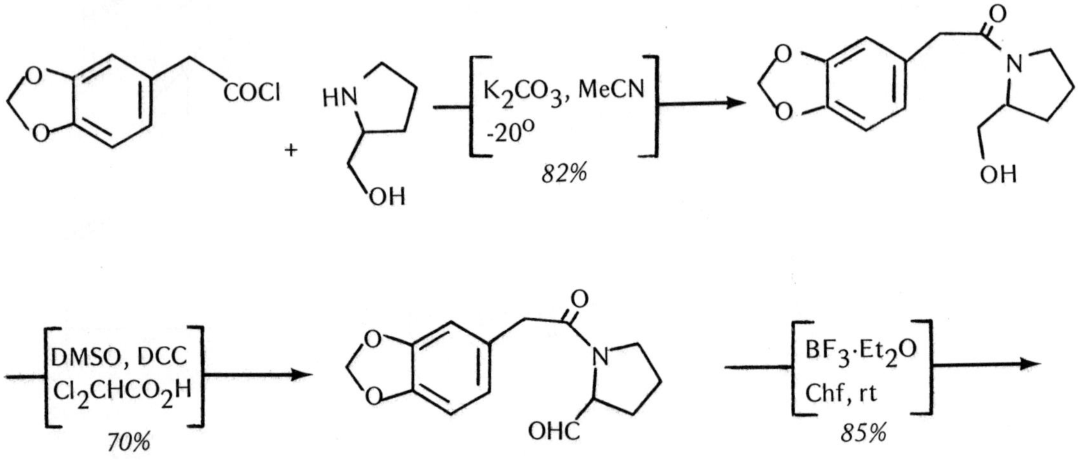

COCl + HN ... OH — [K_2CO_3, MeCN, -20°] *82%* →

— [DMSO, DCC, Cl_2CHCO_2H] *70%* → ... OHC — [$BF_3{\cdot}Et_2O$, Chf, rt] *85%* →

1. R. G. Powell, Phytochemistry, **11**, 1467 (1972).

2. J. Auerbach and S. M. Weinreb, J. Amer. Chem. Soc., **94**, 7172 (1972).

LAH, THF, Δ

100%

(A)[3]

OAc

COCl

$NaHCO_3$, MeCN, Δ

75%

AcO

O

1. K_2CO_3, aq MeOH
2. PbO_2, Tol, Δ

80%

O

O

(B)[4]

$Mg(OMe)_2$, MeOH[5]

52%

HO

O

Desmethylcephalotaxinone

---

3. This tricyclic enamine (A) has also been synthesized by L. J. Dolby, S. J. Nelson and D. Senkovich, J. Org. Chem., **37**, 3691 (1972).

4. Direct conversion of the enamine (A) to the α-dicarbonyl compound (B) could also be accomplished in a single step by alkylation with the mixed anhydride prepared from pyruvic acid and ethyl chloroformate, cf., E. W. Colvin, J. Martin, W. Parker, R. A. Raphael, B. Shroot and M. Doyle, J. Chem. Soc. Perkin 1, 860 (1972).

5. Cf. H. Muxfeldt, M. Weigee and V. Van Rheenen, J. Org. Chem., **30**, 3573 (1965).

[TsOH, (MeO)$_2$CMe$_2$, MeOH-Diox, Δ] 40%

[$NaBH_4$, MeOH, rt] 80%

dl-Cephalotaxine

The efficient convergent synthesis developed by Semmelhack *et al.*[6,7] is based on construction of the two sections of cephalotaxinone molecule as dissected in representation (i), and their combination for creation of the central 7-membered azepine ring. Several variations of the key cyclization step have been developed, involving (a) a nucleophilic attack on a transient benzyne derivative,[7] (b) nucleophilic aromatic substitution *via* an aryl nickel species,[8] and (c) an intramolecular $S_{RN}1$ reaction.[9]

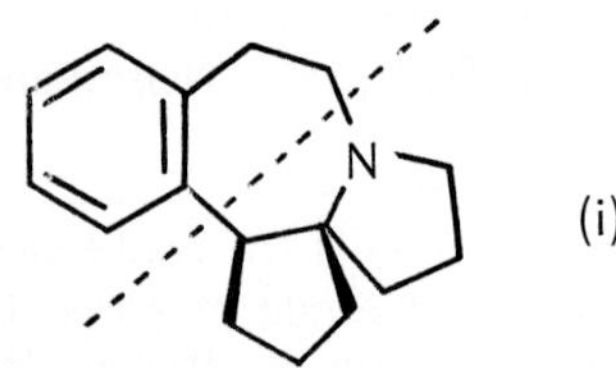

(i)

6. M. F. Semmelhack, B. P. Chong and L. D. Jones, J. Amer. Chem. Soc., **94**, 8629 (1972).

7. M. F. Semmelhack, R. D. Stauffer and T. D. Rogerson, Tetrahedron Lett., 4519 (1973).

Preparation of alicyclic part (A):

[Et$_2$O, 25°, 18 hr] (allyl-MgBr) 78%

HN

[^{t}BuOCON$_3$, aq THF; MgO, 50°, 18 hr]

^{t}BuOC(=O)-N

[1. O_3, MeOH, -78°; 2. Diox-H_2O, 80°; 3. Ag_2O, KOH; 4. $CH(OMe)_3$, MeOH, HCl]

61%

HN

MeO_2C CO_2Me

[TMSCl, Na-K; C_6H_6, 25°]

TMS-N

TMSO OTMS

[Br_2, CH_2Cl_2, -78°]

$H_2\overset{+}{N}$

O O$^-$

[CH_2N_2, EtOH; 0°, CH_2Cl_2]

HN

O OMe

(A)

60% for three steps

Preparation of aromatic part (B):

OH → Cl → CN → CO_2H —[LAH]→ 96%

—[NaH, THF → NsCl]→ 92%

ONs

(B)

Ns=$-SO_2-C_6H_4-NO_2$

Preparation of cephalotaxine:

ONs + HN ... O OMe —[R_3N, MeCN, 50°, 15 hr]→ 80%

(B) (A)

$(C_6H_5)_3C^-K^+$
DME, 50°, 2 hr

15%

H^+

OMe

OMe

Dibal, C_6H_6
25°, 1 hr

HO H OMe

Cephalotaxinone

Cephalotaxine

Improved preparation of cephalotaxinone:[7]

CO_2H

LAH

OH

I_2, CF_3CO_2Ag

OH

I

NsCl, Py
Et_2O, 25°

ONs

I

[(A), R_3N, MeCN, 25°] →
61%

$Ph_3C^-Li^+$, THF → $Ni(COD)_2$, 25° (30%)[8]

KO^tBu, $h\nu$ (94%)[9]

KNH_2, NH_3 → Na-K (45%)[10]

Cephalotaxinone

8. This reaction takes advantage of the unique reactivity of aryl halides in oxidative-addition to low-valent transition metals, providing a means of activating aryl halides toward carbon-carbon coupling, as exemplified below:

PhBr + NiL_4 → PhNi(L)(L)-Br —$PhCOCH_2^-$→ [$PhCOCH_2$-Ni(L)(L)-Ph] → Ph-CH_2COPh

L=Ph_3; NiL_4=tetrakis(triphenylphosphine)nickel(0)

9. This photostimulated aromatic $S_{RN}1$ reaction is typified by the reaction of halobenzenes and related substrates with enolate anions under irradiation, a reaction which does not otherwise proceed in the dark, R. A. Rossi and J. F. Bunnett, J. Org. Chem., **38**, 1407 (1973).

$$C_6H_5{-}X + CH_3COCH_2^- \xrightarrow[\text{liq } NH_3]{h\nu} C_6H_5{-}CH_2COCH_3 + X^-$$

10. Cf. R. A. Rossi and J. F. Bunnett, J. Amer. Chem. Soc., **94**, 683 (1972).

CHELIDONINE

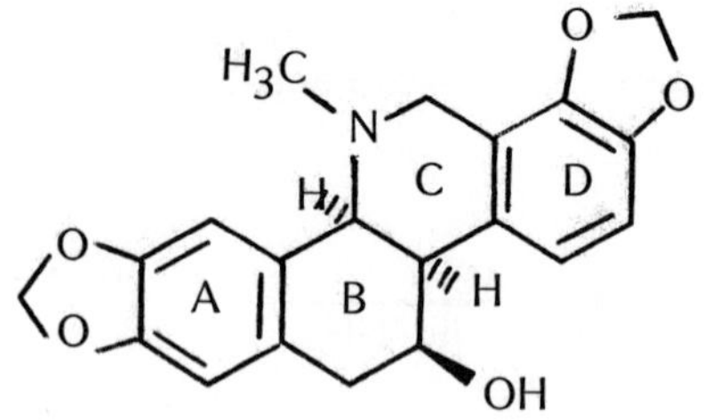
H3C
N
H
C
D
A
B
H
OH
O
O
O
O

The synthesis of *dl*-chelidonine[1] exploits a versatile approach to annelated heterocycles based on intramolecular cycloaddition of o-quinodimethanes, which are readily derived by thermolysis of benzocyclobutenes (B→C).[2]

Preparation of ring D component:

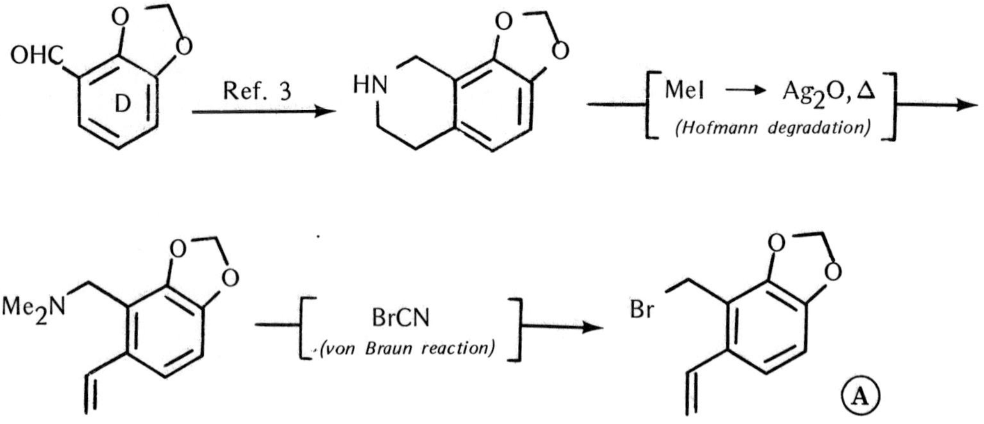
OHC
D
Ref. 3
HN
MeI → Ag2O, Δ
(Hofmann degradation)
Me2N
BrCN
(von Braun reaction)
Br
A

---

1. W. Oppolzer and K. Keller, J. Amer. Chem. Soc., **93**, 3836 (1971).

2. Benzocyclobutenes such as (i) readily undergo thermal rearrangements to o-quinodimethanes (ii), which can further participate in intramolecular cycloadditions to give heterocyclic systems (iii), W. Oppolzer, J. Amer. Chem. Soc., **93**, 3836 (1971).

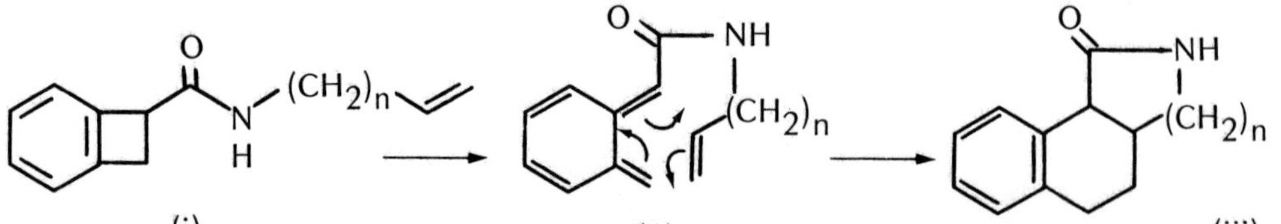
O
N
H
(CH2)n
NH
O

(i) (ii) (iii)

3. Cf. W. J. Gensler, K. T. Shamsundar and M. Marburg, J. Org. Chem., **33**, 2861 (1968).

Preparation of chelidonine:

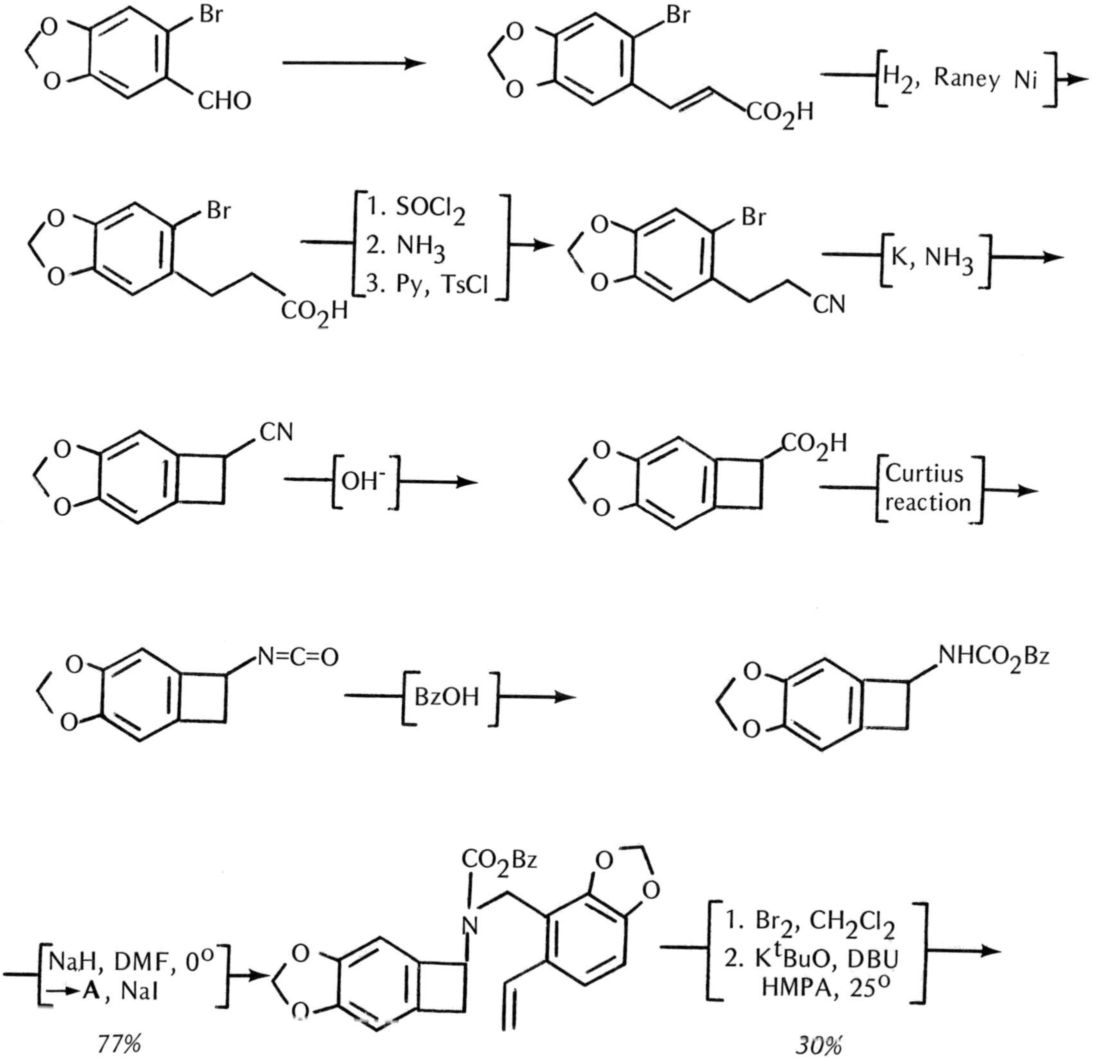
Br
CHO
Br
CO2H
H2, Raney Ni
Br
CO2H
1. SOCl2
2. NH3
3. Py, TsCl
Br
CN
K, NH3
CN
OH-
CO2H
Curtius reaction
N=C=O
BzOH
NHCO2Bz
NaH, DMF, 0°
→A, NaI
77%
CO2Bz
N
1. Br2, CH2Cl2
2. KtBuO, DBU
HMPA, 25°
30%

B

[Xy, 120°]
73%

C

[B_2H_6, THF ⟶ H_2O_2]
68%

4b
Note cis addition

D[4]

[CrO_3, Me_2CO]
32%

[$NaBH_4$, MeOH-Diox, 0°]

---

4. C-4b epimer separated by chromatography.

CO_2Bz ... H ... H ... OH

[H_2, Pd-C, EtOH] →
90%

H ... N ... H ... H ... OH

dl-Norchelidonine

[N-Methylation] →

H_3C ... N ... H ... H ... OH

dl-Chelidonine

α-COPAENE

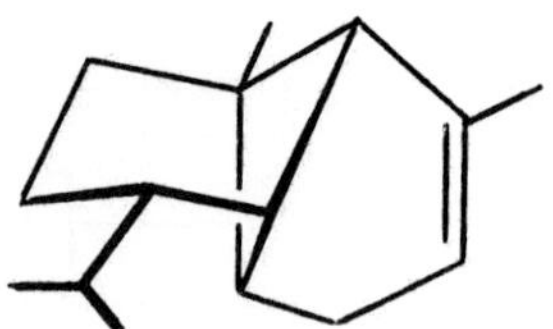

The unusual tricyclic ring system of sesquiterpenes copaene and ylangene offers an interesting synthetic challenge.[1] The relatively simple and direct approach to α-copaene outlined below[2] involves construction of a *cis*-decalin structure (A) using the Diels-Alder reaction and subsequent generation of the tricyclic system by an internal S_N2 cyclization establishing the four-membered ring.

[150°, 24 hr][3]

25%

---

1. For an earlier synthesis of α-copaene, see C. H. Heathcock, J. Amer. Chem. Soc., **88**, 4110 (1966); C. H. Heathcock, R. A. Badger and J. W. Patterson, J. Amer. Chem. Soc., **89**, 4113 (1967); Art Org. Synth., 117 (1970).

2. E. J. Corey and D. S. Watt, J. Amer. Chem. Soc., **95**, 2303 (1973).

3. The remarkable selectivity of the cycloaddition reaction appears to be due to intervention of the enol form of the cyclohexenone as the true dienophilic partner:

(A)

[ethylene glycol, TsOH]

[m-$ClC_6H_3CO_3H$, CH_2Cl_2] 72%

[LAH, Et_2O, 0°] 87%

CO_2Me H OH O

[Py, $ClCO_2Et$, 0°]* 75%

CH_2OH

*Note greater reactivity of primary allylic alcohol vs. secondary alcohol.

CH_2OCO_2Et

[Ac_2O, DMSO, 24 hr] 73%

[Li, NH_3, EtOH] 92%

[TsCl, Py, 0°]

OTs

[HCl, AcOH, THF] →

OTs

H

[$MeS(O^-)=CH_2(Na^+)$
DMSO, 75°] →
36%

H

OTs

[$(EtO)_2P(O)$-CH(CH_3)-CN
NaH, DME] →
94%

(B)

CN

[Mg, MeOH
→ H_3O^+] →
76%

CN

[Dibal, Hex, -78°] →

CHO

[1. $TsNHNH_2$,THF*
2. LAH, Diox] →
53%
**(-CHO → -CH=NNHTs)*

+

α-Copaene
(25%)

α-Ylangene
(25%)

CORYNANTHEINE

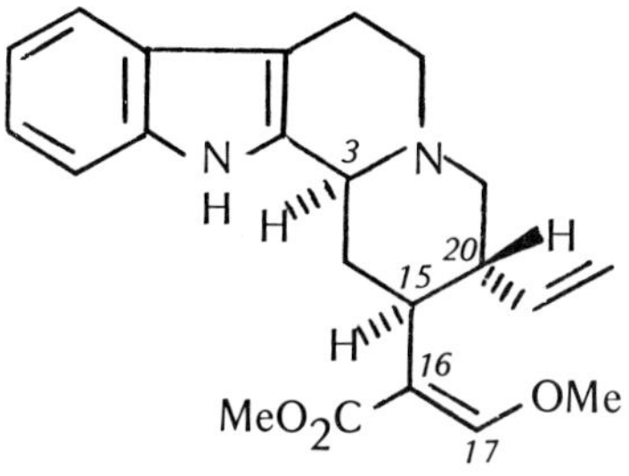

While the four centers of dissymmetry in corynantheine[1] (C-3, C-15, C-17 and C-20) do not pose a serious problem, a major obstacle in the synthesis of corynantheine consists in introduction of the C-20 vinyl group.[2] In the synthesis of corynantheine by Autrey and Scullard[3] this has been elegantly accomplished by a novel sulfur directed Beckmann fragmentation[6] of the pentacyclic *trans* fused D/E ring system in yohimbone. The resulting thioenol ether thus generated serves as a progenitor of the vinyl group.

---

1. The curious history of isolation and characterization of corynantheine has been reviewed by Autrey (ref. 3) and by J. E. Saxton, The Alkaloids, **7**, 37 (1960).

2. The relatively simpler objective of synthesizing dihydrocorynantheine has been accomplished earlier: C. Sźantay and L. Toke, Acta Chim. Hung., **34**, 249 (1963); J. A. Weisbach, J. L. Kirkpatrick, K. R. Williams, E. L. Anderson, N. C. Yim and B. Douglas, Tetrahedron Lett., 3457 (1965); E. E. van Tamelen and J. B. Hester, J. Amer. Chem. Soc., **81**, 3805 (1959); **91**, 7342 (1969).

3. R. L. Autrey and P. W. Scullard, J. Amer. Chem. Soc., **90**, 4917 (1968).

HCO$_2$Et, NaOMe
C$_6$H$_6$, rt
98%

CHOH

Yohimbone[4]

TsSCH$_3$, NaOMe[5]
95% EtOH
62%

17
18
SMe

NH$_2$OH, OH$^-$
98%

---

4. For a total synthesis of yohimbone, see G. A. Swan, J. Chem. Soc., 1534 (1964).

5. The desired 18(S)-methioxyyohimban-17-one was accompanied by a bismethioxy product, formation of which could be suppressed by increasing proton availability and increasing concentration of the reaction mixture.

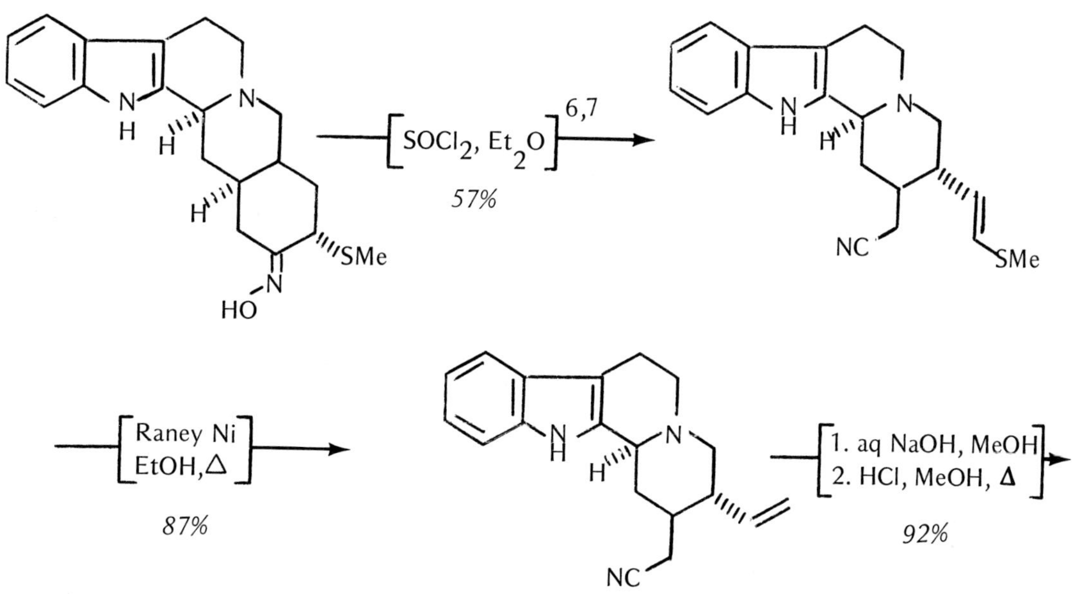

[$SOCl_2$, Et_2O] [6,7] 57%

[Raney Ni, EtOH,Δ] 87%

[1. aq NaOH, MeOH; 2. HCl, MeOH, Δ] 92%

---

6. The novel Beckmann fragmentation of an α-methioxyketoxime was developed by Autrey to provide a method for unsymmetrical cleavage of the bond between a carbonyl and an adjacent methylene such that the termini of the cleaved bond are left in different oxidation states to permit their selective modification, R. L. Autrey and P. W. Scullard, J. Amer. Chem. Soc., **90**, 4924 (1968).

7. While any substituent on the carbon atom α to the oxime is expected to stabilize the carbonium ion at the site of fragmentation [R. K. Hill, J. Org. Chem., **27**, 29 (1962)], the fragmentation governed by sulfur has been rationalized as proceeding more likely by the relatively strain-free 1,2-thiazetine intermediate (i). Loss of a proton β to the sulfur leads to the fragmented product.

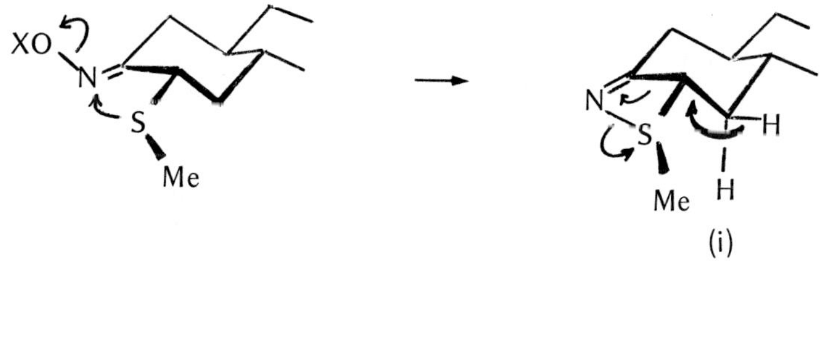

(i)

—[HCO_2Me, $Ph_3C^-Na^+$, THF-Et_2O]→

MeO_2C CHO + MeO_2C OH

—[MeOH, HCl, Δ]→

MeO_2C OMe

Corynantheine

In a biogenetically patterned synthesis of *dl*-corynantheine carried out by van Tamelen and Wright,[8] the essential features of the alkaloid have been assembled at the very outset and incorporated into the key tetracyclic intermediate (A) previously utilized for the synthesis of ajmalicine.[9]

---

8. E. E. van Tamelen and I. G. Wright, Tetrahedron Lett., 295 (1964); J. Amer. Chem. Soc., **91**, 7333 (1969).

9. E. E. van Tamelen and C. Placeway, J. Amer. Chem. Soc., **83**, 2594 (1961); E. E. van Tamelen, C. Placeway, I. G. Wright and G. P. Schiemenz, J. Amer. Chem. Soc., **91**, 7359 (1969).

Preparation of ketotriester:[9]

CO_2Me CO_2Me + CO_2Me —[NaOMe]→ CO_2Me CO_2Me CO_2Me

Preparation of corynantheine:[8]

N H + NH_2 CO_2Me CO_2Me CO_2Me —[HCHO, ^{t}BuOH]→ N H O N CO_2Me CO_2Me

—[1. $POCl_3$ 2. H_2, Pd-C, EtOH]→ N H H N H CO_2Me CO_2Me (A) —[dil HCl, Δ]→

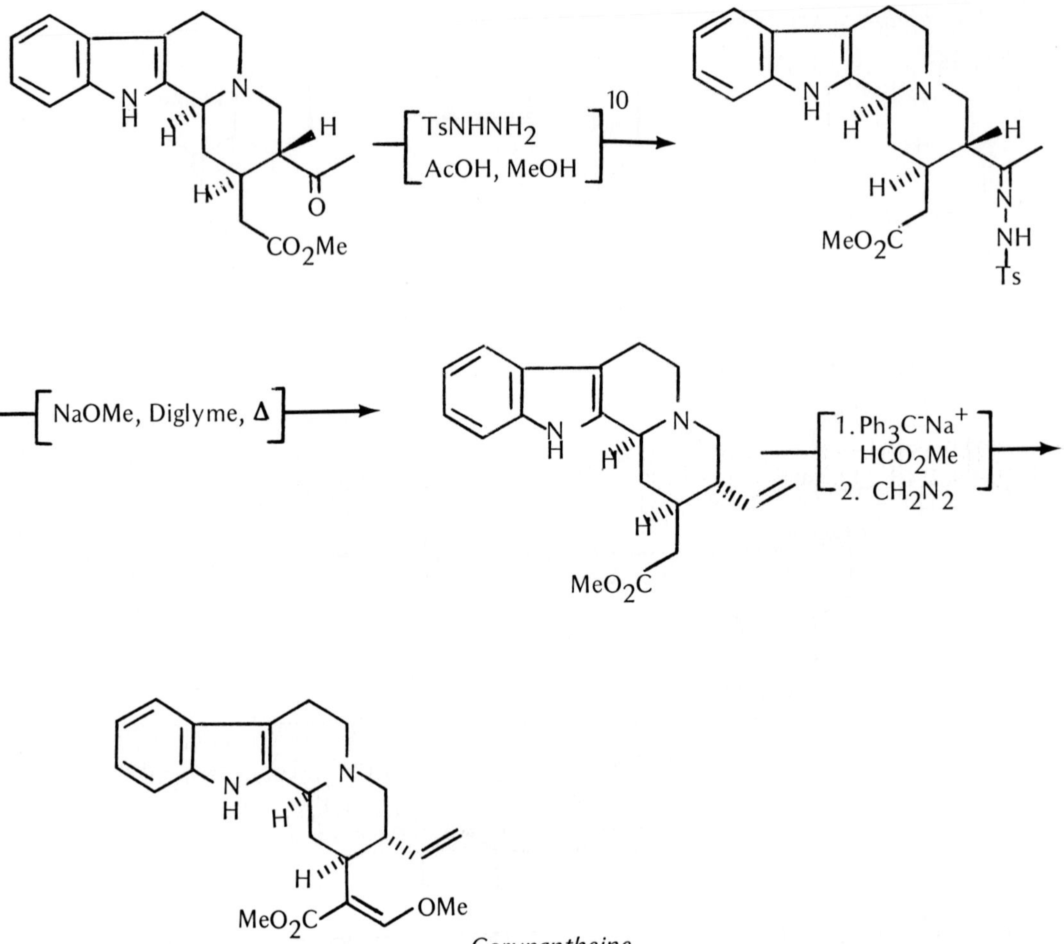

$TsNHNH_2$, AcOH, MeOH [10]

NaOMe, Diglyme, Δ

1. $Ph_3C^-Na^+$, HCO_2Me
2. CH_2N_2

Corynantheine

10. The mixture of tosyl hydrazones could be separated into pure *cis* and *trans* compounds. The *trans* isomer was transformed into corynantheine.

CYCLOARTENOL

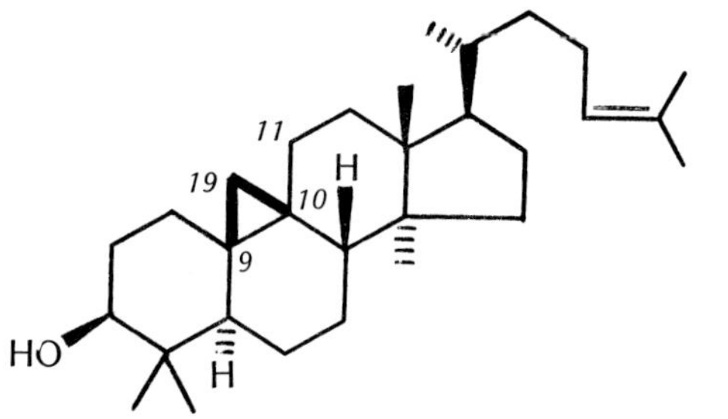

Cycloartenol is the parent compound of a family of triterpenoids characterized by a 9 ,10—cyclopropane ring, and has been suggested as an important intermediate in the biosynthesis of tetracyclic triterpenoids and steroids in plants. A key feature of the synthesis of cycloartenol by Barton *et al.*[1] is functionalization of the unactivated C-19 methyl group in lanosterol[2] by photolysis of an 11β-nitrite (A),[3] followed by an intramolecular alkylation for closure of the cyclopropane ring.

---

1. D. H. R. Barton, D. Ranganathan (neé, Kumari), P. Welzel, L. J. Danks and J. F. McGhie, Chem. Commun., 643 (1968); J. Chem. Soc. C., 332 (1969).

2. The total synthesis of lanosterol has been accomplished earlier by R. B. Woodward, A. A. Patchett, D. H. R. Barton, D. A. J. Ives and R. B. Kelly, J. Chem. Soc., 1131 (1957). See also Art Org. Synth., 230 (1970).

3. The photolysis of nitrite esters for intramolecular hydrogen abstraction reactions, a procedure that has become known as the Barton reaction, is a useful process for functionalization of unactivated sites. The carbon radical thus generated may be quenched by a variety of reagents (i→ii), D. H. R. Barton, J. M. Beaton, L. E. Geller, M. M. Pechet, J. Amer. Chem. Soc., **83**, 4076 (1961); M. Akhtar, D. H. R. Barton and P. G. Sammes, J. Amer. Chem. Soc., **87**, 4601 (1965).

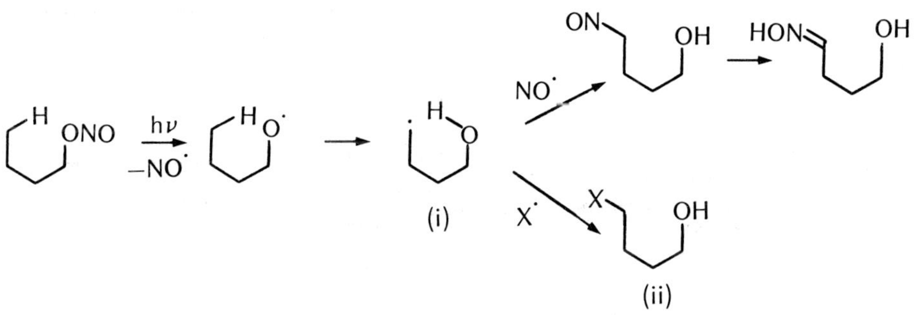

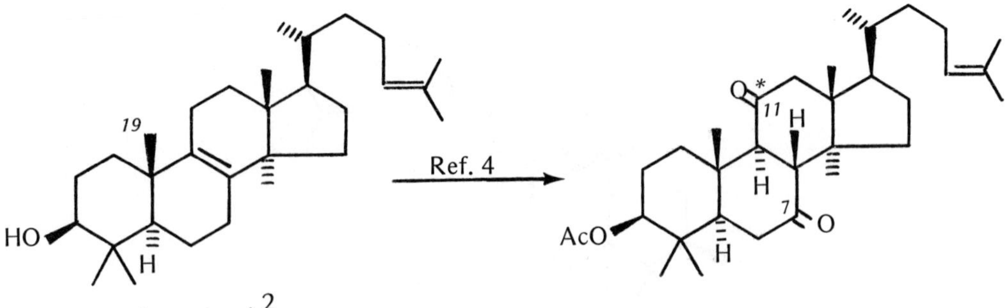

19

HO H

Lanosterol 2

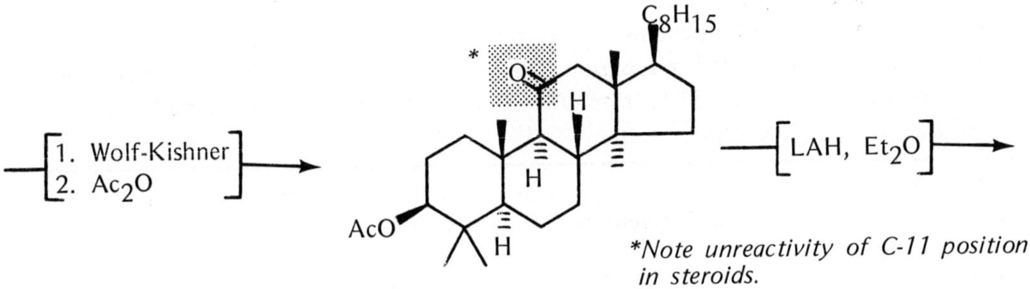

Ref. 4

AcO H H 11 H * O 7 O

[1. Wolf-Kishner 2. Ac_2O]

C_8H_{15}

* O H H AcO H

[LAH, Et_2O]

*Note unreactivity of C-11 position in steroids.

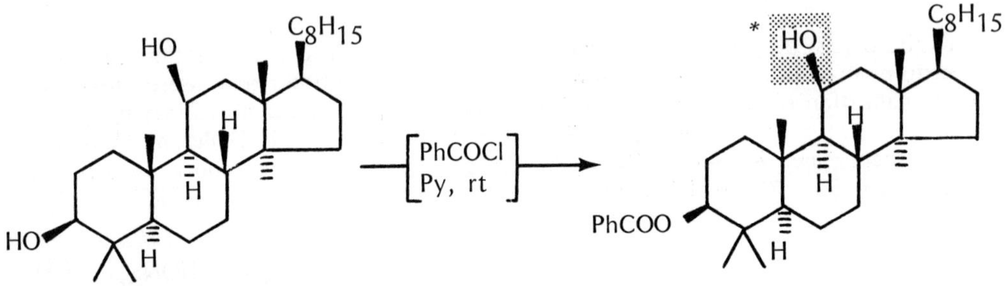

HO C_8H_{15} H H HO H

[PhCOCl Py, rt]

* HO C_8H_{15} H H PhCOO H

4. W. Lawrie, F. S. Spring and H. S. Watson, Chem. Ind., 1458 (1956).

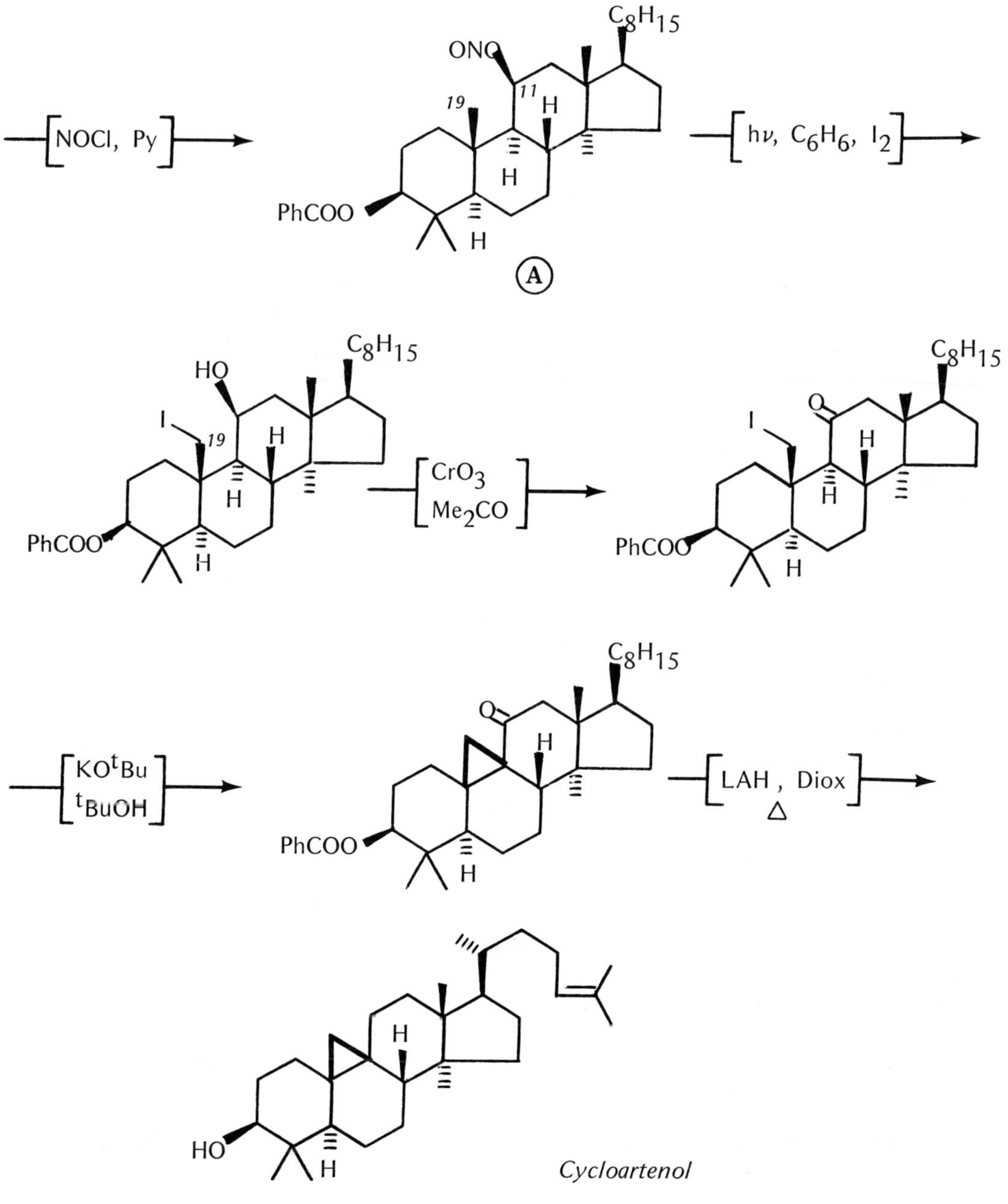

[NOCI, Py] →

C_8H_{15} ONO 19 11 H H PhCOO H

(A)

[hν, C_6H_6, I_2] →

C_8H_{15} HO I 19 H H PhCOO H

[CrO_3, Me_2CO] →

C_8H_{15} O I H H PhCOO H

[KO^tBu, ^tBuOH] →

C_8H_{15} O H PhCOO H

[LAH, Diox, Δ] →

H HO H

Cycloartenol

DENDROBINE

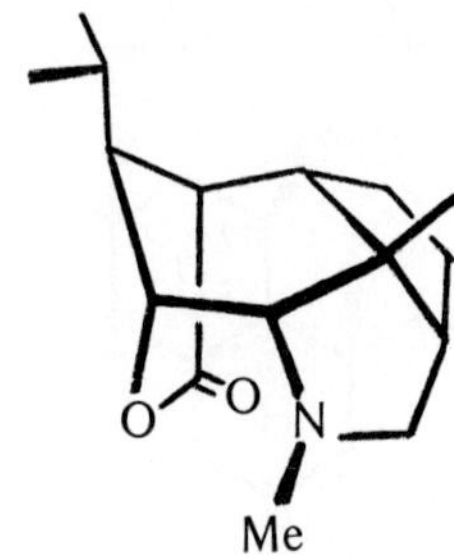

Dendrobine, a congener of the powerful convulsant picrotoxinin, was first isolated from the Chinese tonic 'Chin Shih Hu' in 1932, but its complete structure was not unravelled until 1964.[1] Three groups of workers have reported the synthesis of dedrobine since that time.[2-4] The synthesis outlined below was developed by Kende and his group at Rochester[4], and is based on the *cis*-hydroindan nucleus (B) derived from the Diels-Alder adduct (A) by cleavage and aldol cyclization.[5] The hydroindan derivative (B) already carrying the isopropyl unit, bears appropriate functionality for elaboration of the N-methylpyrrolidine ring and the γ-lactone moiety.

OAc, AcO, OAc → [1. Saponify; 2. $FeCl_3$] 100% → O, HO, O

→ [diene, EtOH, 110°] 95% → O, H, HO, O (A) → [1. MeI, K_2CO_3, Me_2CO; 2. OsO_4, $Ba(ClO_3)_2$, aq Diox] →

1. Review: L. A. Porter, Chem. Rev., **67**, 441 (1967).

2. Y. Inubushi, T. Kikuchi, T. Ibuka, K. Tanaka, I. Saji and K. Tokane, Chem. Commun., 1252 (1972); Chem. Pharm. Bull., **22**, 349 (1974).

3. K. Yamada, M. Suzuki, Y. Hayakawa, K. Aoki, H. Nakamura, H. Nagase and Y. Hirata, J. Amer. Chem. Soc., **94**, 8278 (1972).

4. A. S. Kende, T. J. Bentley, R. A. Mader and D. Ridge, J. Amer. Chem. Soc., **96**, 4332 (1974).

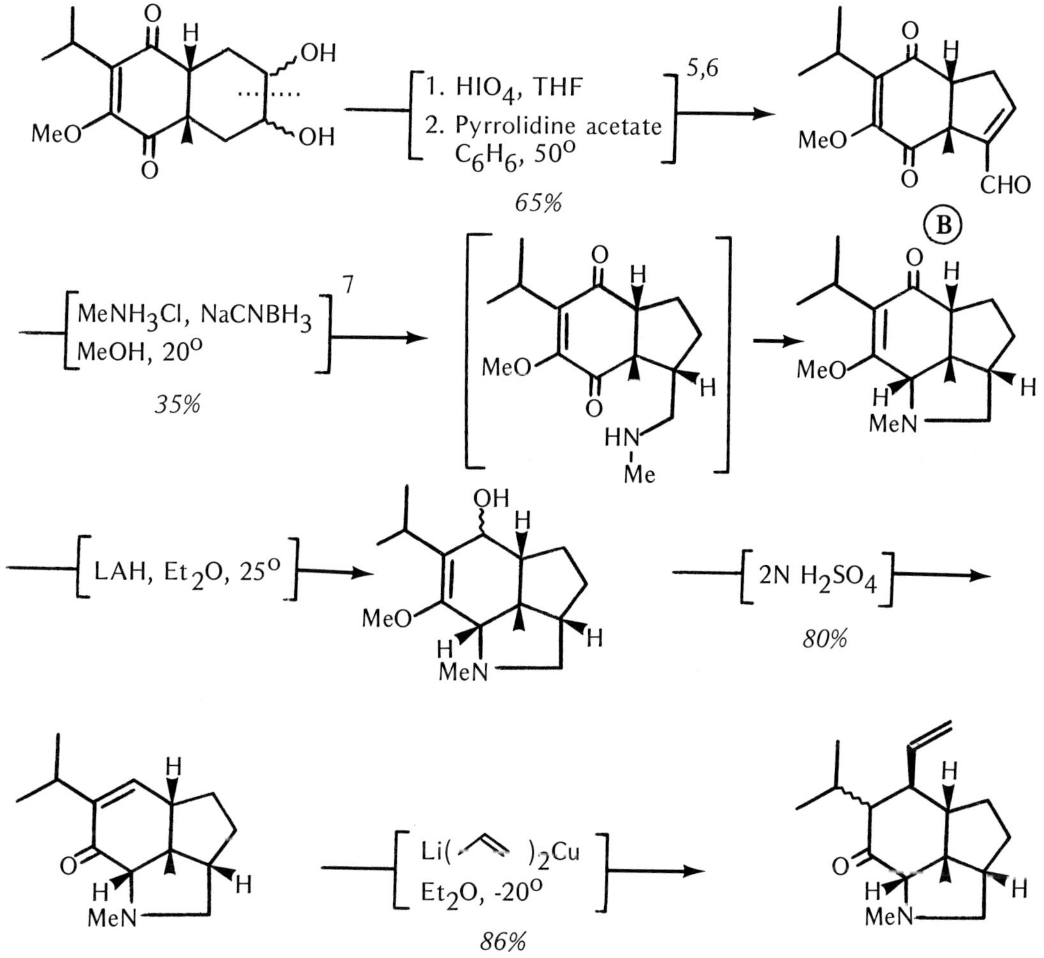

1. HIO_4, THF
2. Pyrrolidine acetate C_6H_6, 50° [5,6]

65%

(B)

$MeNH_3Cl$, $NaCNBH_3$
MeOH, 20° [7]

35%

LAH, Et_2O, 25°

2N H_2SO_4

80%

$Li(CH{=}CH_2)_2Cu$
Et_2O, -20°

86%

---

5. For a related approach to the dendrobine skeleton, see K. Yamamoto, I. Kawaski and T. Kaneko, Tetrahedron Lett., 4859 (1970).

6. An equal amount of the isomer in which cyclization had occurred in the wrong direction was formed and separated on Florisil.

7. Reductive amination: R. F. Borche, M. Bernstein and H. D. Durst, J. Amer. Chem. Soc., **93**, 2897 (1971).

1. RuO_4, 60% AcOH
2. CH_2N_2

49%

CO_2Me H O H MeN H

NaOMe, MeOH, Δ

43%

CO_2Me H O H MeN H

$NaBH_4$, ^iPrOH, 25°
3 days

CO_2Me H HO H MeN H

Chromatography on silica

O H O H MeN H

Dendrobine

ELEMOL

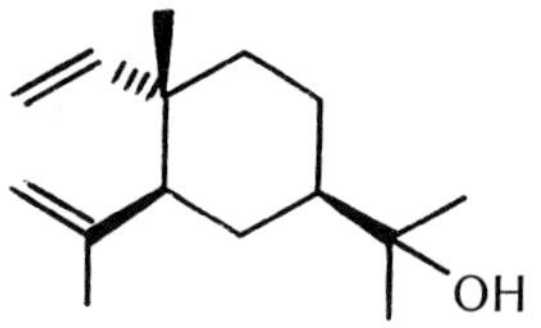

Corey's synthesis of the sesquiterpene elemol[1] illustrates a novel method for construction of 1,2-divinylcyclohexanes involving the reaction of acyclic allylic dibromides and nickel carbonyl.

Preparation of malonate:

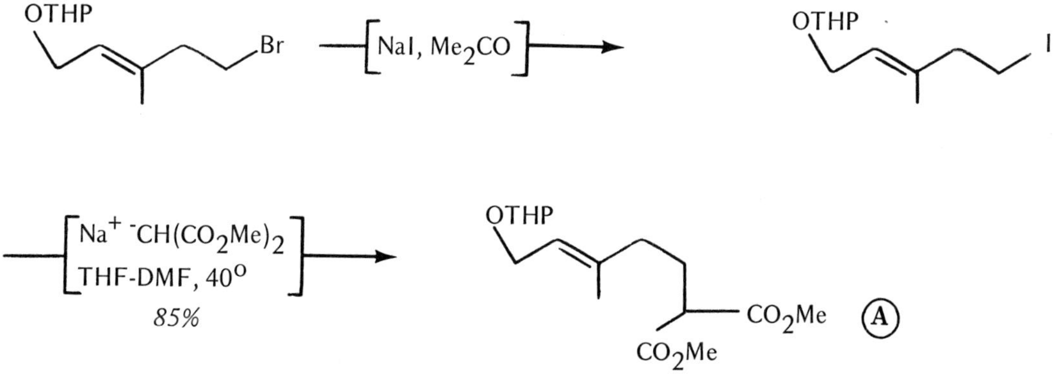

OTHP ... Br —[NaI, Me_2CO]→ OTHP ... I

—[Na^+ $^-CH(CO_2Me)_2$ THF-DMF, 40°]→ *85%* OTHP ... CO_2Me CO_2Me (A)

1. E. J. Corey and E. A. Broger, Tetrahedron Lett., 1779 (1969).

2. For a review on the application of π-allylmetal derivatives in organic synthesis, see R. Baker, Chem. Rev., **73**, 487 (1973).

3. The formation of cycloolefins from allylic dihalides (i) and nickel carbonyl has previously been shown to provide an unusually efficient route for the formation of large rings, and is exemplified in the synthesis of humulene (11-membered ring formation) by E. J. Corey and E. Hamanaka, J. Amer. Chem. Soc., **89**, 2758 (1967). However, in the case of 1,10-dibromo-2,8-decadiene (i, n = 4), the predominant course of cyclization was formation of the 6-membered ring structure (ii) by 3,8 coupling rather than the expected 1,5-cyclodecadiene (iii), E. J. Corey and E. K. W. Watt, J. Amer. Chem. Soc., **89**, 2757 (1967).

$BrCH_2CH{=}CH(CH_2)_nCH{=}CH_2Br$ · · · · · · · · (i)

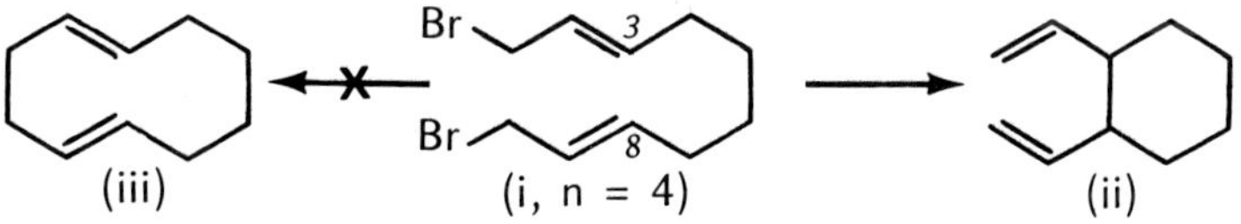

(iii) ←✗— Br 3 Br 8 (i, n = 4) → (ii)

Preparation of bromide:

Cl, OAc — [o-TolCO$_2^-$Me$_3$(Bz)N$^+$, EtOH, 80°] → OCOTol, OAc

— [NaOH, aq EtOH, 25°] → OCOTol, OH — [PBr$_3$, Et$_2$O, 0°] →

OCOTol, Br (B)

Preparation of elemol:

A* + B — [THF-DMF, 80°] → THPO, TolCO (O), CO$_2$Me, CO$_2$Me

*As sodio derivative

— [KOH, MeOH, 25°] → THPO, HO, CO$_2$Me, CO$_2$Me — [DHP, Diox, TsOH] →

THPO, THPO, CO$_2$Me, CO$_2$Me — [(^{i}Bu)$_2$AlH, Pentane-Tol, -70°][4] (79%) → THPO, THPO, CHO, CO$_2$Me

[NaOMe, MeOH, Tol 100°] [4] →

THPO THPO H CO_2Me

[H_3O^+] →

HO HO H CO_2Me

[PBr_3, Et_2O, -5°] →

Br Br H CO_2Me

[$Ni(CO)_4$, N-Methyl pyrrolidone, 50°] [5] → 32%

CO_2Me

[MeMgBr] →

OH

Elemol

---

4. Since hydrolysis and decarboxylation of the malonate function under acidic conditions was precluded due to presence of acid sensitive groups in the molecule, the desired transformation was effected by selective reduction followed by base catalyzed deformylation of the resulting aldehyde.

5. Cyclization resulted in a mixture of isomeric products, which could be separated by chromatographic methods.

The synthesis of β-elemene outlined below[6] is also based on the coupling of 1,10-dibromo-2,8-decadienes under influence of nickel carbonyl.[3] The key allylic dibromide (B) has been built up from the unsaturated ester (A) utilizing a Claisen rearrangement and two successive phosphonate Wittig reactions.[7]

(A) CO_2Et — 1. $(CH_2OH)_2$ 2. LAH → OH — Hg(OAc)$_2$ →

— 195° → CHO — $(EtO)_2\overset{O}{\overset{\|}{P}}-\overset{CH_3}{\underset{Na^+}{C}}-CO_2Et$ →

CO_2Et — H_3O^+ → CO_2Et

6. O. P. Vig, K. L. Matta, J. C. Kapur and B. Vig, J. Indian Chem. Soc., **45**, 973 (1968).

7. J. Boutagy and R. Thomas, Chem. Rev., **74**, 87 (1974).

$\left[\text{(EtO)}_2\overset{\text{O}}{\overset{\|}{\text{P}}}\text{–}\bar{\text{C}}\text{HCO}_2\text{Et}\ \ \text{Na}^+\right]$ →

CO_2Et CO_2Et ≡ EtO_2C EtO_2C

[1. LAH; 2. PBr_3, Et_2O] →

Br Br

(B)

[$Ni(CO)_4$][5] →

β-Elemene
(as mixture of isomers)

ELAEOCARPINE

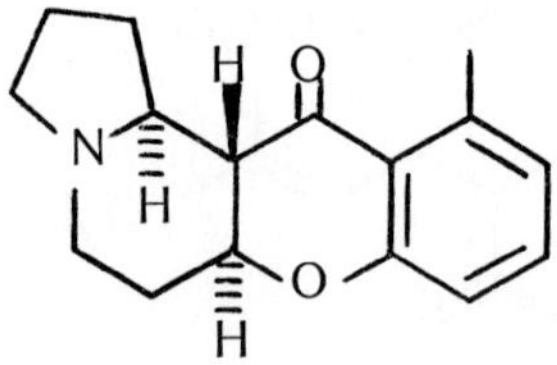

Plants of Elaeocarpus species growing in New Guinea yield a number of alkaloids characterized by the unique chromanoindolizidine skeleton.[1] An efficient two-step synthesis of these alkaloids developed by Onaka[2] is based on the known reaction of enamines with salicylaldehyde[3].

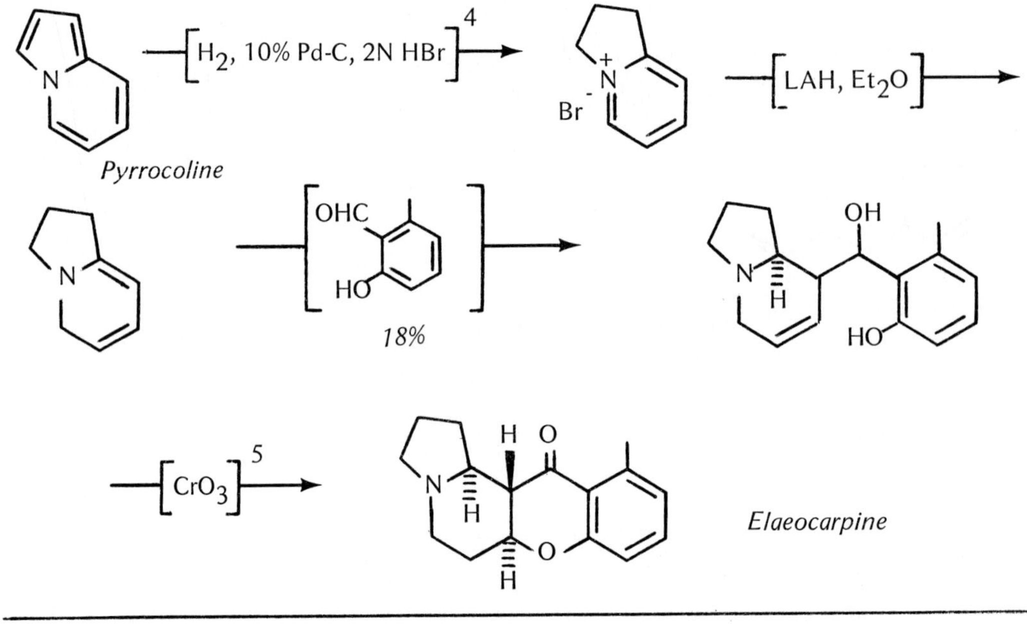

[H_2, 10% Pd-C, 2N HBr][4]

Pyrrocoline

[LAH, Et_2O]

18%

[CrO_3][5]

Elaeocarpine

---

1. S. R. Johns, J. A. Lamberton and A. A. Sioumis, Chem. Commun., 290 (1968); Austral. J. Chem., **22**, 775 (1969).

2. T. Onaka, Tetrahedron Lett., 4395 (1971).

3. L. A. Paquette and H. Stucki, J. Org. Chem., **31**, 1232 (1966).

4. O. G. Lowe and L. C. King, J. Org. Chem., **24**, 1200 (1959).

5. A different synthesis of elaeocarpine has been reported by T. Tanaka and I. Iijima, Tetrahedron Lett., 3963 (1970); Tetrahedron, **29**, 1285 (1973).

ESTRONE

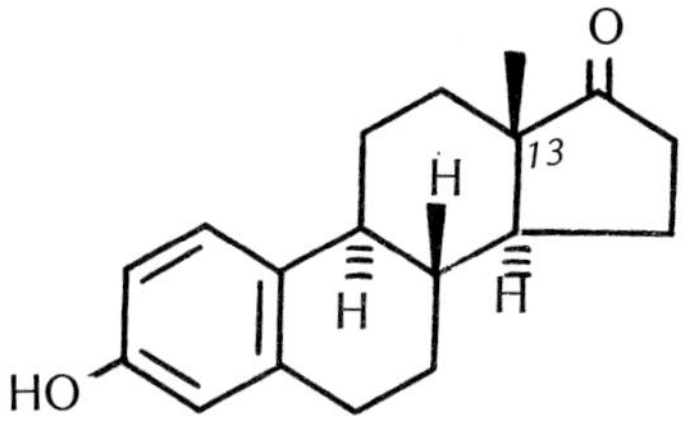

For nearly 25 years the synthesis of estrone has held a special appeal for the organic chemist, and a variety of approaches have been devised towards this natural sex hormone.[1] In recent years interest has focussed on developing asymmetric total syntheses of estrone and estrone intermediates.[2] One such approach, selected from the prodigious and elegant work by the Hoffmann-La Roche group on total synthesis of 19-norsteroids,[3] is portrayed below.[4] The basic scheme utilizes a novel asymmetric induction step involving condensation of the optically active Mannich base (A) with 2-methyl-1,3-cyclopentanedione. The resulting ketol (B) bearing the natural configuration at C-13, can then be readily transformed to the desired 19-norsteroids.

---

1. Estrone is an important intermediate in the production of 19-nor steroids, and is one of the few steroids produced commercially by total synthesis. For a review of the earlier synthetic approaches,see Art Org. Synth., 177 (1970).

2. C. Rufer, E. Schroder and H. Gibian, Annalen, **701**, 206 (1967); R. Bucourt, M. Vignau and J. Weil-Raynal, Comptes Rendus C, **265**, 834 (1967); U. Eder, G. Sauer and R. Wiechert, Angew. Chem. Int. Ed., **10**, 496 (1971).

3. For papers in this series, see G. Saucy, R. Borer and A. Furst, Helv., **54**, 2034 (1971); G. Saucy and R. Borer, Helv., **54**, 2857 (1971); J. W. Scott and G. Saucy, J. Org. Chem., **37**, 1652 (1972); J. W. Scott, R. Borer and G. Saucy; J. Org. Chem., **37**, 1659 (1972); M. Rosenberger, A. J. Duggan and G. Saucy, Helv., **55**, 1333 (1972); M. Rosenberger, A. J. Duggan, R. Borer, R. Muller and G. Saucy, Helv., 2663 (1972); N. Cohen, B. Banner, R. Borer, R. Mueller, R. Yang, M. Rosenberger and G. Saucy, J. Org. Chem., **37**, 3385 (1972).

4. N. Cohen, B. L. Banner, J. F. Bount, T. Tsai and G. Saucy, J. Org. Chem., **38**, 3229 (1973).

OMe

[CH_2O, $Me_2NH{\cdot}HCl$] 80%

OMe, NMe_2, O

[aq KCN] 58%

OMe, CN, O

[$NaBH_4$, EtOH, 0°] 94%

OMe, CN, H, OH

[phthalic anhydride, Py, 100°] 94%

OMe, CN, CO_2H, H, OCO

[1. Resolve, 2. NaOH → HCl]

MeO, O, O, H

[Dibal, Tol, -70°] 77%

OH, MeO, O, H

[vinyl-MgCl, THF, -50°] 99%

MeO, OH, OH, H

[MnO_2, Et_2NH, C_6H_6]* 69%

**Note selective oxidation of allylic alcohol in presence of a benzylic alcohol.*

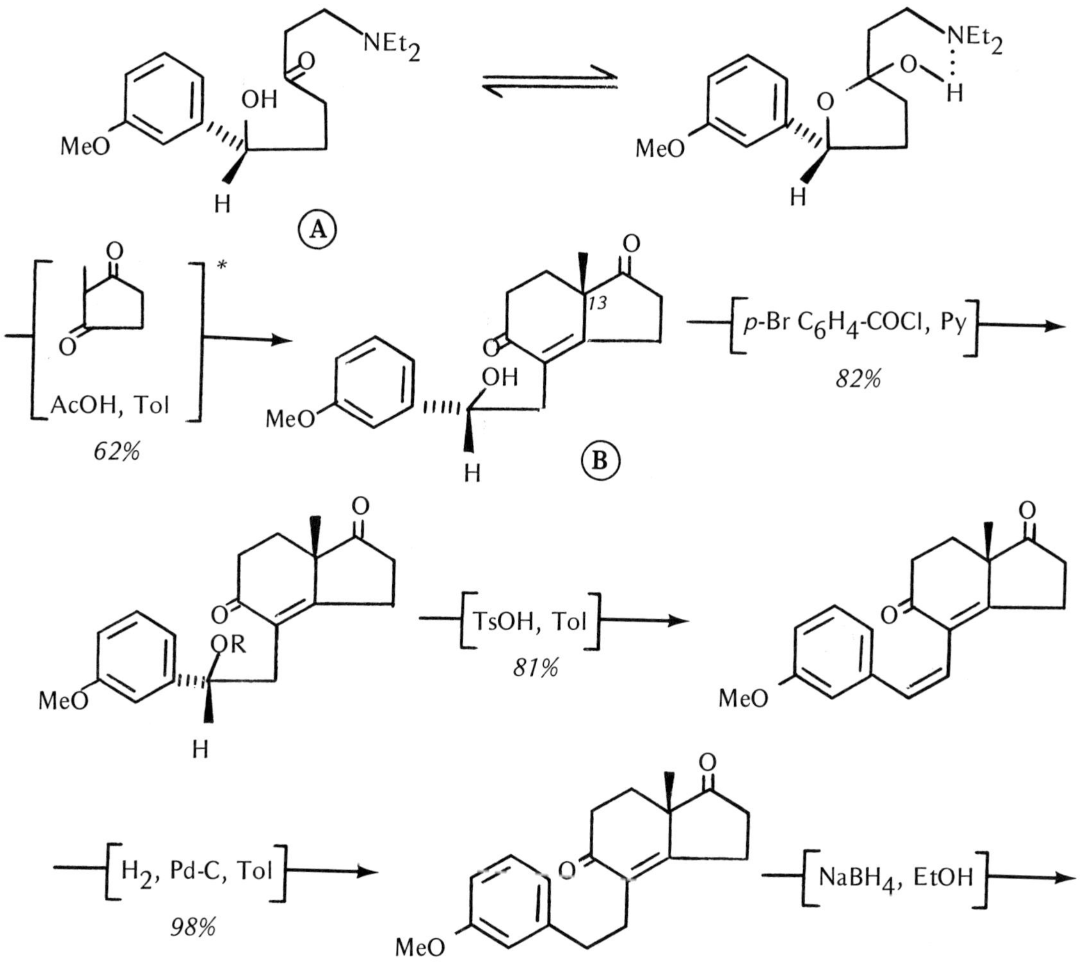

MeO OH NEt$_2$ O H (A) ⇌ MeO O O H NEt$_2$ H

[AcOH, Tol]* 62% → (B)

[p-Br C_6H_4-COCl, Py] 82% →

[TsOH, Tol] 81% →

[H_2, Pd-C, Tol] 98% →

[$NaBH_4$, EtOH] →

---

*When R=alkyl, the reaction furnishes a dienol ether (i), which can be readily transformed into optically active 19-nor steroids or related BCD-tricyclic intermediates (Ref. 3).

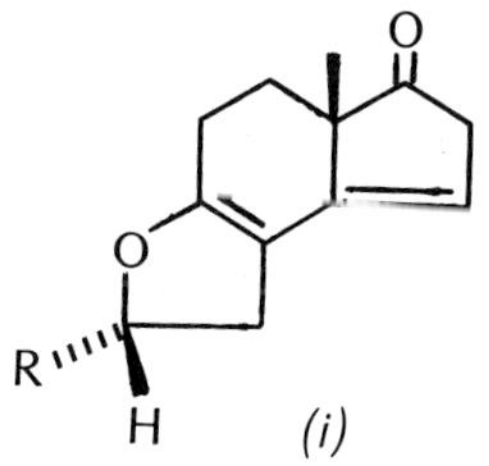

(i)

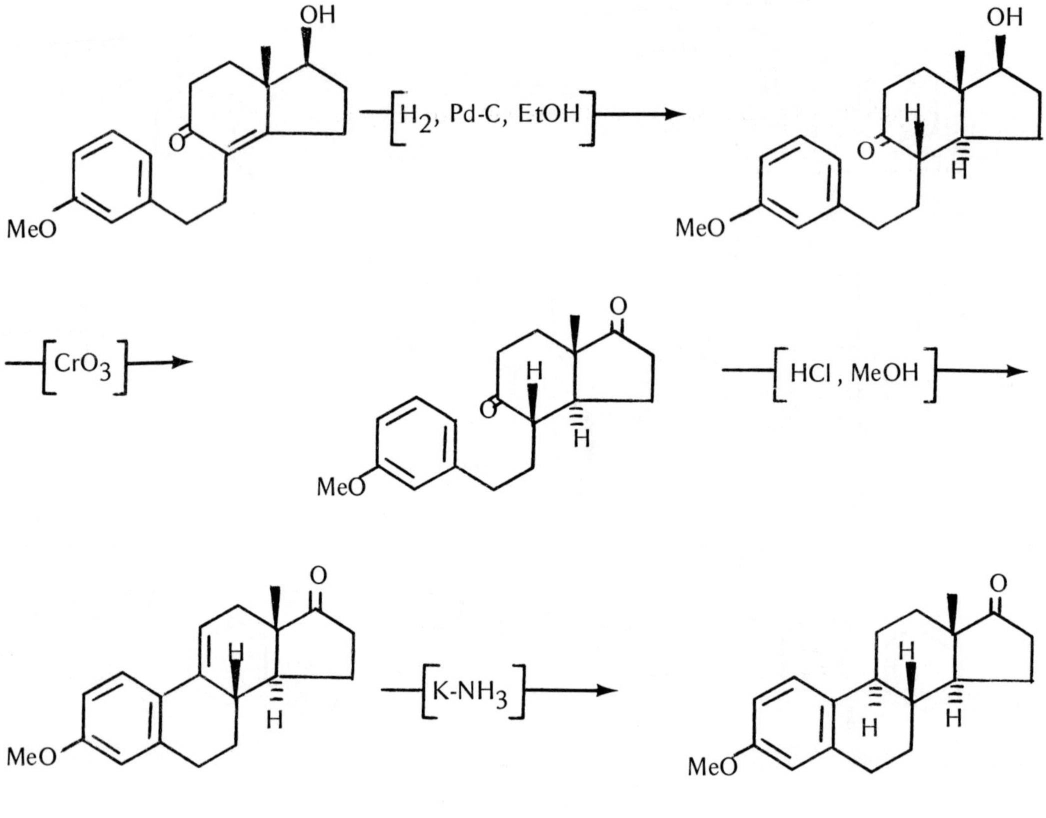

OH

MeO — [H_2, Pd-C, EtOH] → OH, H, O, H, MeO

— [CrO_3] → O, H, H, MeO — [HCl, MeOH] →

O, H, H, MeO — [K-NH_3] → O, H, H, H, MeO

dl-Estrone methyl ether

An ingenious new approach to estrone, by Bartlett and Johnson[5], based on a highly efficient stereospecific cationic olefinic cyclization is outlined below.

Preparation of phosphonium salt (A):

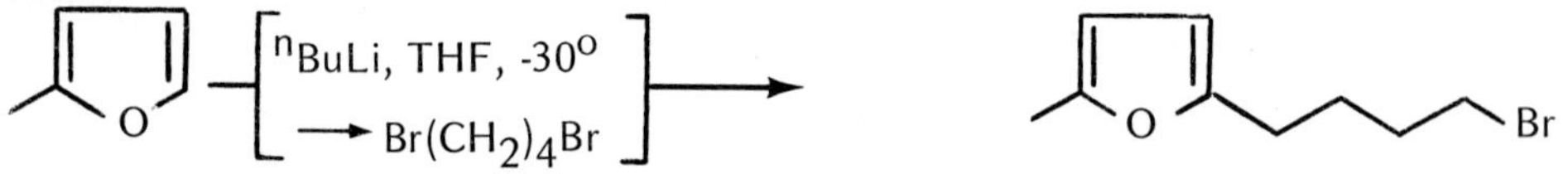

O — [^nBuLi, THF, -30°; → $Br(CH_2)_4Br$] → O, Br

5. P. A. Bartlett and W. S. Johnson, J. Amer. Chem. Soc., **95**, 7501 (1973).

[ethylene glycol, TsOH, C_6H_6, Δ] →

[NaI, MEK, 80°] →

[Ph_3P, C_6H_6, Δ] → (A)

Preparation of estrone:

[Red-Al*, THF, -75°] →

*$*NaAlH_2(OCH_2CH_2OCH_3)_2$

[(A), PhLi, Et_2O, -70°] →

[H_2SO_4, EtOH → 0·1M NaOH] → *83% overall*

[Red-Al, 0°] →

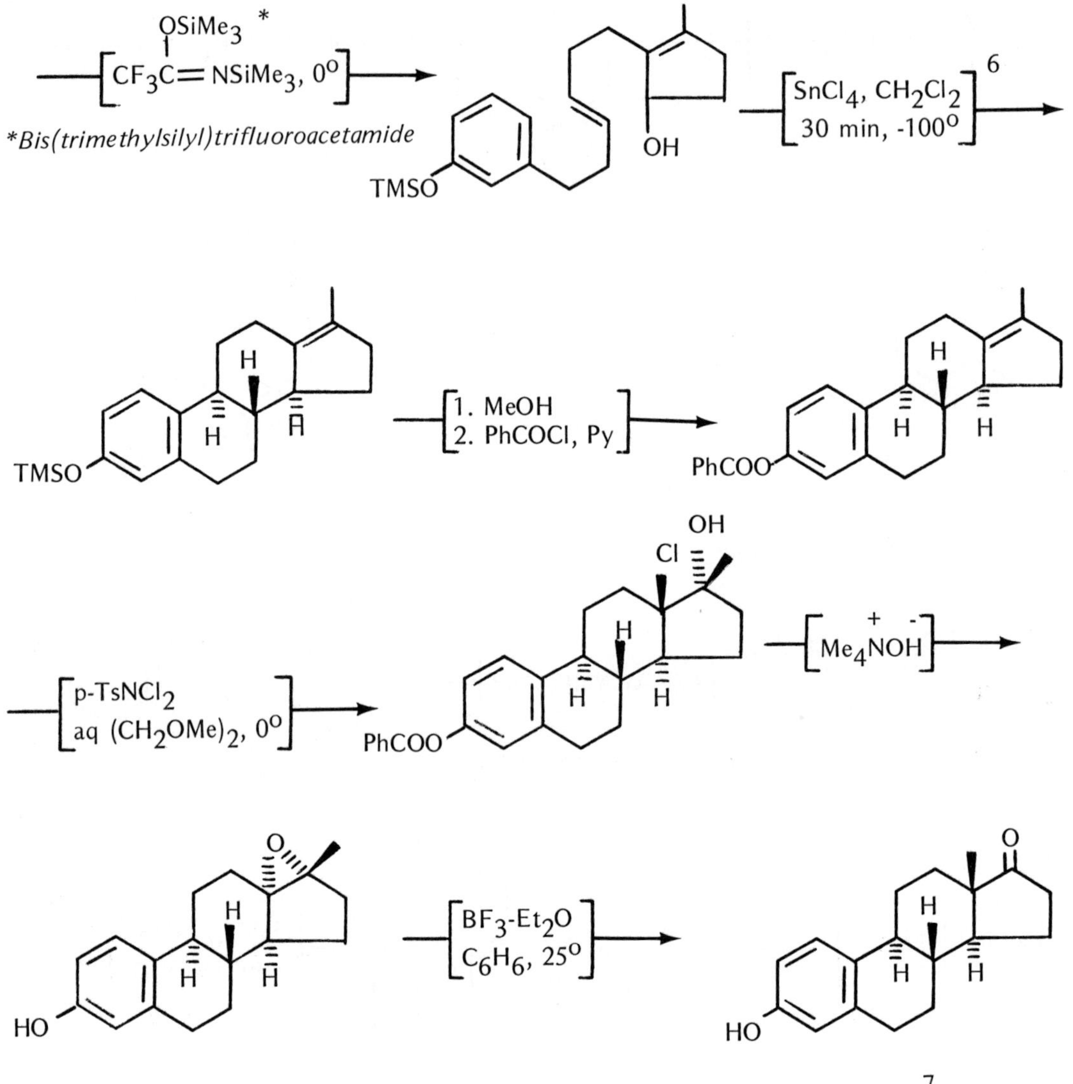

$\left[CF_3\overset{OSiMe_3}{C}{=}NSiMe_3,\ 0^o\right]$ *

*Bis(trimethylsilyl)trifluoroacetamide

TMSO OH

$\left[SnCl_4,\ CH_2Cl_2;\ 30\ min,\ -100^o\right]$ 6

TMSO H H H

[1. MeOH; 2. PhCOCl, Py]

PhCOO H H H

$\left[p\text{-}TsNCl_2;\ aq\ (CH_2OMe)_2,\ 0^o\right]$

PhCOO OH Cl H H H

$\left[Me_4\overset{+}{N}\overset{-}{O}H\right]$

HO O H H H

$\left[BF_3\text{-}Et_2O;\ C_6H_6,\ 25^o\right]$

HO O H H H

dl-Estrone 7

---

6. The cyclization appears to proceed with concerted formation of both rings directly to give the tetracyclic product, P. A. Bartlett, J. I. Brauman, W. S. Johnson and R. A. Volkmann, J. Amer. Chem. Soc., **95**, 7502 (1973).

7. A novel approach to estrone *via* D-homoestrone by S. Danishefsky and A. Nagel [Chem. Commun., 373 (1972)] is based on a novel bisannelation sequence in which the 6-vinyl-2-picoline moiety serves as a masked equivalent of two moles of methyl vinyl ketone (i→ii) [S. Danishefsky and R. Cavanaugh, J. Amer. Chem. Soc., **90**, 520 (1968)].

1. $K^{t}PentO$
2. H_3O^+
70%

1. $NaBH_4$
2. H_2, Pd-C
3. Ketalize
28%

(i)

$Na\text{-}NH_3$
EtOH

NaOH
aq EtOH

(ii)

H_3O^+
90% overall from (i)

1. CrO_3
2. NaOEt
45%

D-Homo-estrone

FUMAGILLIN

H
H
OMe
O
H
O
$CO(CH=CH)_4CO_2H$

Amongst the metabolic products of *Aspergillus fumigatus* is a crystalline antibiotic, fumigallin, having antiparasitic and carcinolytic activity. This molecule, characterized by a structural combination of no less than six asymmetric centers and highly reactive functionality, defied attempts at its synthesis for nearly a decade until it was finally secured by Corey and Snider[1] in 1972. A key feature of this synthesis is the application of Diels-Alder reaction for simultaneous introduction of elements of the eight carbon side-chain (borne on the diene, A) in the molecule and a bromoaldehyde functionality suitable for elaboration to the terminal epoxide.

Preparation of the diene component:

Br + O CO_2Me → → CO_2Me

SeO_2, DME
Δ, 8 hr
41%

CHO CO_2Me

$P(Ph)_3$
THF
84%

CO_2Me

(A)

1. E. J. Corey and B. B. Snider, J. Amer. Chem. Soc., **94**, 2549 (1972).

Preparation of fumagillin:

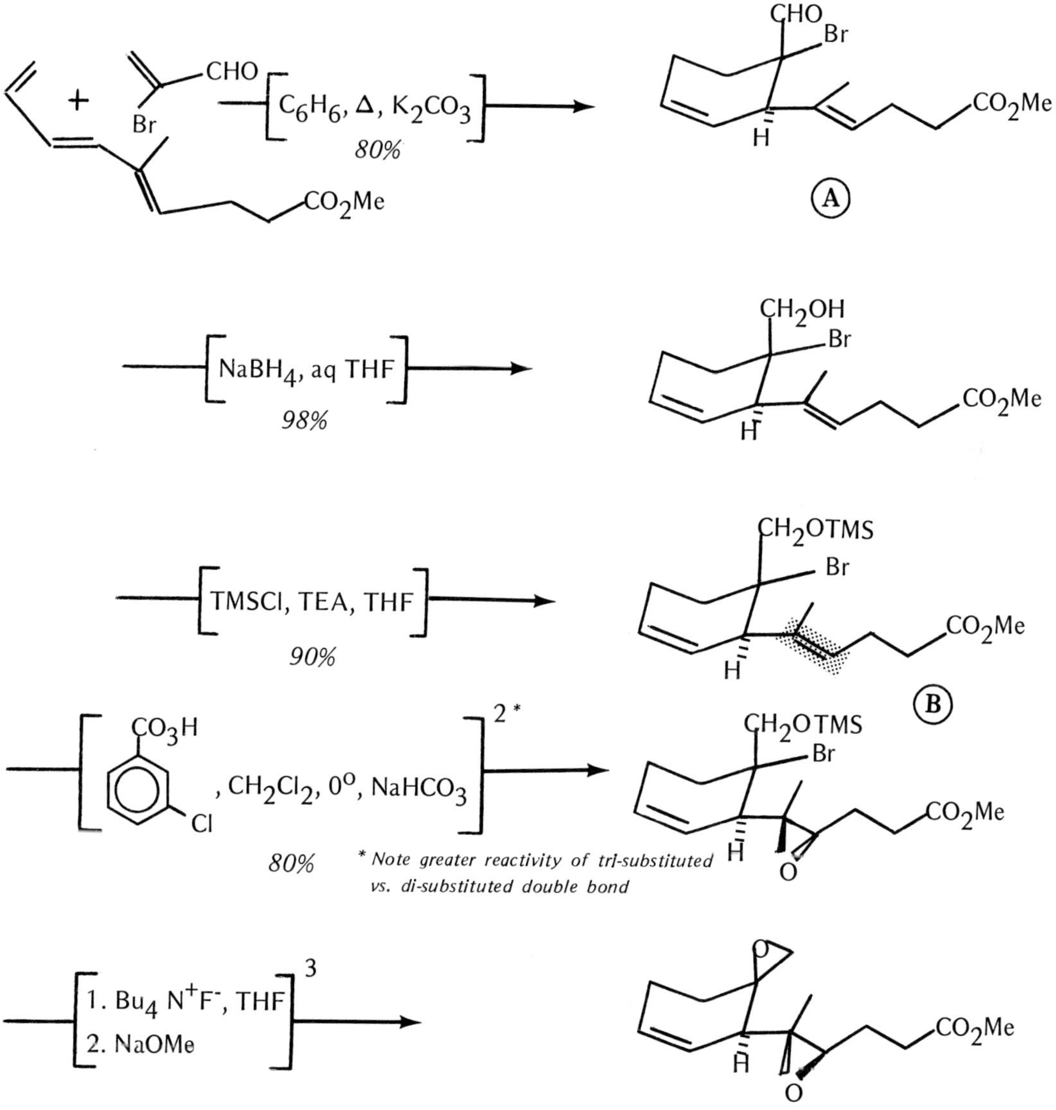

CHO, Br, + , C_6H_6, Δ, K_2CO_3, 80%, CO_2Me, CHO, Br, H, CO_2Me, (A)

$NaBH_4$, aq THF, 98%, CH_2OH, Br, H, CO_2Me

TMSCl, TEA, THF, 90%, CH_2OTMS, Br, H, CO_2Me, (B)

CO_3H, Cl, , CH_2Cl_2, 0°, $NaHCO_3$, 2*, 80%, CH_2OTMS, Br, H, O, CO_2Me

* *Note greater reactivity of tri-substituted vs. di-substituted double bond*

1. $Bu_4N^+F^-$, THF
2. NaOMe

3, O, H, O, CO_2Me

---

2. Stereoselectivity of the epoxidation process was anticipated on basis of the six-carbon chain assuming the most stable conformation as depicted in (B).

3. Hydrolysis of the silyl ether linkage under the usual conditions in protic solvents led to a rapid intramolecular attack on the epoxide ring by the liberated primary hydroxyl group.

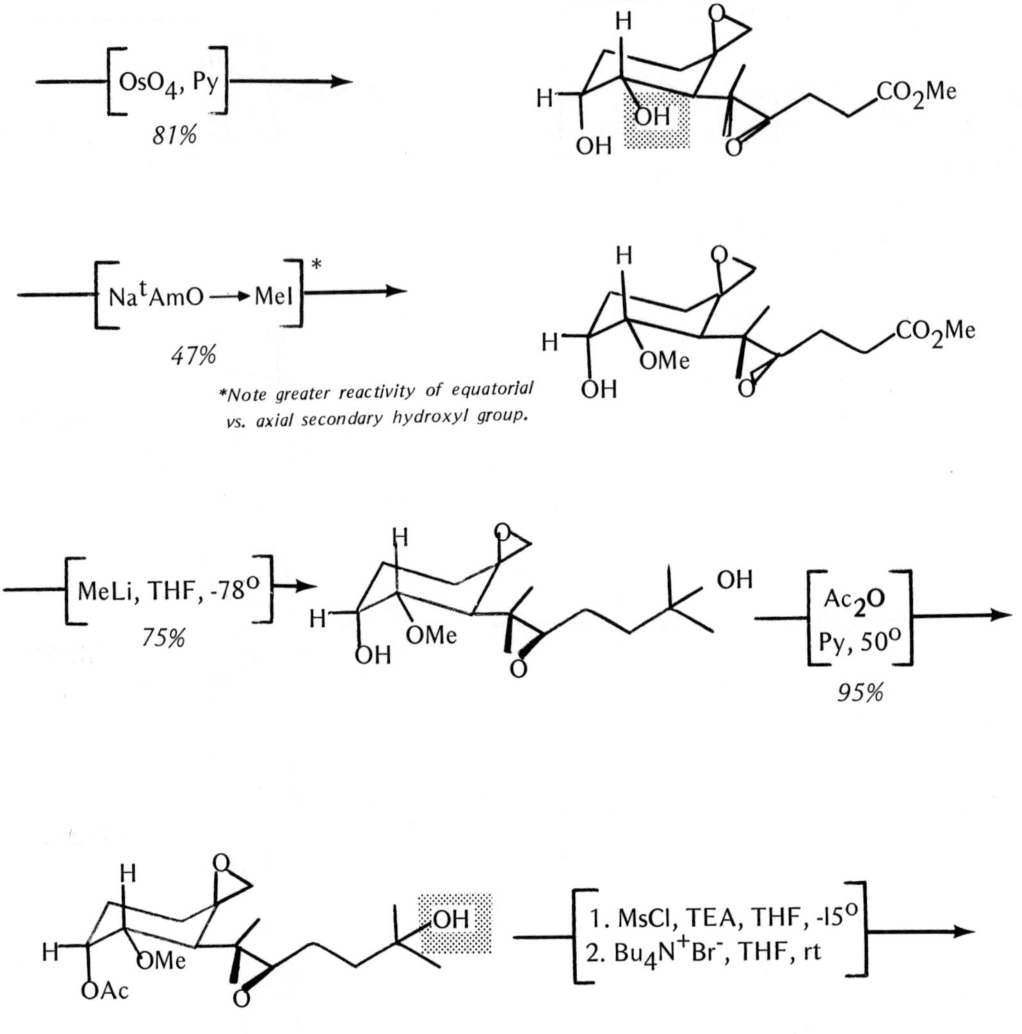

[OsO_4, Py] 81%

H, H, OH, OH, CO_2Me, O, O

[$Na^tAmO \rightarrow MeI$]* 47%

*Note greater reactivity of equatorial vs. axial secondary hydroxyl group.

H, H, OH, OMe, CO_2Me, O, O

[MeLi, THF, -78°] 75%

H, H, OH, OMe, OH, O, O

[Ac_2O Py, 50°] 95%

H, H, OAc, OMe, OH, O, O

[1. MsCl, TEA, THF, -15° 2. $Bu_4N^+Br^-$, THF, rt]

Therefore, hydrolysis was carried out under aprotic conditions using tetrabutylammonium fluoride; the use of fluoride ion being suggested by the high value of Si-F bond energy (ca. 135 KCal/Mol).

H OMe OAc O + H OMe OAc O

—[K_2CO_3, MeOH]→

H OMe OH O

—[1. Separation by HPLC
2. MeLi→ClCO(CH=CH)$_4$CO$_2$H
THF, -78°]→

H OMe O O
CO(CH=CH)$_4$CO$_2$H

Fumagillin

GALANTHAMINE

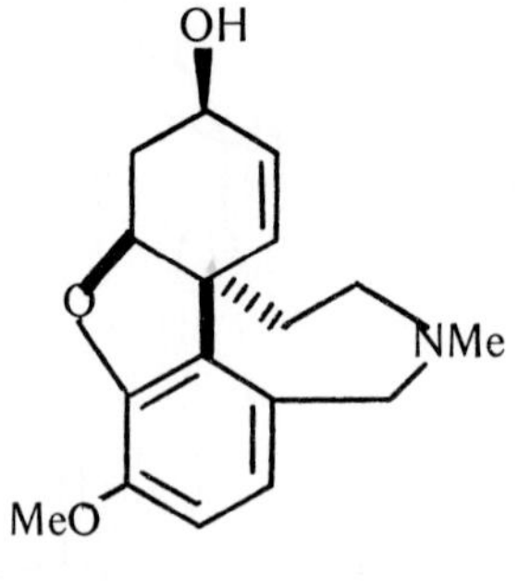

The synthesis of galanthamine through phenolic oxidative coupling along the lines of Barton's biogenetic hypothesis[1] has now been executed in the laboratory several times. In the example[2] presented below a bromine atom has been introduced in one of the phenolic rings of a norbelladine precursor (A) in order to direct coupling *ortho* to the hydroxy group.[3] The resulting dienone (B) readily undergoes intramolecular ether formation to the desired tetracyclic system.

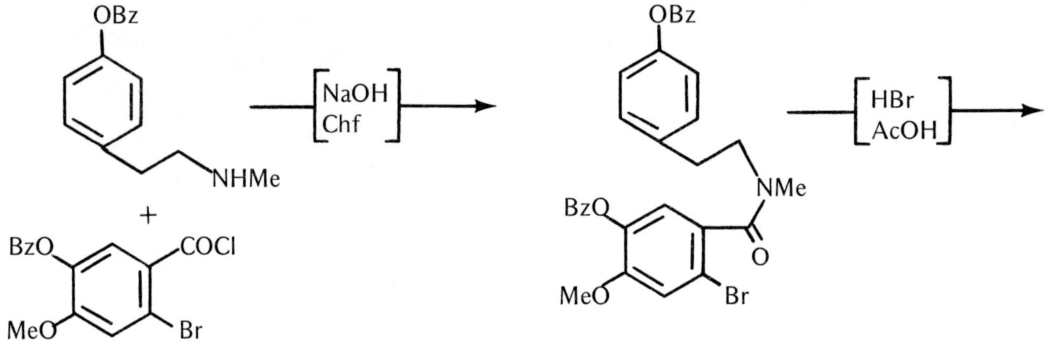

---

1. D. H. R. Barton, Proc. Chem. Soc., 293 (1963); D. H. R. Barton and G. W. Kirby, J. Chem. Soc., 806 (1962).

2. T. Kametani, K. Yamaki, H. Yagi and K. Fukomuto, Chem. Commun., 425 (1969); J. Chem. Soc. C., 2602 (1969).

[aq K_2FeCN_6, $NaHCO_3$, Chf, 60°][3]

40%

[LAH, THF]

Epigallanthamine
(50%)

Gallanthamine
(40%)

---

3. In the absence of a blocking group cyclization proceeds preferentially to give *para-para* coupled products.

GERMANICOL

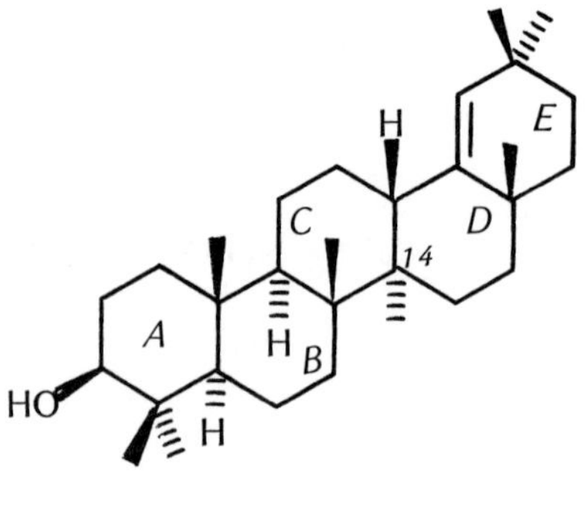

The key tricyclic ketone (A) which embodies the characteristics common to the ABC rings of many triterpenoids has been designed to pave the way for the synthesis of a variety of unsymmetrically substituted pentacyclic triterpenes. The application of this intermediate in the synthesis of germanicol is outlined below. Noteworthy features of this synthesis include conjugate delivery of elements of rings D and E, introduction of the angular methyl group at C-14 (B→C) and cyclodehydration to the desired pentacyclic nucleus (D).[1]

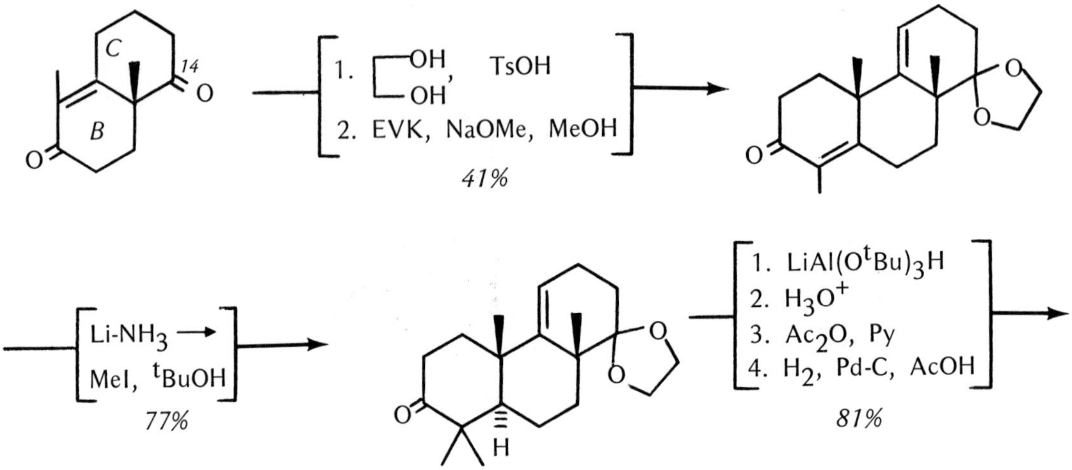

1. $\text{HOCH}_2\text{CH}_2\text{OH}$, TsOH
2. EVK, NaOMe, MeOH

41%

Li-NH_3 → MeI, ^{t}BuOH

77%

1. LiAl(O^{t}Bu)$_3$H
2. H_3O^+
3. Ac_2O, Py
4. H_2, Pd-C, AcOH

81%

1. R. E. Ireland, S. W. Baldwin, D. J. Dawson, M. I. Dawson, J. E. Dolfini, J. Newbould, W. S. Johnson, M. Brown, R. J. Crawford, P. F. Hurdlik, G. H. Rasmussen and K. K. Schmiegel, J. Amer. Chem. Soc., **92**, 5743 (1970).

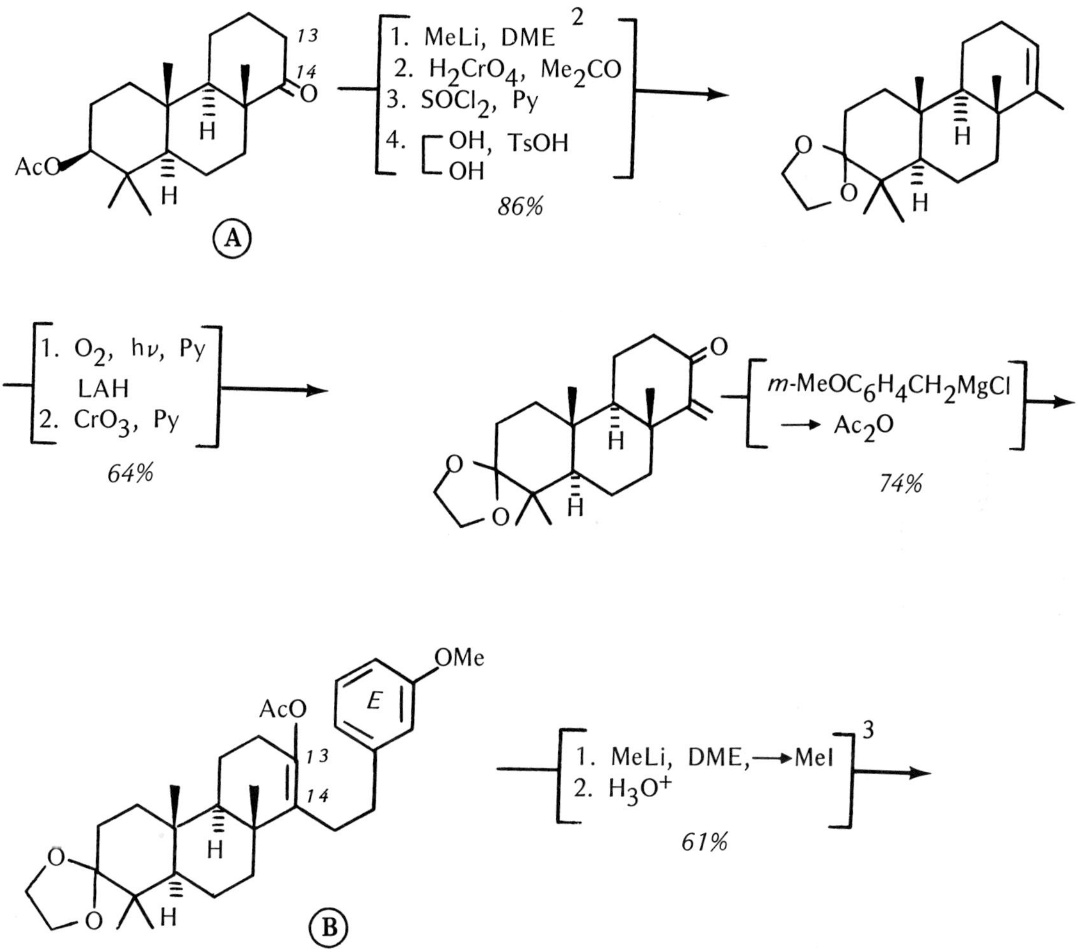

13

14

AcO

H

H

(A)

1. MeLi, DME [2]
2. H_2CrO_4, Me_2CO
3. $SOCl_2$, Py
4. (CH₂OH)₂, TsOH

86%

1. O_2, $h\nu$, Py
LAH
2. CrO_3, Py

64%

m-$MeOC_6H_4CH_2MgCl$
→ Ac_2O

74%

OMe

AcO

E

13

14

H

H

(B)

1. MeLi, DME, → MeI [3]
2. H_3O^+

61%

2. Excess methyl lithium used in this reaction cleaves the O-acetate.

3. The "trapped" enolate generated during conjugate Grignard addition serves to direct alkylation to C-14.

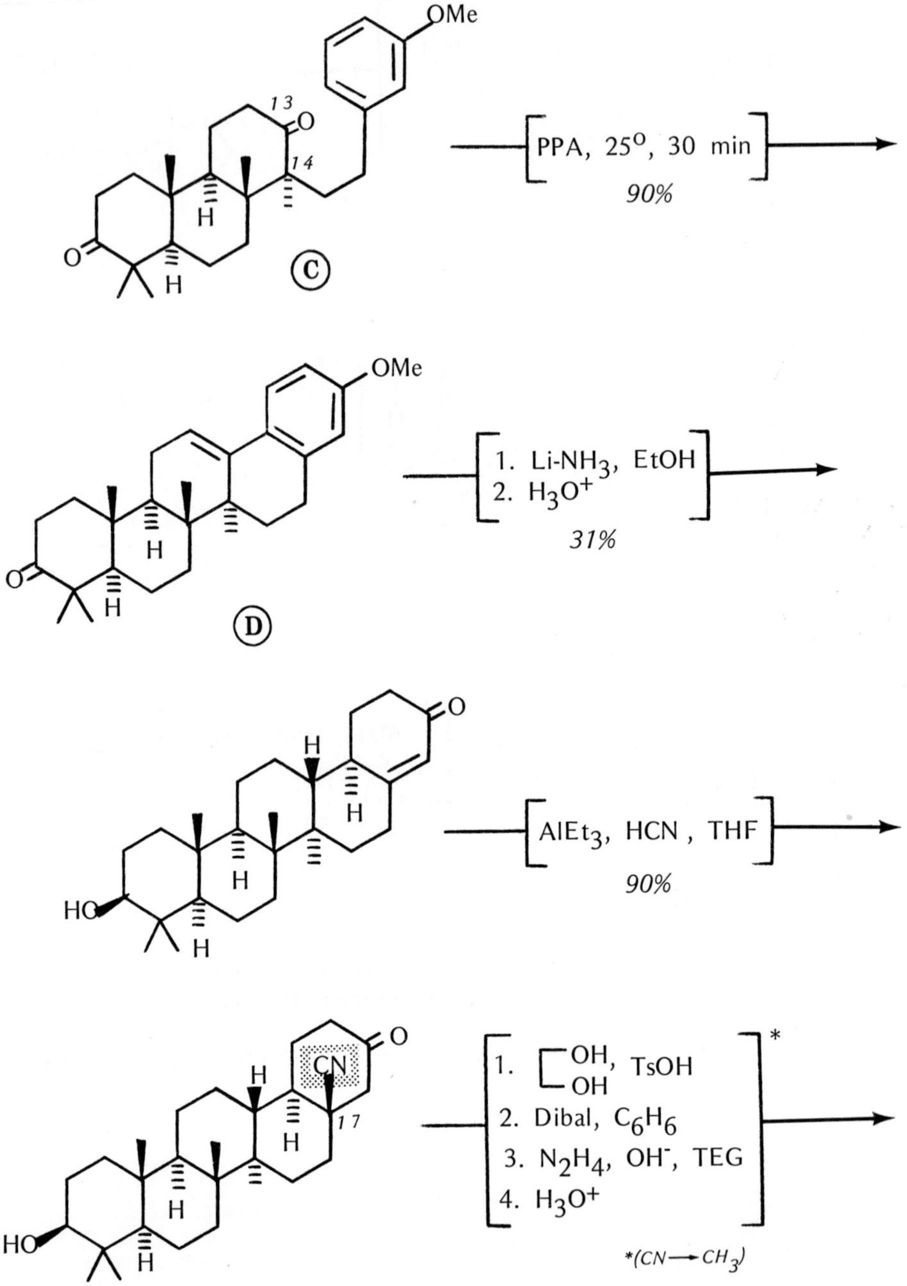

OMe

13

14

H

H

C

[PPA, 25°, 30 min]

90%

OMe

H

H

D

[1. Li-NH_3, EtOH
2. H_3O^+]

31%

O

H

H

H

HO

H

[$AlEt_3$, HCN , THF]

90%

O

CN

H

17

H

H

HO

H

[1. (ethylene glycol) OH, OH, TsOH
2. Dibal, C_6H_6
3. N_2H_4, OH^-, TEG
4. H_3O^+]*

*($CN \longrightarrow CH_3$)

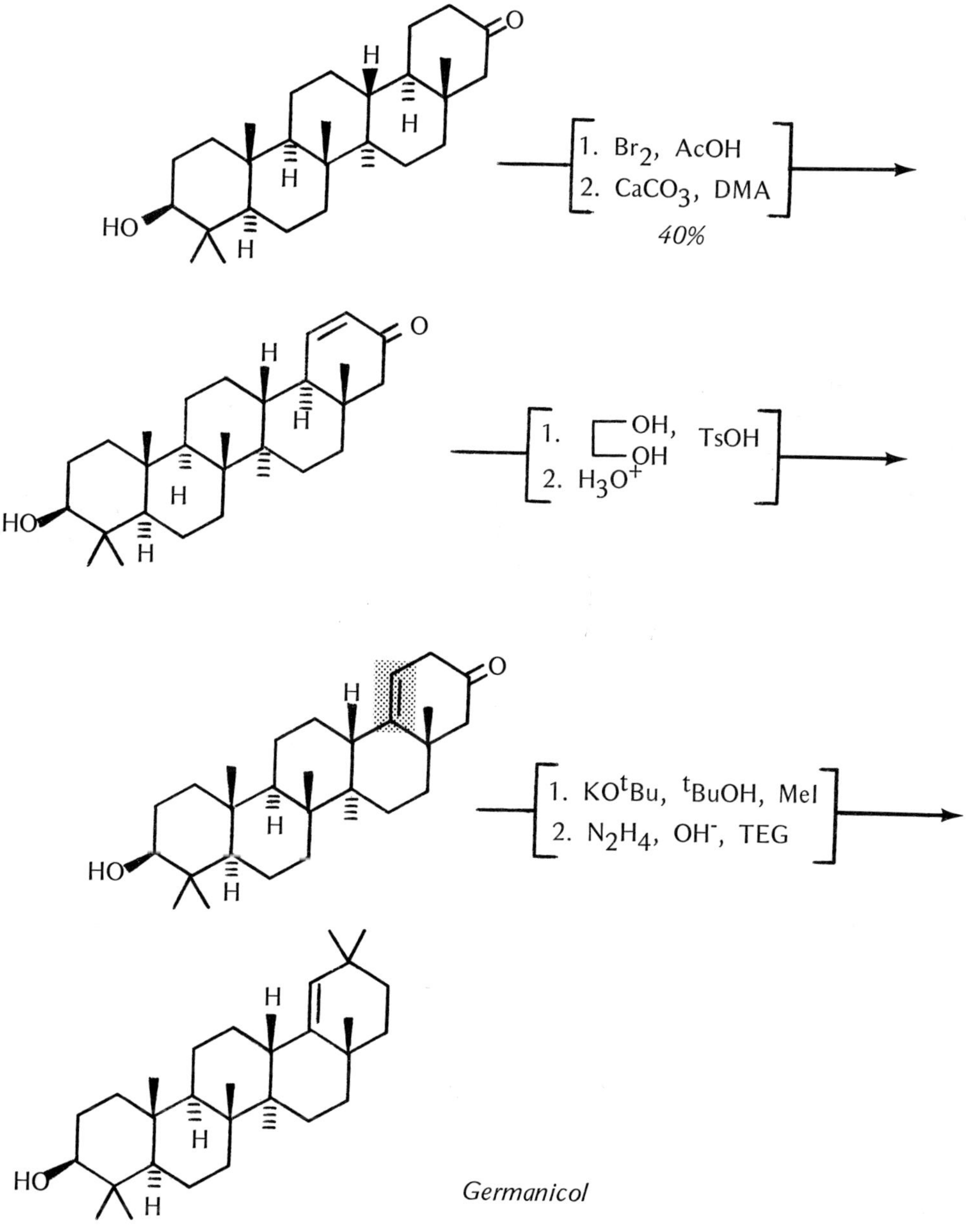

1. Br_2, AcOH
2. $CaCO_3$, DMA

40%

1. (HOCH2CH2OH), TsOH
2. H_3O^+

1. KO^tBu, ^tBuOH, MeI
2. N_2H_4, OH^-, TEG

Germanicol

GRANDISOL

One aspect of the current work on chemical control of insects centers on the use of sex pheromones as lures.[1] Recently, chemists at the U.S. Department of Agriculture's Boll Weevil Research Laboratory isolated and synthesized components of the sex attractant of the male boll weevil,[2] a pest responsible for a prodigious loss of cotton crop in the United States. A stereoselective synthesis of the active pheromone was later carried out at the Zoecon Corporation, and is outlined below.[3] A key feature of this synthesis is the photochemical cycloaddition of ethylene to 3-methylcyclohex-2-enone, followed by oxidative disconnection of the resulting *cis*-fused bicyclo[4.2.0]octane (A) thereby generating the *cis* disposed side-chains.

$CH_2{=}CH_2$ + [3-methylcyclohex-2-enone] —[$h\nu$][4]→ (55%) (A) —[$PhN^+(Me)_3Br_3^-$, THF, 20°]→ (85%) [bromo ketone] —[Li_2CO_3, DMA, 125°]→ [enone] + [enone]; B^-

---

1. M. Berozo, Ed., "Chemicals Controlling Insect Behavior," Academic Press, New York (1970).

2. J. H. Tumlinson, D. D. Hardee, R. C. Gueldner, A. C. Thompson, P. A. Hedin and J. P. Minyard, Science, **166**, 1010 (1969); J. Org. Chem., **36**, 2616 (1971).

3. R. Zurfluh, L. L. Dunham, V. L. Spain and J. B. Siddall, J. Amer. Chem. Soc., **92**, 425 (1970).

4. Cf. Y. Yamada, H. Uda and K. Nakanishi, Chem. Commun., 423 (1966).

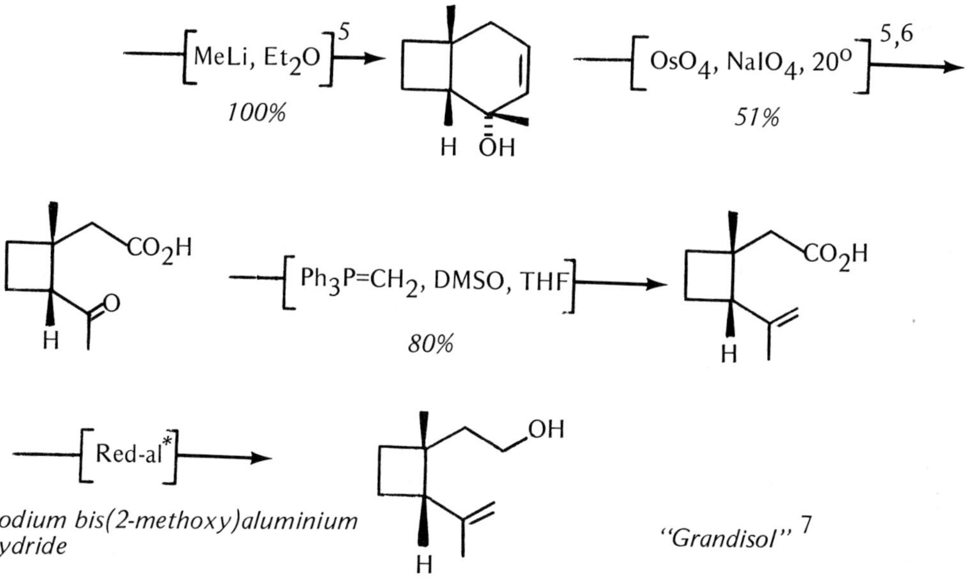

[MeLi, Et_2O][5] 100%

H OH

[OsO_4, $NaIO_4$, 20°][5,6] 51%

CO_2H H

[$Ph_3P{=}CH_2$, DMSO, THF] 80%

CO_2H H

[Red-al*]

*Sodium bis(2-methoxy)aluminium hydride

OH H

"Grandisol"[7]

---

5. The double bond cleavage is straightforward but nonetheless subtle, involving *addition* of a carbon followed by *removal* of one carbon to generate the desired functional group.

6. Cf. R. Pappo, D. S. Allen, R. V. Lemieux and W. S. Johnson, J. Org. Chem., **21**, 478 (1956).

7. For a related synthesis of the boll weevil sex pheromone, see R. C. Gueldner, A. C. Thompson and P. A. Hedin, J. Org. Chem., **37**, 1854 (1972). A key feature of this synthesis is the 2+2 cycloaddition of ethylene to 5,6-dihydro-4-methyl-2H-pyran-2-one:

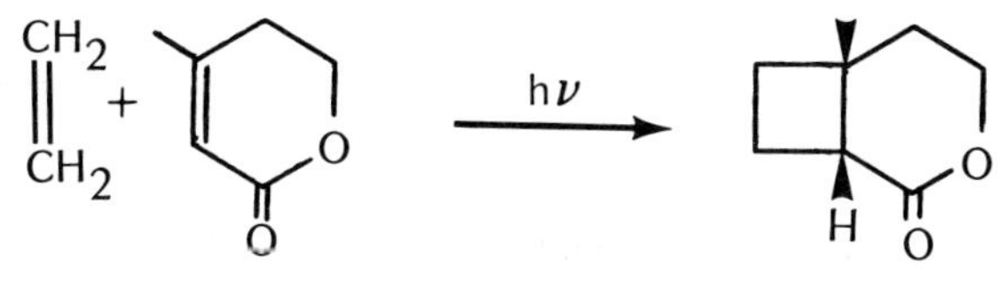

$CH_2{=}CH_2$ + (O, O) $\xrightarrow{h\nu}$ (O, H, O)

Recently, Billups *et al.*[8] reported an unusually short, nonphotochemical synthesis of grandisol based on the production of *cis*-1,2-divinyl cyclobutanes from 1,3-dienes under influence of zero-valent nickel complexes.[9]

2 —[Ni(COD)$_2$, tris(2-biphenylyl)phosphite]^{10,11}→

12-15%

—[Sia$_2$BH, THF, 0°→alkaline H_2O_2]^{12}→ OH

52%

Grandisol

---

8. W. E. Billups, J. H. Cross and C. V. Smith, J. Amer. Chem. Soc., **95**, 3438 (1973).

9. P. Heimbach and W. Brenner, Angew. Chem. Int. Ed., **6**, 800 (1967).

10. COD = 1,5-cyclooctadiene.

11. The *cis*-1,2-divinylcyclobutane is probably formed from the bis-σ-allyl form (i).

Lg—Ni

(i)

Lg = P- [O-C$_6$H$_4$(C$_6$H$_6$)]$_3$

12. Above room temperature the divinylcyclobutane undergoes a Cope rearrangement (ii→iii).

(ii) → (iii)

The synthesis of grandisol by Stork and Cohen[13] illustrates the application of epoxynitrile cyclization, a facile general ring forming reaction developed at Columbia.[14]

CN + Br OTHP → [$LiN(^iPr)_2$ / THF-HMPA, rt] 50% → CN, OTHP

→ [m-$ClC_6H_4CO_3H$ / CH_2Cl_2, rt] 95% → CN, O, OTHP → [$LiN(SiMe_3)_2$ / C_6H_6, 0°][15] →

Li, :N, O, OTHP, H → OH, H, CN, OTHP

13. G. Stork and J. F. Cohen, J. Amer. Chem. Soc., **96**, 5272 (1974).

14. G. Stork, L. D. Cama and D. R. Coulson, J. Amer. Chem. Soc., **96**, 5268 (1974).

15. Cyclobutanes are formed in preference to cyclopentanes during the epoxynitrile cyclization since the transition state for formation of a four-membered ring (i) seems to allow easy attainment of collinearity as opposed to the transition state for five-membered ring formation (ii). Geometric constraints imposed by the oxirane ring also explain the easier formation of cyclohexanes (iii) compared to cyclopentanes during this cyclization.

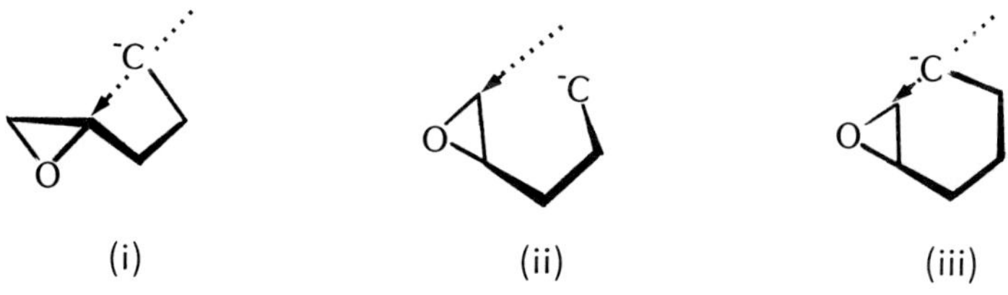

[$({}^{i}Bu)_2AlH$] → 52% for two steps

H, OH, CHO, OTHP

[1. Wolff-Kishner 2. CrO_3] →

H, O, OTHP

[$Ph_3\overset{+}{P}CH_3$ Br^-, BuLi, THF] → 77%

H, OTHP

[0.05N $HClO_4$, THF, rt] →

H, OH

Grandisol

GRASS-HOPPER KETONE

HO H O

The large, flightless grasshopper, *Romalea microptera* excretes a novel sesquiterpenoid allenic ketone in its defensive secretions.[1] Meinwald and Hendry[2] confirmed the proposed structure and stereochemistry of this ant-repellant substance by stereospecific synthesis from isophorone.

Isophorone — [MeMgBr, $FeCl_3$]* → — [1. LAH, 2. Ac_2O, Py] → OAc

*M. S. Kharasch and P. O. Towney, J. Amer. Chem. Soc., **63**, 2308 (1941)

— [m-$ClC_6H_4CO_3H$] → (99%) OAc — [aq $HClO_4$, Me_2CO][3] → (96%) HO, OH, OAc

---

1. For a review of chemical defense against predation in anthropods, see T. Eisner in "Chemical Ecology", Eds. E. Sondheimer and J. B. Simeone, Academic Press, New York (1970) p. 157.

2. J. Meinwald and L. Hendry, Tetrahedron Lett., 1657 (1969).

3. Epoxidation occurs from the less hindered side, and *trans* diaxial epoxide opening results in formation of diol with the stereochemistry shown:

H :OH_2 AcO O H → H OH AcO H OH

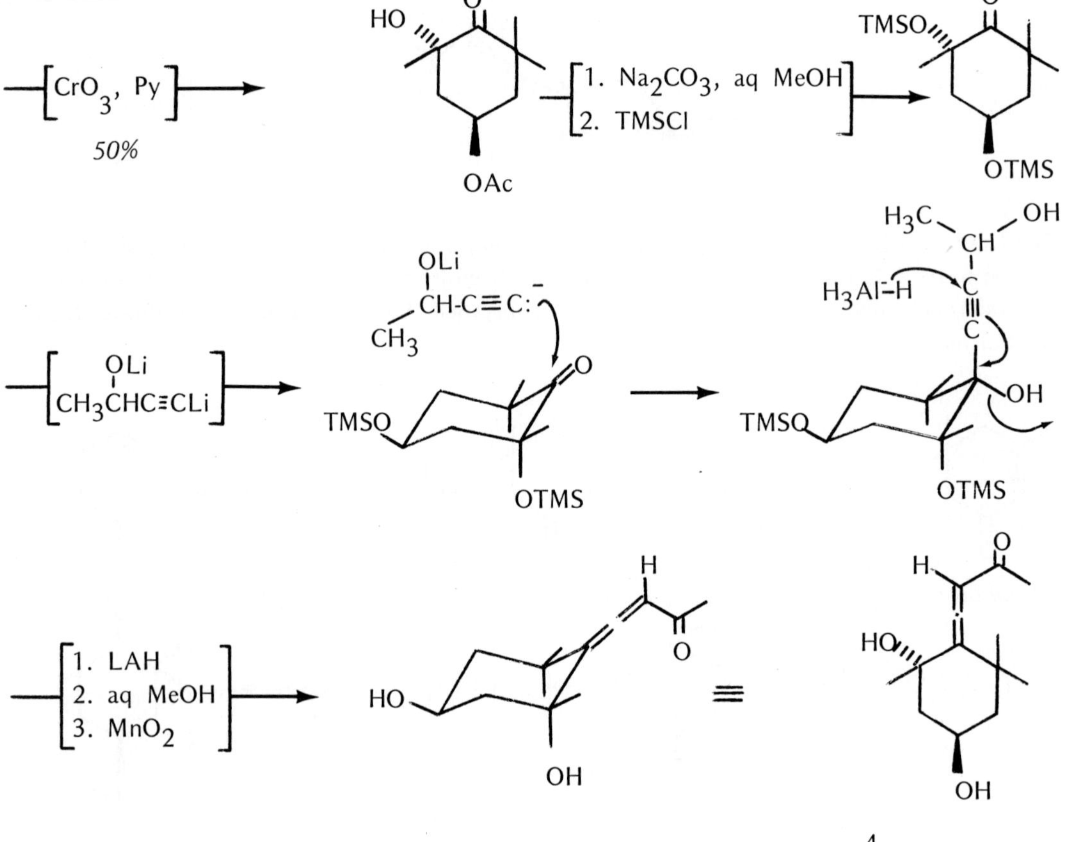

[CrO_3, Py] 50%

[1. Na_2CO_3, aq MeOH; 2. TMSCl]

[OLi $CH_3CHC{\equiv}CLi$]

[1. LAH; 2. aq MeOH; 3. MnO_2]

Grass-Hopper Ketone[4]

4. For other syntheses of the grasshopper ketone, see S. W. Russell and B. C. L. Weedon, Chem. Commun., 85 (1969); S. Isoe, S. Katsumura, S. B. Hyeon and T. Sakan, Tetrahedron Lett., 1089 (1971); K. Mori, Tetrahedron Lett., 723 (1973).

HAEMANTHIDINE

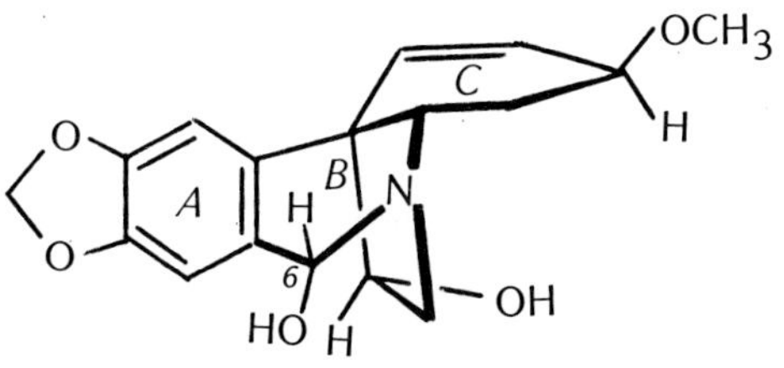

Synthesis of the complex *Amaryllidaceae* alkaloid, haemanthidine,[1,2] proceeds along lines laid down earlier by Hendrickson during the synthesis of dihydroxycrinene.[3] A noteworthy feature of this synthesis is the application of Diels-Alder reaction for construction of the key B/C *trans* lactam (A)[3] bearing a sterically dominant axial carboxyl for steric control of ring-C functionalization as well as construction of the ethano bridge.

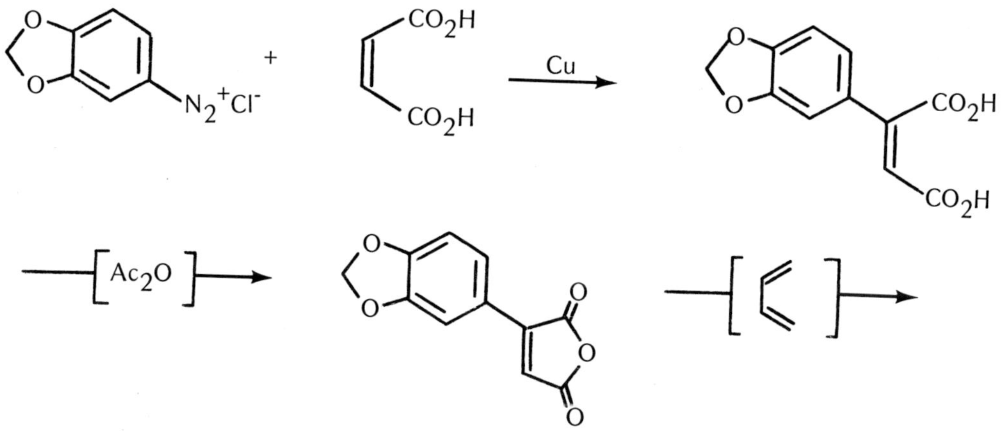

---

1. Haemanthidine, one of the more complex *Amaryllidaceae* alkaloids characterized by the crinine skeleton bears a C-6 hydroxyl substitution unique among the 5,10b-ethanophenanthridine group. For a review of *Amaryllidacea* alkaloids, see W. C. Wildman, The Alkaloids, **6**, 290 (1960).

2. J. B. Hendrickson, T. L. Bogard and M. E. Fisch, J. Amer. Chem. Soc., **92**, 5538 (1970).

3. J. B. Hendrickson, C. Foote and N. Yoshimura, Chem. Commun., 165 (1965).

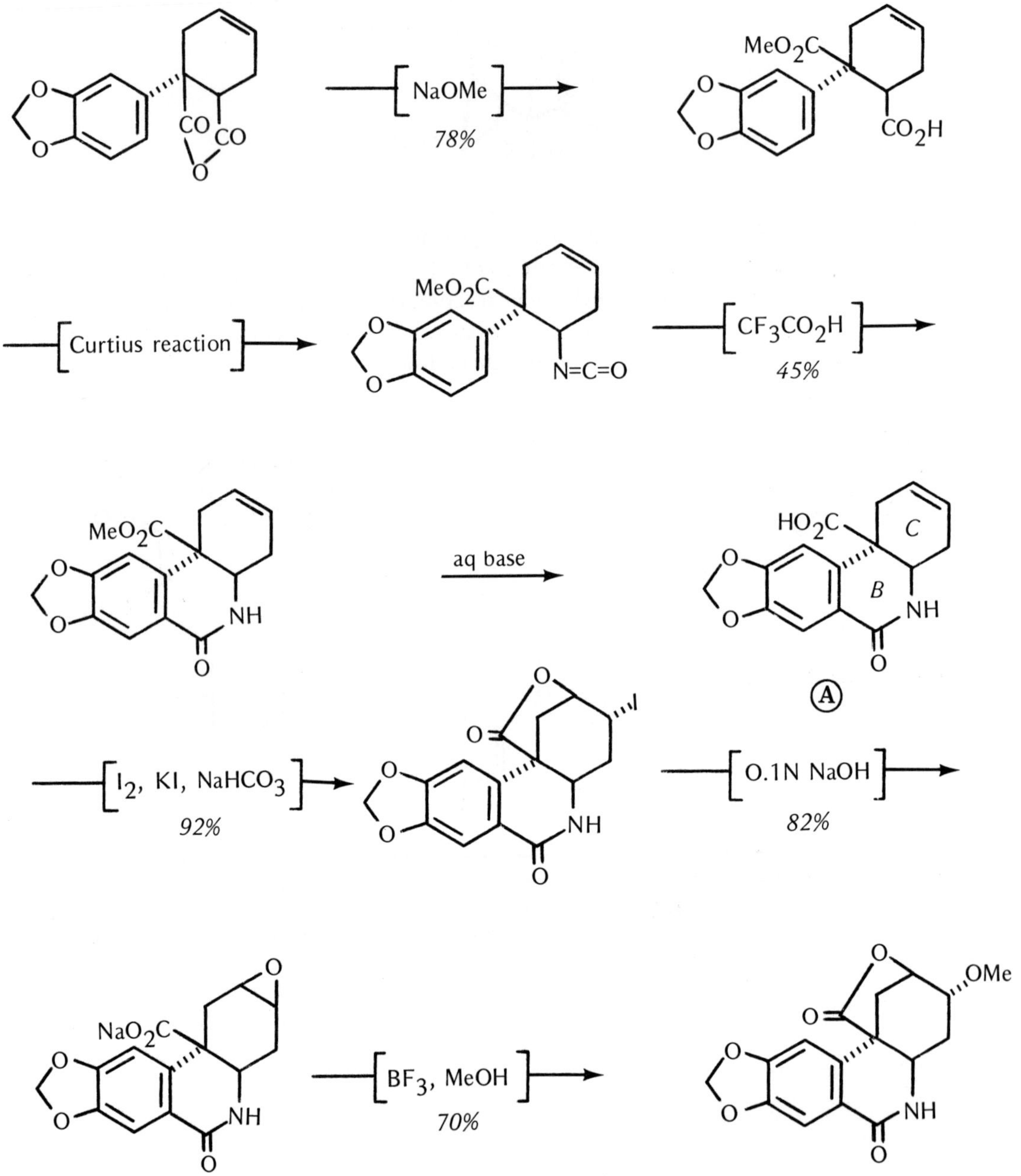

[NaOMe] 78%

MeO_2C CO_2H

[Curtius reaction]

MeO_2C N=C=O

[CF_3CO_2H] 45%

MeO_2C NH O

aq base

HO_2C C B NH O

(A)

[I_2, KI, $NaHCO_3$] 92%

O NH O I

[0.1N NaOH] 82%

NaO_2C NH O

[BF_3, MeOH] 70%

OMe O NH O

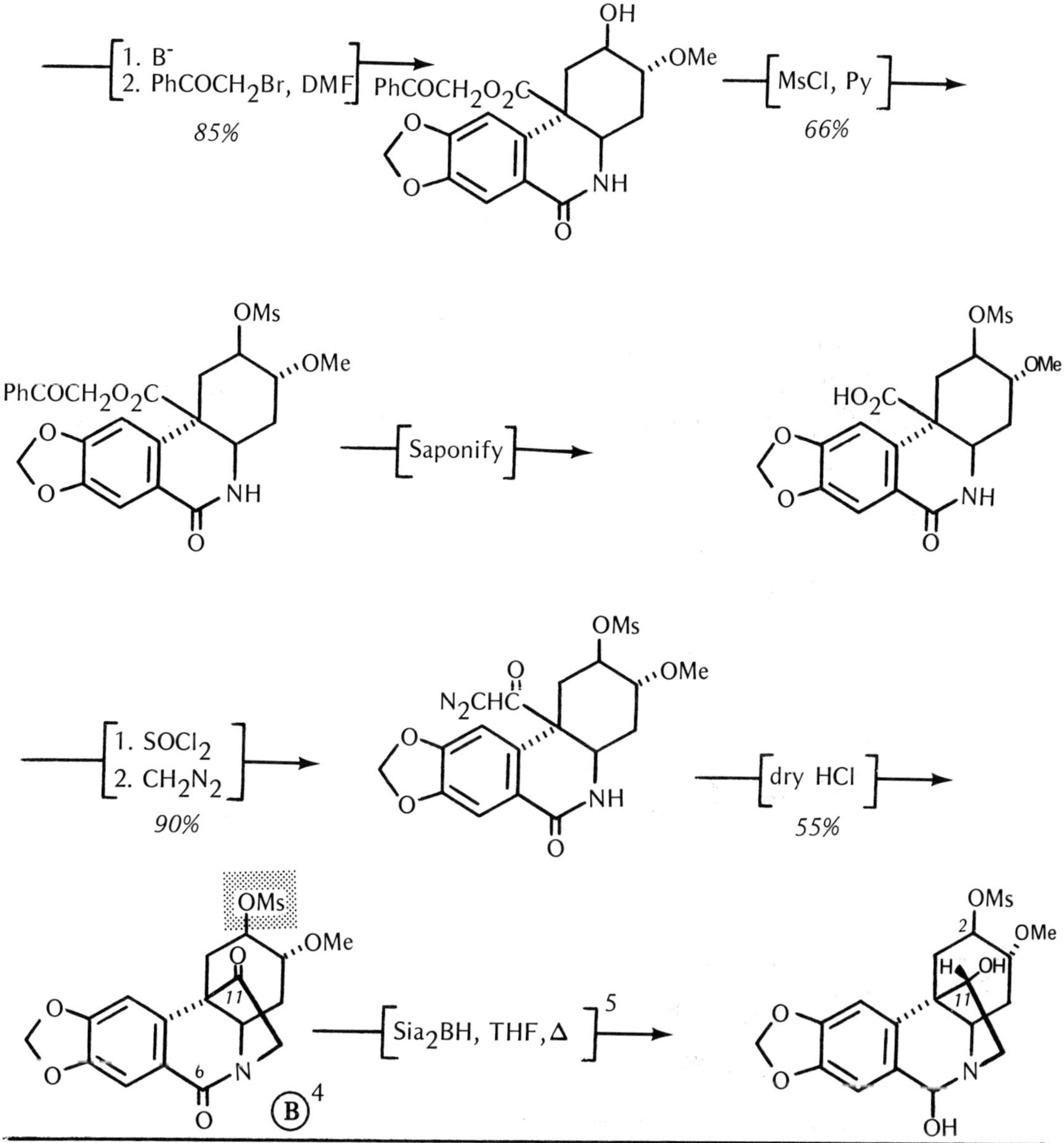

[1. B^- 2. $PhCOCH_2Br$, DMF] 85% → [MsCl, Py] 66% →

[Saponify] →

[1. $SOCl_2$ 2. CH_2N_2] 90% → [dry HCl] 55% →

(B)[4] → [Sia_2BH, THF, Δ][5] →

4. Since the carbonyl at C-6 in (B) does not exhibit normal amide resonance owing to Bredt's Rule prohibition of the canonical form bearing a π-bond between the carbonyl and bridge-head nitrogen, reduction of the amide proceeds readily to a carbinolamine with $NaBH_4$ or hindered boranes.

5. Since reduction at C-11 in natural 11-oxo derivatives is known to result predominantly in

1. Acetylate
2. DBN
3. LAH

Haemanthidine

The synthesis of haemanthamine illustrates another approach to the ethano–phenanthridine nucleus and is based on the cycloaddition reaction of butadienes and Δ^2-pyrroline-4,5-diones to give tetrahydroisatins. Transformation of the key intermediate diol (B) to the desired nucleus has been achieved *via* an unusual Bischler-Napieralski reaction involving the bridge-head nitrogen.[6-8]

NaOEt

95%

H_2, Raney Ni
EtOH, 70°

25-30%

the wrong 11-epimer, retention of the axial mesylate at C-2 so late in the synthesis was intended to reverse this trend by steric hindrance.

6. Y. Tsuda, K. Isobe and A. Ukai, Chem. Commun., 1554 (1971).

7. Y. Tsuda, A. Ukai and K. Isobe, Tetrahedron Lett., 3153 (1972).

8. Y. Tsuda and K. Isobe, Chem. Commun., 1555 (1971).

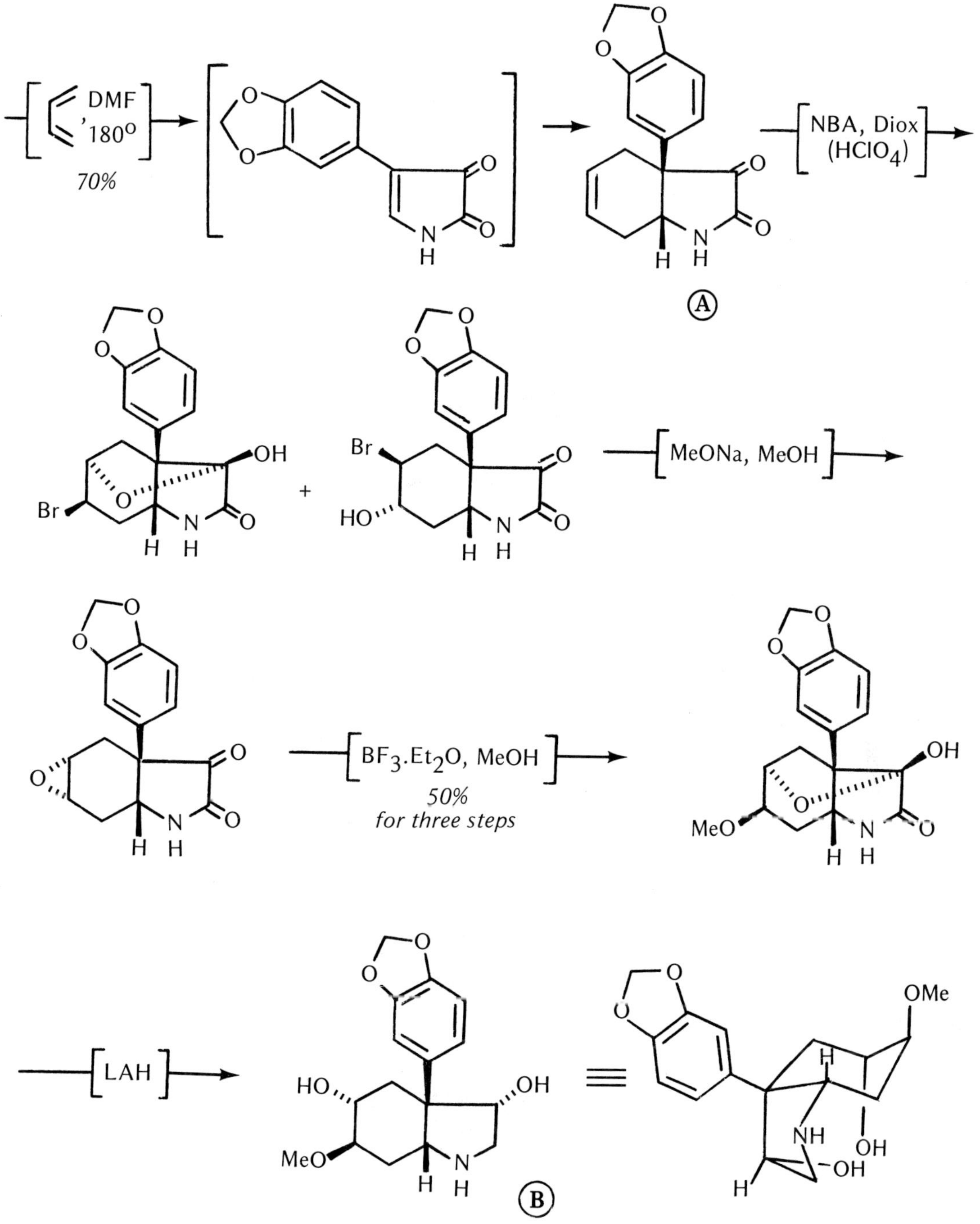

DMF, 180°

70%

A

NBA, Diox ($HClO_4$)

MeONa, MeOH

$BF_3.Et_2O$, MeOH

50% for three steps

LAH

B

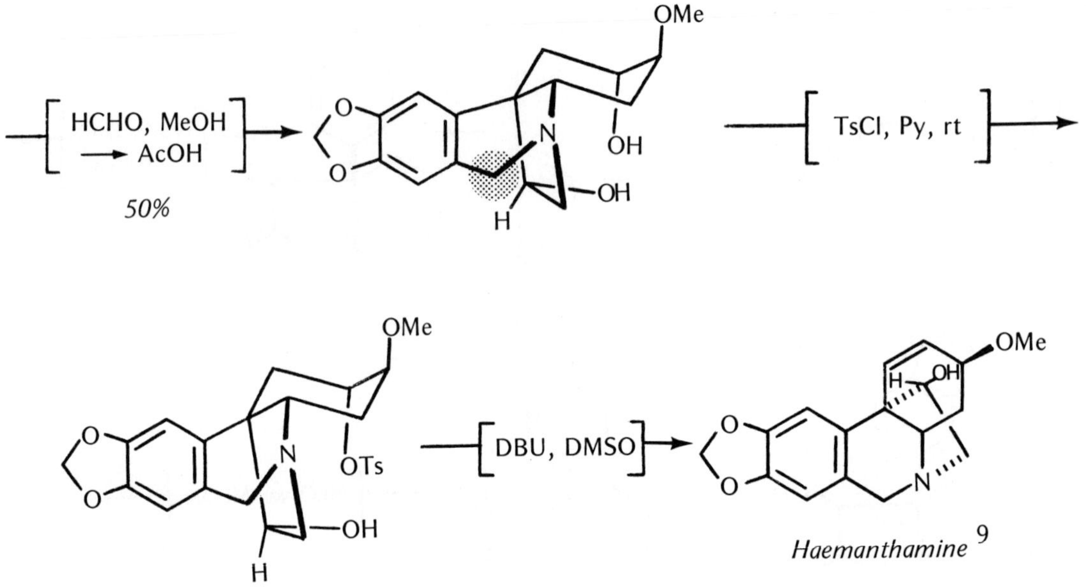

[HCHO, MeOH → AcOH] 50% → [TsCl, Py, rt] →

→ [DBU, DMSO] → *Haemanthamine*[9]

9. For syntheses of haemanthidine and tazettine using this approach, see Y. Tsuda, A Ukai and K. Isobe, Tetrahedron Lett., 3153 (1972).

ILLUDIN-S

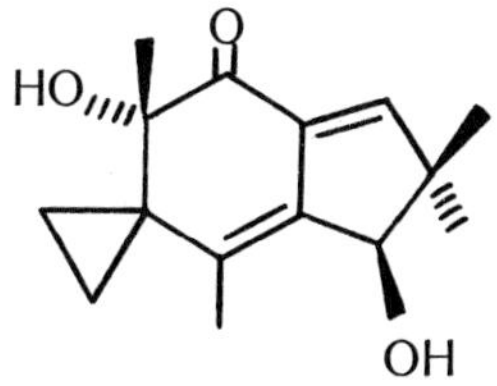

The beautiful saffron colored Jack-o'-Lantern (*Clitocybe illudens*), a poisonous bioluminescent mushroom growing in large clusters during late summer and autumn particularly in the eastern United States, yields an antitumor substance illudin-S shown to possess a unique sesquiterpenoid structure.[1] A total synthesis of Illudin-S has been reported by Matsumoto *et al.*[2], and involves combination of cyclopropane (A) and cyclopentane (B) moieties of the molecule by Michael addition of a β-keto sulfoxide, followed by an aldol condensation for completion of the six-membered ring.[3]

Preparation of cyclopropane (A)[4]*:*

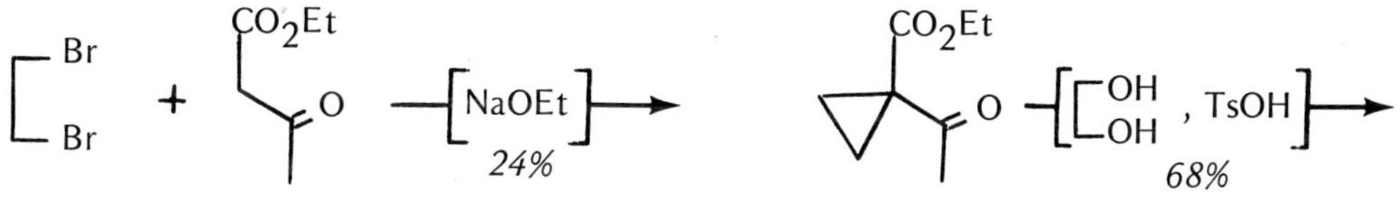

NaOEt 24%

CO_2Et , TsOH 68%

---

1. Illudin-S has also been isolated from the moon-night-mushroom (*Lampteromyces Japonicus*), a poisonous bioluminescent mushroom growing in Japan, K. Nakanishi, M. Ohashi, M. Tada and Y. Yamada, Tetrahedron, **21**, 1231 (1965).

2. T. Matsumoto, H. Shirahama, A. Ichihara, H. Shin, S. Kagawa, F. Saken and K. Miyano, Tetrahedron Lett., 2049 (1971).

3. Preliminary work on the synthesis of illudane and functionalized illudane skeleton has been discussed by T. Matsumoto *et al.*, Tetrahedron Lett., 4097 (1967); 1925 (1968).

4. For synthesis of cyclopropane intermediates, see T.Matsumoto *et al.*, Bull. Chem. Soc., Japan, **45**, 1136 (1972).

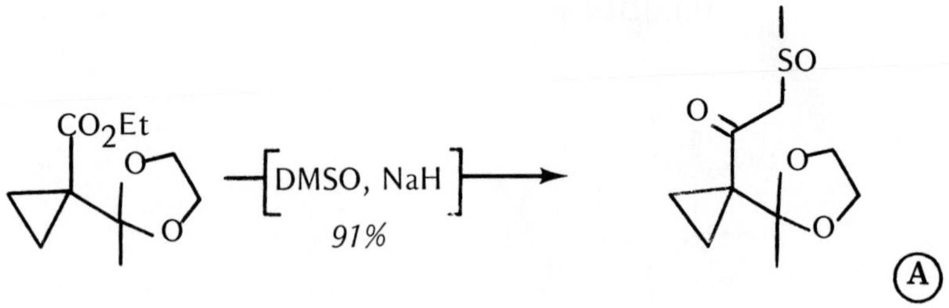

SO

CO_2Et O O —[DMSO, NaH]→ 91% O O O Ⓐ

Preparation of cyclopentenone (B)[5]:

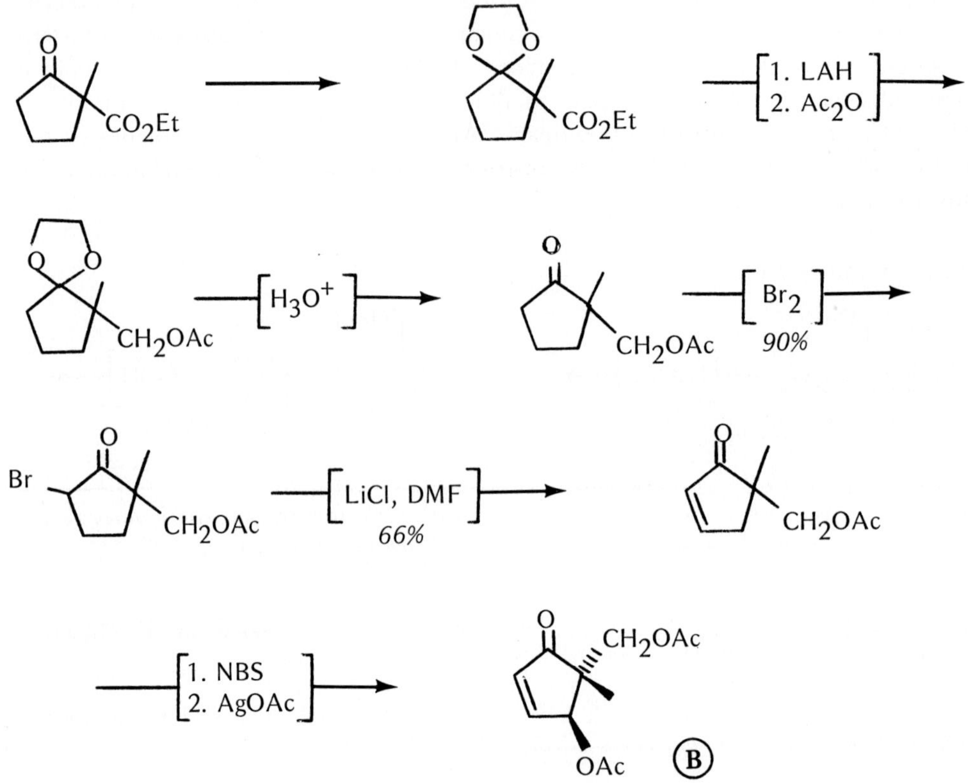

O CO_2Et ——→ O O CO_2Et —[1. LAH 2. Ac_2O]→

O O CH_2OAc —[H_3O^+]→ O CH_2OAc —[Br_2]→ 90%

Br O CH_2OAc —[LiCl, DMF]→ 66% O CH_2OAc

—[1. NBS 2. AgOAc]→ O CH_2OAc OAc Ⓑ

5. For details of synthesis of the cyclopentenone intermediates for illudin syntheses, see T. Matsumoto *et al.*, Bull. Chem. Soc. Japan, **45**, 1140, 1144 (1972).

Preparation of illudin-S:

Ⓐ + Ⓑ —[KO^tBu, ^tBuOH, rt] 70% →

—[$(MeOCH_2CO)_2O$, Py]* 100% →

**Pummerer rearrangement: For a recent review, see G. A..Russell and G. J. Mikol, Mech. Mol. Migrations, 1, 157 (1968).*

—[aq EtOH, rt][6] →

—[KO^tBu, rt → Ac_2O] →

---

6. This novel intramolecular transketalization may be rationalized in terms of an eight membered intermediate (i):

HO^- → (i) →

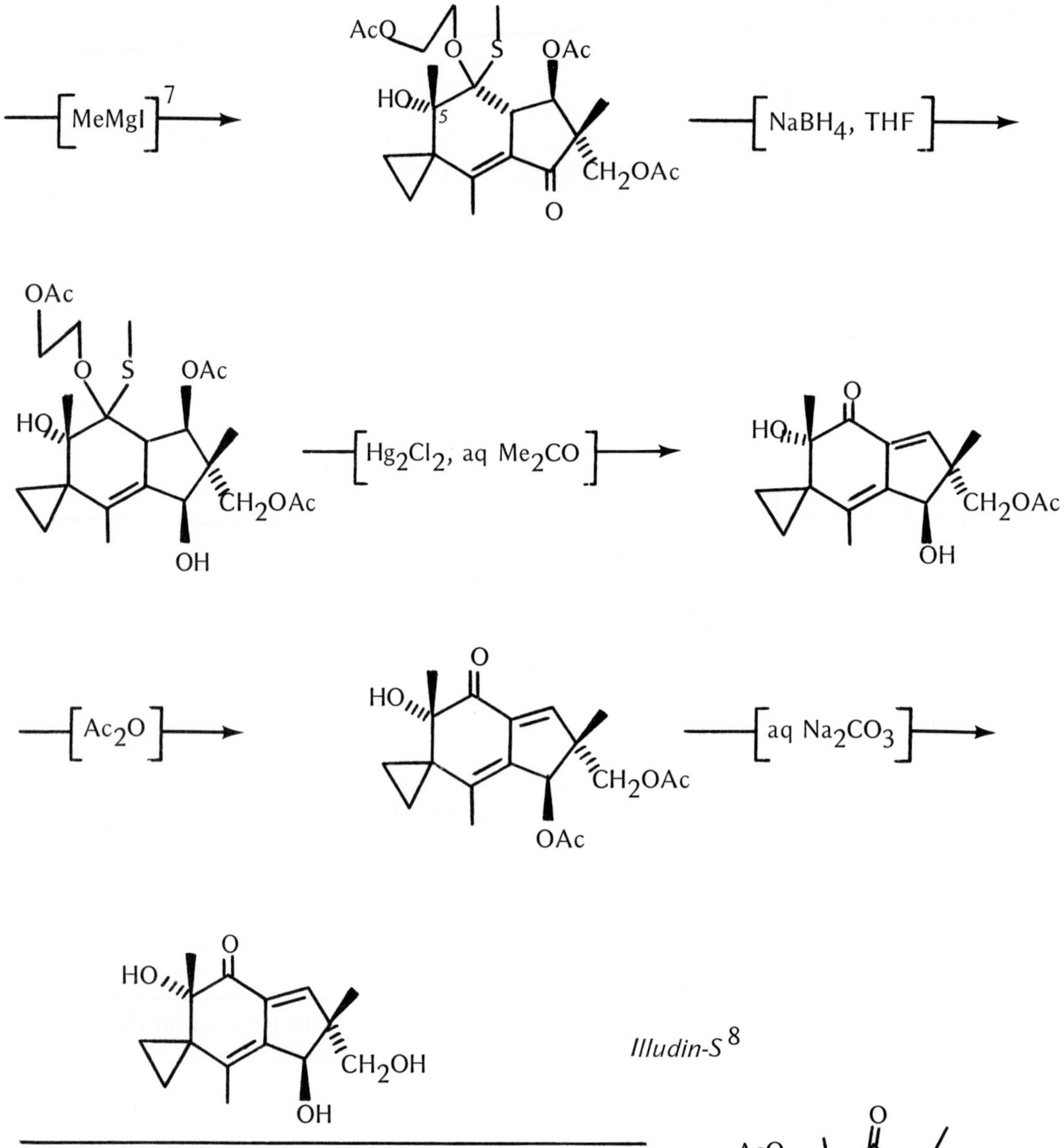

[$MeMgI$][7] →

[$NaBH_4$, THF] →

[Hg_2Cl_2, aq Me_2CO] →

[Ac_2O] →

[aq Na_2CO_3] →

Illudin-S[8]

---

7. Attack of Grignard reagent to the C-5 carbonyl group occurs from the less hindered convex face of the molecule.

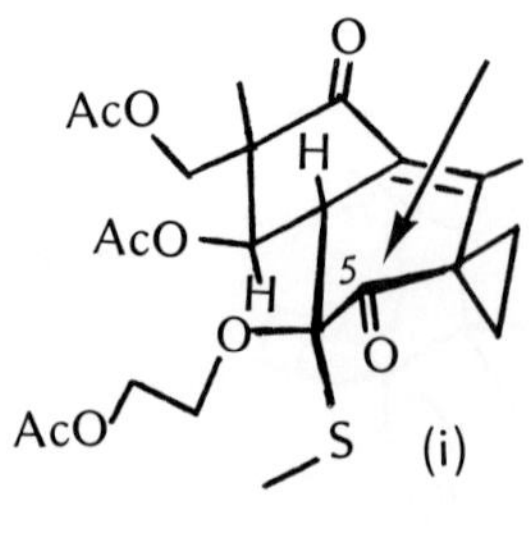

(i)

8. For the closely related total synthesis of Illudin-M see, T. Matsumoto, H. Shirahama, A. Ichihara, H. Shin, S. Kagawa, F. Sakan, S. Matsumoto and S. Nishida, J. Amer. Chem. Soc., **90**, 3280 (1968); Tetrahedron Lett., 1171 (1970).

ISHWARANE

The synthetic route to ishwarane,[1] a novel tetracyclic sesquiterpene, proceeds through the bicyclic intermediate (A) bearing not only the C-4, C-5 *cis* methyl groups, but also possessing an α,β-unsaturated ketone functionality suitable for development of the tricyclooctane system. This has been accomplished by photo-addition of allene to (A), followed by skeletal rearrangement of the resulting cyclobutanol (B) into intermediates possessing a bicyclo[2·2·2]octane system (C) and, finally, the transformation of the bicyclooctane system into the desired tricyclo[$3{\cdot}2{\cdot}1{\cdot}0^{2,7}$]octane system.

+ [NaOEt, EtOH] →

cis-2,3-Dimethylcyclohexanone[2]

OH

→ [TsOH] →

(A)[3]

[CH_2=C=CH_2, hν][4] →

---

1. R. B. Kelly, J. Zamecnik and B. A. Beckett, Chem. Commun., 479 (1971); Canad. J. Chem., **50**, 3455 (1972).

2. For a preparation of the *cis*-2,3-dimethylcyclohexanone by acid-catalyzed cyclization of acyclic dienes, see H. E. Ulery and J. H. Richards, J. Am. Chem. Soc., **86**, 3113 (1964).

3. The octalone may also be prepared by the method of E. Piers, R. W. Britton and W. de Waal, Canad. J. Chem., **47**, 4307 (1969).

[OH, OH (ethylene glycol), TsOH, C_6H_6] →

[$C_6H_5CO_3H$, $NaHCO_3$] →

[LAH] →

[aq HCl][5] →
97%

(B)

(C)

---

4. Stereochemistry of the photo-addition of allene to the α,β-unsaturated ketone is controlled by tendency of the excited state to exist preferentially in the most stable configuration (i). The excited state is assumed to have carbonium character in the position α to the carbonyl group and an orbital with an electron pair in the β-position, wherein the orbital and the adjacent methyl are *trans*-axial, cf., K. Wiesner, L. Poon, I. Jirkovsky and M. Fishman, Canad. J. Chem., **47**, 433 (1969).

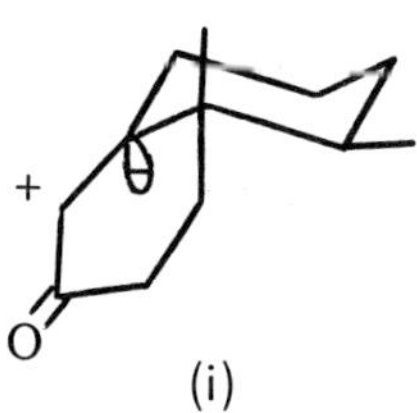
(i)

5. This rearrangement proceeds *via* deacetalization and retro-aldolization of the resulting keto cyclobutanol to an intermediate diketone (ii) which undergoes aldol condensation to the ketoalcohol (C). A similar transformation has been accomplished earlier in the synthesis of atisine by R. W. Guthrie, Z. Valenta and K. Wiesner, Tetrahedron Lett., 4645 (1966).

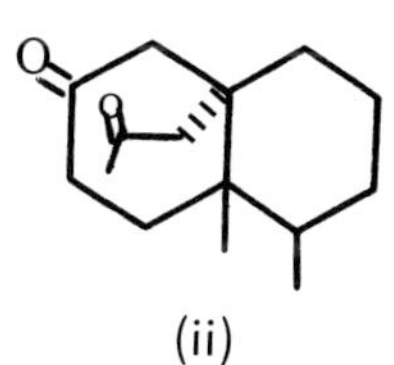
(ii)

[TsOH] 89%

≡

[LAH, Et_2O] 76%

* 1:1 mixture of epimeric alcohols

HO

*

[NaH, C_6H_6 → $C_6H_5CH_2Cl$]

OBz

[B_2H_6, THF] 71-82%

OBz

OH

[CrO_3] 80-90%

OBz

O

[H_2, Pd-C] 80-90%

[TsCl, Py]

[$MeS(O^-)=CH_2$, Na^+, DMSO]
60-70%

[NH_2NH_2; Na, HO~~OH, Δ]
67%

Ishwarane[6]

6. For a promising approach to the synthesis of ishwarone, see C. J. V. Scanio and D. L. Lickei, Tetrahedron Lett., 1363 (1972).

LOGANIN

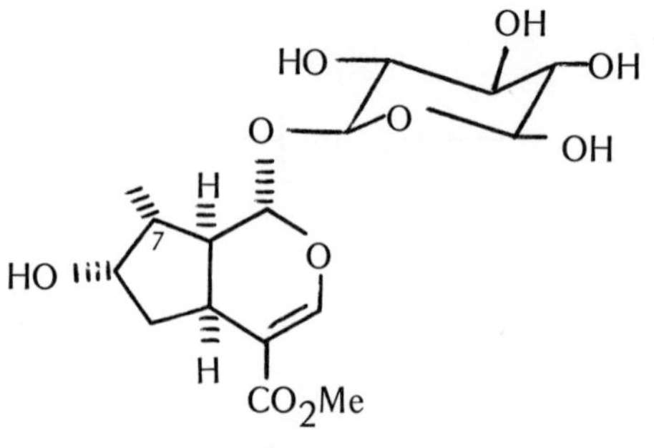

The synthesis of loganin[1] by Büchi *et al.*[2] illustrates the tactical use of the enone photoannelation technique, developed by de Mayo,[3] for synthesis of δ-diketones *via* photochemical cycloaddition of enolized β diketones to olefins. Thus, construction of the bicyclic iridoid nucleus is accomplished in a single photochemical operation followed by introduction of the C-7 methyl group and glucosidation to afford loganin penta-acetate.[2]

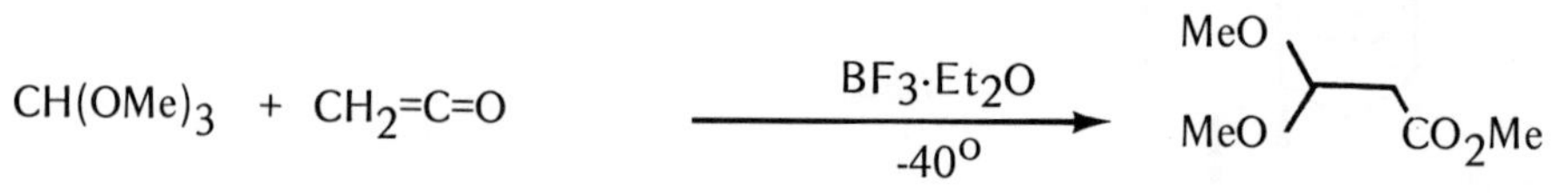

$CH(OMe)_3$ + CH_2=C=O $\xrightarrow[-40^{o}]{BF_3\cdot Et_2O}$ $(MeO)_2CHCH_2CO_2Me$

1. Loganin is an iridoid glucoside which occupies a central position in the biosynthesis of *Corynanthe, Aspidosperma, Iboga, Ipecacuanha, Cinchona,* and structurally simpler monoterpene alkaloids. For a review of the extensive researches that led to confirmation of the key role of loganin in alkaloid biosynthesis, see A. R. Battersby, Biochem. Soc. Symp., **29**, 157 (1970); A. R. Battersby, Chem. Soc. Spec. Period. Rep., **1**, 31 (1971); A. I. Scott, Accts. Chem. Res., **3**, 151 (1970). For a useful account of the chemistry of iridoid glucosides see, J. M. Bobbitt and K. -P. Segebarth in "Cyclopentanoid terpene Derivatives," Eds., W. I. Taylor and A. R. Battersby, Marcel Dekker, New York, 1969, p 1.

2. G. Büchi, J. A. Carlson, J. E. Powell and L. -F. Tietze, J. Amer. Chem. Soc., **92**, 2165 (1970); **95**, 540 (1973).

3. P. deMayo, Accts. Chem. Res., **4**, 41 (1971); B. D. Challand, H. Hikino, G. Kornis, G. Lange and P. deMayo, J. Org. Chem., **34**, 794 (1969).

HCO$_2$Me, Na, Et$_2$O
68%

CO$_2$Me

hν OTHP

THPO CHO OH CO$_2$Me

THPO CHO CHOH CO$_2$Me

4

OH THPO O CO$_2$Me

Amberlite IR-120
MeOH, rt
59%

OMe HO O CO$_2$Me

CrO$_3$, Py, CH$_2$Cl$_2$
95%

OMe O O CO$_2$Me

1. HCO$_2$Me, Na^{t}AmO
2. TsCl, Py → ^{n}BuSH

^{n}BuS OMe O O CO$_2$Me

Raney Ni, MeOH, rt [5]
86%

OMe 7 O O CO$_2$Me

---

4. The primary photoproduct, a cyclobutane formed by cis 2+2 addition, presumably undergoes retroaldol cleavage to form the corresponding 1,5-dialdehyde which then cyclizes to the tetrahydrocoumalate.

5. The preparation of α-ketones by reductive desulfurization of *n*-butylthiomethylene derivatives was pioneered by R. E. Ireland and J. A. Marshall, J. Org. Chem., **27**, 1615 (1962).

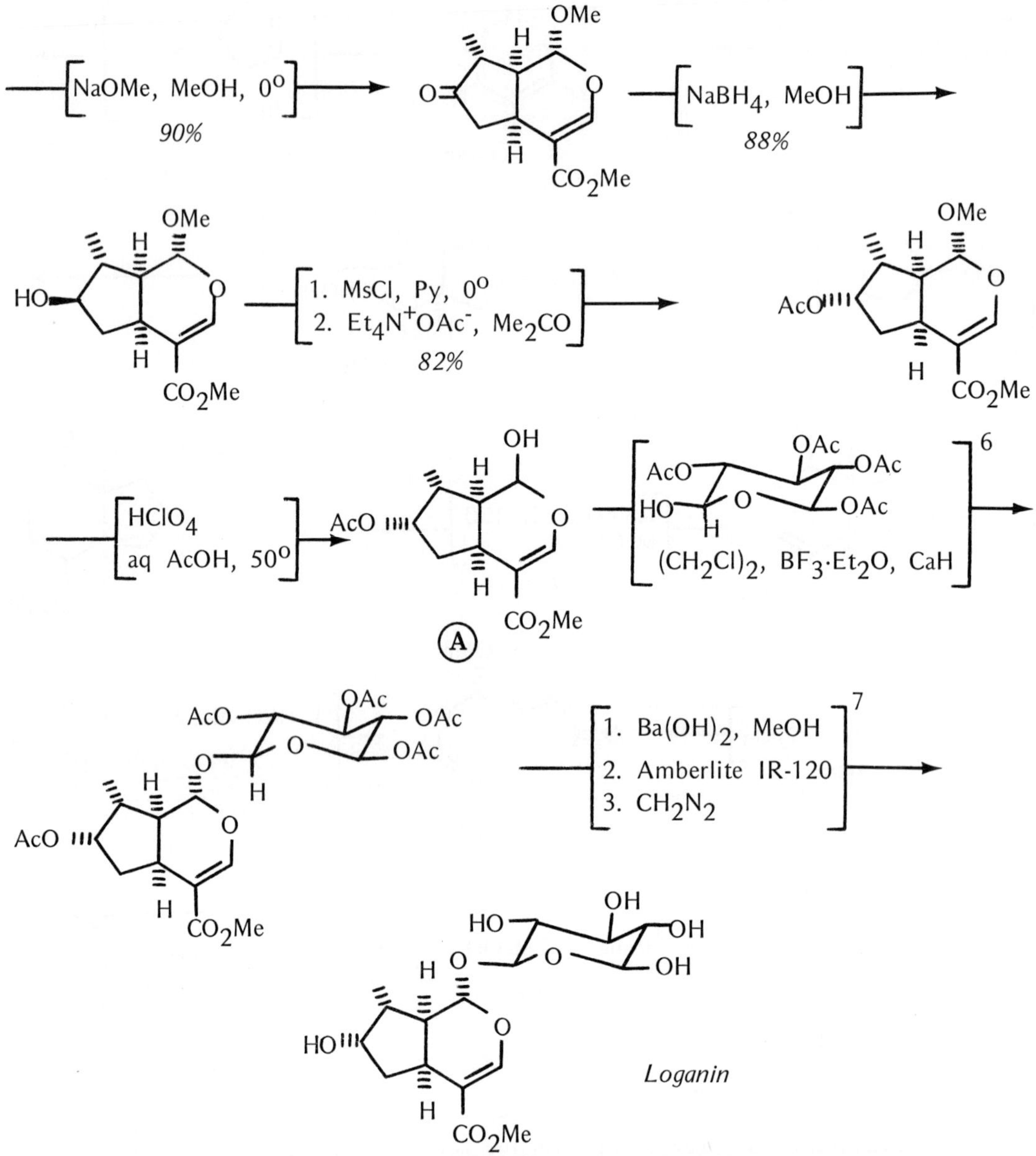

[NaOMe, MeOH, 0°] 90%

[$NaBH_4$, MeOH] 88%

[1. MsCl, Py, 0°; 2. $Et_4N^+OAc^-$, Me_2CO] 82%

[$HClO_4$, aq AcOH, 50°]

(A)

[$(CH_2Cl)_2$, $BF_3{\cdot}Et_2O$, CaH][6]

[1. $Ba(OH)_2$, MeOH; 2. Amberlite IR-120; 3. CH_2N_2][7]

Loganin

6. This method for the synthesis of β-glucopyranosides was developed in the laboratories of Sandoz A.G. by M. Kuhn and A. von Wartburg, Helv. Chim. Acta, **51**, 1631 (1968); **52**, 948 (1969).

7. A. R. Battersby, E. S. Hall and R. Southgate, J. Chem. Soc. C, 721 (1969).

Recently, workers at Hoffman-La Roche[8] described a direct asymmetric synthesis of loganin penta-acetate in which the C-7 methyl group is already incorporated on the cyclopentane nucleus. The synthesis begins with the prochiral reagent 5-methylcyclopentadiene in which a high degree ofasymmetrywas induced by asymmetric hydroboration thus avoiding the need for a resolution step. The photoannelation reaction proceeds regioselectively with asymmetric induction to give the loganin aglucone derivative.

Na^+

MeI, THF, -78°

(+)-[ ]$_2$B (9)

THF, -78° → NaOH, H_2O_2

30%

R, R, HO

MsCl, Py (10)

MsO, 1, 7

$Me_4N^+OAc^-$, HMPA, Me_2CO, 65°

63%

AcO, S, R

H, CHO, HO, CO_2Me

hν

22%

OH, CHO, H, OAc, CO_2Me

11

8. J. J. Partridge, N. K. Chadha and M. R. Uskokovic, J. Amer. Chem. Soc., **95**, 532 (1973).

9. Di-3-pinanylborane is a versatile chirial reagent for selective hydroboration and exhibits a remarkable asymmetric stereoselectivity when applied to the hydroboration of *cis*-olefins, H. C. Brown, N. R. Ayyangar, and G. Zweifel, J. Amer. Chem. Soc., **86**, 397 (1964). See also J. D. Morrison and H. S. Mosher in "Asymmetric Organic Reactions," Prentice-Hall, Englewood, 1971, p. 220-241.

10. Inversion of the C-1 configuration of the alcohol was accomplished using a modification of the method of A. C. Cope and D. L. Nealy, J. Amer. Chem. Soc., **87**, 3122 (1965).

11. The photoannelation reaction would be expected to proceed regioselectively if the enol attacks the less hindered face of the optically active cyclopentenol.

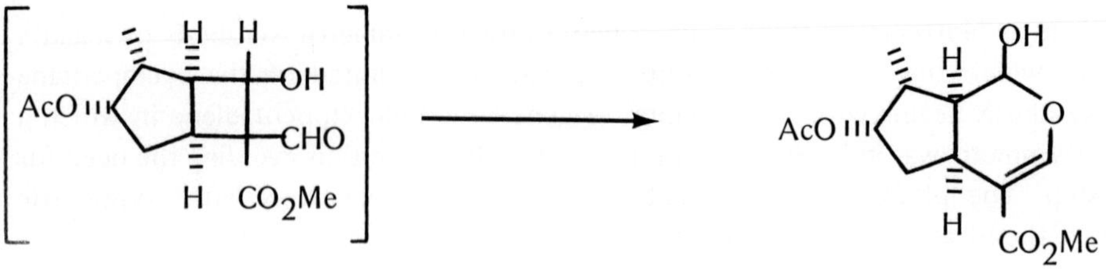

A

LONGIFOLENE

Longifolene, one of nature's most interesting and elegantly constructed sesquiterpenes, was first synthesized by Corey *et al.*[1] using an ingenious intramolecular cyclization reaction of a homodecalin derivative (i→ii). Recently, McMurry and Isser[2] reported a second approach for

(i) (ii)

construction of the intricate carbon-network of longifolene based on the intramolecular alkylation of a bicyclic keto epoxide (A) to the tricyclic compound (B). The olefin derived by dehydration of tertiary alcohol function in (B), after a dibromocarbene addition, undergoes silver ion assisted solvolytic ring enlargement to the allylic alcohol (C). Taking advantage of the potential enone system in (C), the remaining *gem*-methyl group has been introduced by a formal conjugate addition involving an unusual reductive process (→D) and fragmentation to generate the dimethylcycloheptane ring of the natural product.

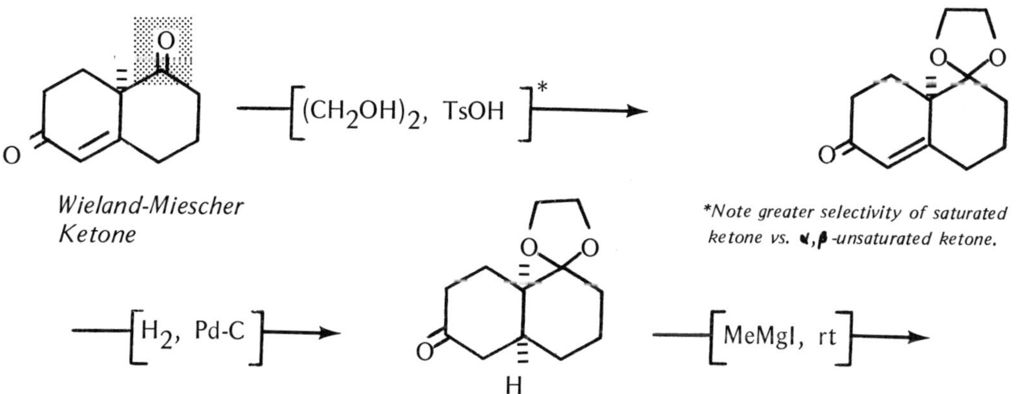

1. E. J. Corey, M. Ohno, R. B. Mitra and P. A. Vatakencherry, J. Amer. Chem. Soc., **86**, 478 (1964); Art Org. Synth., p. 233.

2. J. E. McMurry and S. J. Isser, J. Amer. Chem. Soc., **94**, 7132 (1972).

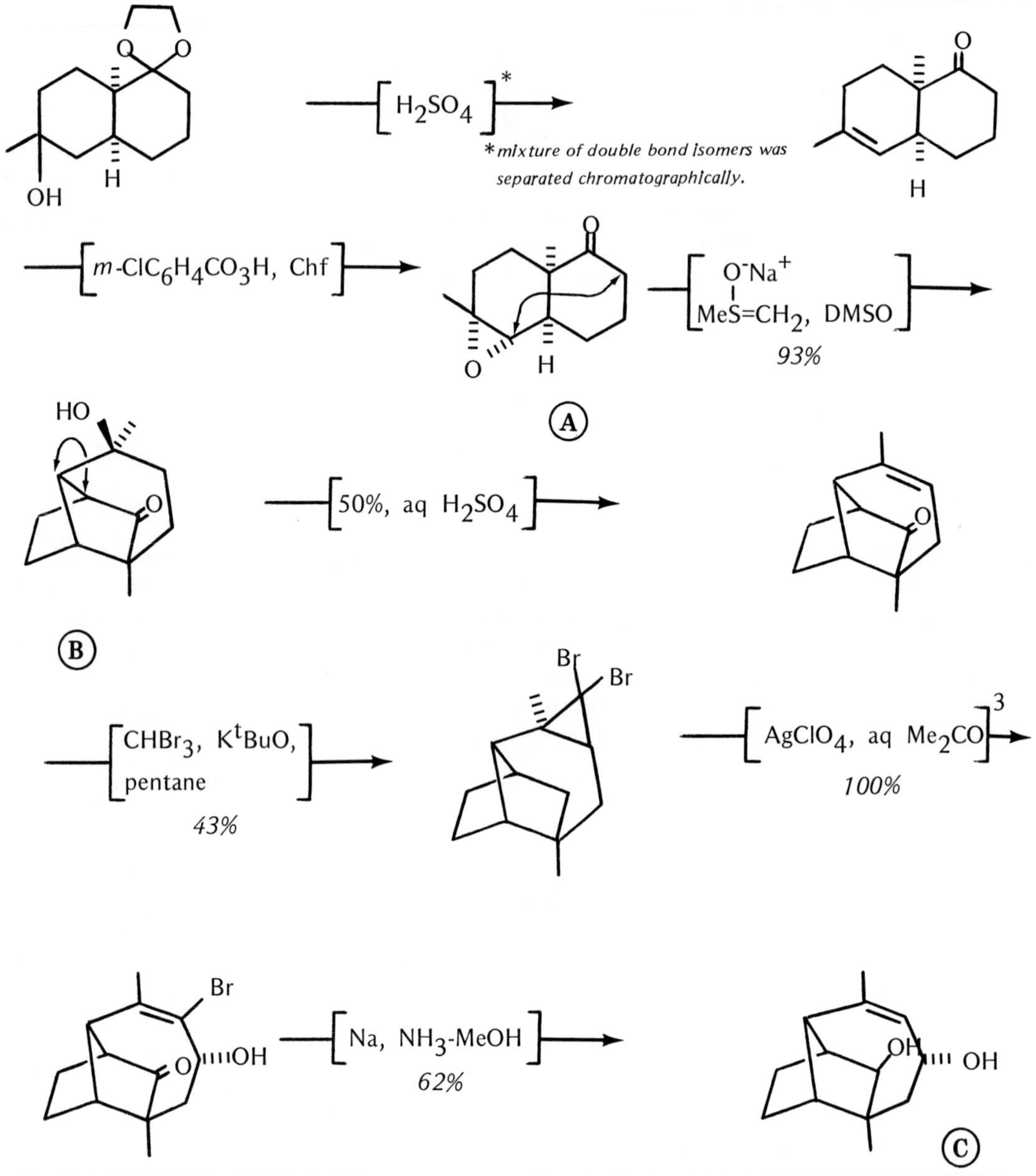

[H_2SO_4]*

*mixture of double bond isomers was separated chromatographically.

[m-$ClC_6H_4CO_3H$, Chf]

(A)

[O^-Na^+ $MeS{=}CH_2$, DMSO] 93%

HO

(B)

[50%, aq H_2SO_4]

[$CHBr_3$, K^tBuO, pentane] 43%

Br Br

[$AgClO_4$, aq Me_2CO][3] 100%

Br

OH

[Na, NH_3-MeOH] 62%

OH OH

(C)

3. Other methods of ring enlargement proved inadequate. However, the sequence involving addition of a dihalocarbene to an olefin followed by silver ion assisted rearrangement and oxidation proceeded smoothly.

[CrO_3, Py] →

[Me_2CuLi, -20°]* → 96%

*This is an intramolecular version of the 'enolate trapping' of an enolate generated via conjugate addition.

O, OH

(D)

[$NaBH_4$] →

OH, H, OH

[MsCl, TEA, CH_2Cl_2] →

OMs, H, OH

[KO^tBu, 45°] → 100%

O

[H_2, $(Ph_3P)_3RhCl$; C_6H_6, MeOH][4] →

O

Longicamphenilone

[1. MeLi; 2. $SOCl_2$, Py] →

Longifolene[5]

4. Wilkinson catalyst: J. F. Young, J. A. Osborn, F. H. Jardine and G. Wilkinson, Chem. Commun., 131 (1965).

5. One of the most interesting aspects of the synthetic problem in the case of longifolene, as also with other bridged polycyclic structures, is the extraordinary variety and diversity of pathways by which the carbon network may be assembled. The importance of exhaustive analysis of the topological properties of the carbon network to define the range of possible precursors from which the desired skeleton can be produced by establishing one or two connecting bonds has been discussed in detail by Corey (Ref. 1).

LUCIDULINE

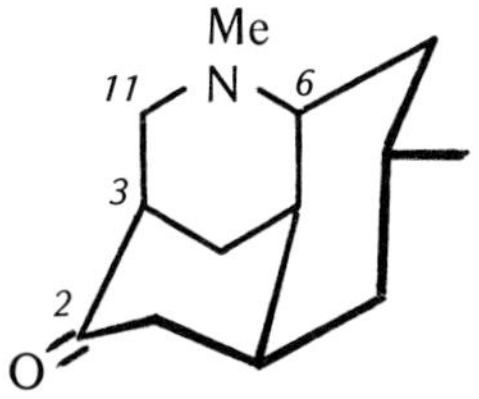

Me
11 N 6
3
2
O

Luciduline is one of the simpler *Lycopodium* alkaloids. In an elegant synthesis of this molecule, Scott and Evans[1] constructed the desired tricyclic skeleton from the key decalin ketoamine (D) by an intramolecular Mannich reaction, C-11 being derived from formaldehyde. The *cis*-decalin-2,6-dione suitable for preparation of (D) was obtained by the oxy-Cope rearrangement[2] of a bicyclo[2·2·2]octene derivative (A) which allowed controlled introduction of the desired sites of asymmetry.

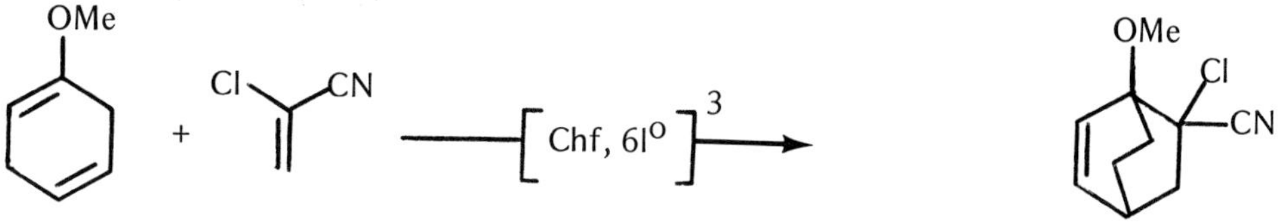

OMe + Cl CN —[Chf, 61°][3]→ OMe Cl CN

---

1. W. L. Scott and D. A. Evans, J. Amer. Chem. Soc., **94**, 4780 (1972).

2. The oxy-Cope rearrangement involves thermal isomerization of 3-hydroxy-1,5-hexadienes (i) to enols (ii), leading to the synthesis of carbonyl compounds [J. A. Berson and M. Jones, J. Amer. Chem. Soc., **86**, 5019 (1964); J. A. Berson and E. J. Walsh, **90**, 4729 (1968)]. For extension of this method to preparation of oxygenated *cis*-decalin carbocycles see, D. A. Evans, W. L. Scott and L. K. Truesdale, Tetrahedron Lett., 137 (1972).

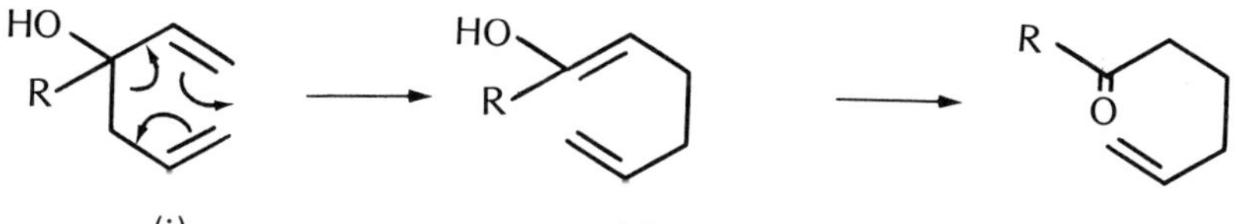

HO R → HO R → R O

(i) (ii)

3. Chloroacrylonitrile serves as a useful ketene equivalent in the Diels-Alder reaction, P. K. Freeman, D. M. Balls and D. J. Brown, J. Org. Chem., **33**, 2211 (1968); D. A. Evans, W. L. Scott and L. K. Truesdale, Tetrahedron Lett., 121 (1972).

[$Na_2S{\cdot}9H_2O$, EtOH] →

OMe

O

≡

O

OMe

[MgBr (isopropenyl)] →

HO

OMe

(A)

[250°]^2 →

65%

H

O

MeO

H

[HOCH$_2$CH$_2$OH, H^+] →

H

O

O

O

H

[$TsNHNH_2$, MeOH] →

H–N–Ts

N

H

O

O

H

[MeLi, Et_2O] →

H

O

O

H

[m-$ClC_6H_4CO_3H$, Chf] →

H

O

O

O

H

[NaSPh, MeOH, Δ] →

85%

[Raney Ni, EtOH, Δ] 82%

[TsCl, Py, 50°]

(B)[4]

[aq HCl, Me_2CO, 50°]

(C)

[$MeNH_2$, C_6H_6, 70° (sealed tube)] 94%

(D)

[$(HCHO)_n$, $C_5H_{11}OH$, Δ]

Luciduline

4. Introduction of the N-methylamine moiety at C-6 by tosylate displacement was thwarted by the sterically congested concave face of the *cis* decalyl system which promoted elimination in addition to substitution. However, intramolecular delivery of nitrogen could be smoothly effected in the ketone (C), probably through intermediacy of the aminal (i).

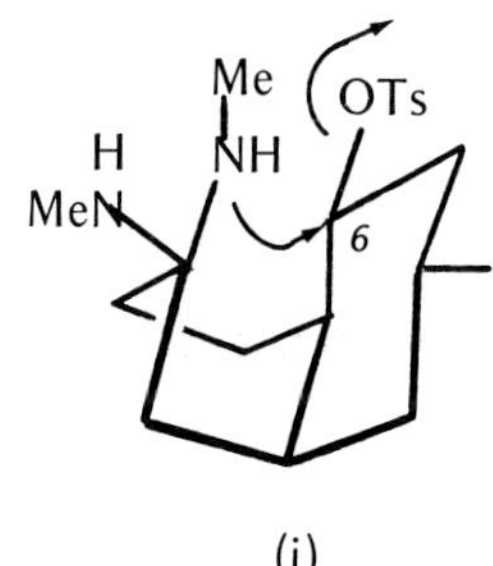

(i)

LUPEOL

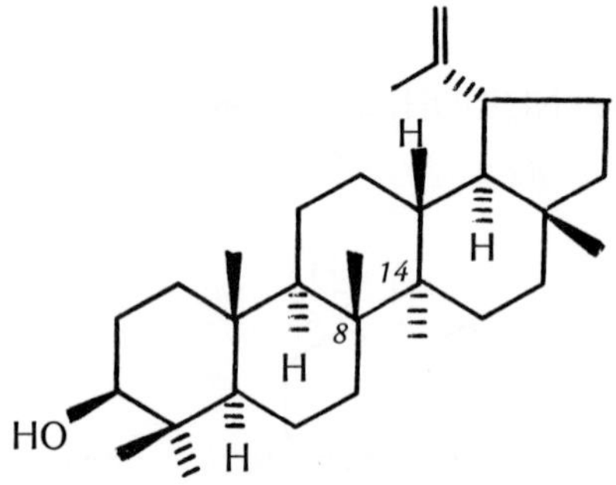

Lupeol, the most abundant of pentacyclic triterpenes, presents a difficult challenge to the synthetic organic chemist, not only because of the presence of ten asymmetric centers in the molecule, but also because of the rather complex structural problems that need to be surmounted for the construction of its non-linear array of rings replete with angular methyl groups. The synthesis of lupeol by Stork and co-workers[1] is based on the tricyclic enone (A), representing rings C, D and E of the pentacyclic system. The tricyclic (A) has been further elaborated to the tetracyclic structure (D) *via* a crucial intermediate (C) already bearing the two vicinal *trans*-methyl groups at C-8 and C-14, together with a substituent at C-8 to serve as a precursor of ring B. The regio- and stereospecific problems attending the establishment of ring A have been efficiently surmounted using the enolate trapping method[2] to afford the pentacyclic system (E) which embodies essentially all the asymmetry of lupeol and can be converted to it by suitable transformations on ring E.

---

1. G. Stork, S. Uyeo, T. Wakamatsu, P. Grieco and J. Labovitz, J. Amer. Chem. Soc., **93**, 4945 (1971); G. Stork, XXIIIrd Int. Congress Pure Appl. Chem., Vol. 2, Butterworths, London (1971) p. 193.

2. The reduction of α,β-unsaturated ketones with lithium in liquid ammonia results in intermediates with nucleophilic character on the β-carbon atom (i). The β-carbon becomes protonated by ammonia, leading to an enolate ion (ii) which can be trapped by a suitable electrophilic reagent, giving rise to a saturated ketone (iii) in which the new group appears on the α-carbon atom. For a definitive account of the synthesis of α-alkylated ketones *via* enolates regiospecifically generated from α,β-unsaturated ketones (i→ii), see G. Stork, P. Rosen, N. Goldman, R. V. Coombs and J. Tsuji, J. Amer. Chem. Soc., **87**, 275 (1965).

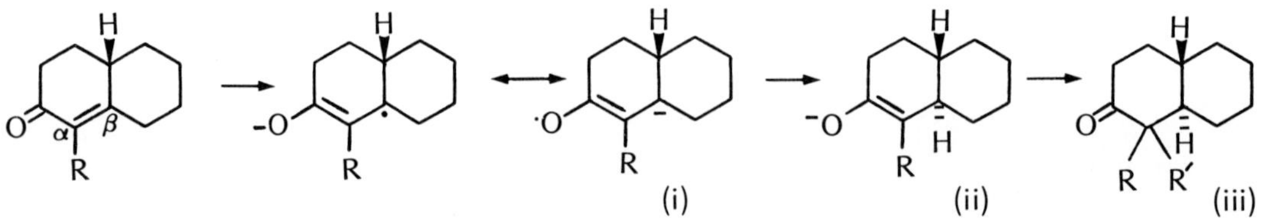

(i) (ii) (iii)

C D

MeO

Footnote 3

MeO CHOH

[MVK→ aq base]

MeO

[OH, OH, TsOH]

MeO

[MeI, B⁻]

MeO

[H^+]

MeO

[$NaBH_4$]

OH

MeO

---

3. The initial stages of the synthesis leading to (A) follow a route developed quite some time ago in connection with the total synthesis of steroids by G. Stork, H. J. E. Loewenthal and P. C. Mukharji, J. Amer. Chem. Soc., **78**, 501 (1956).

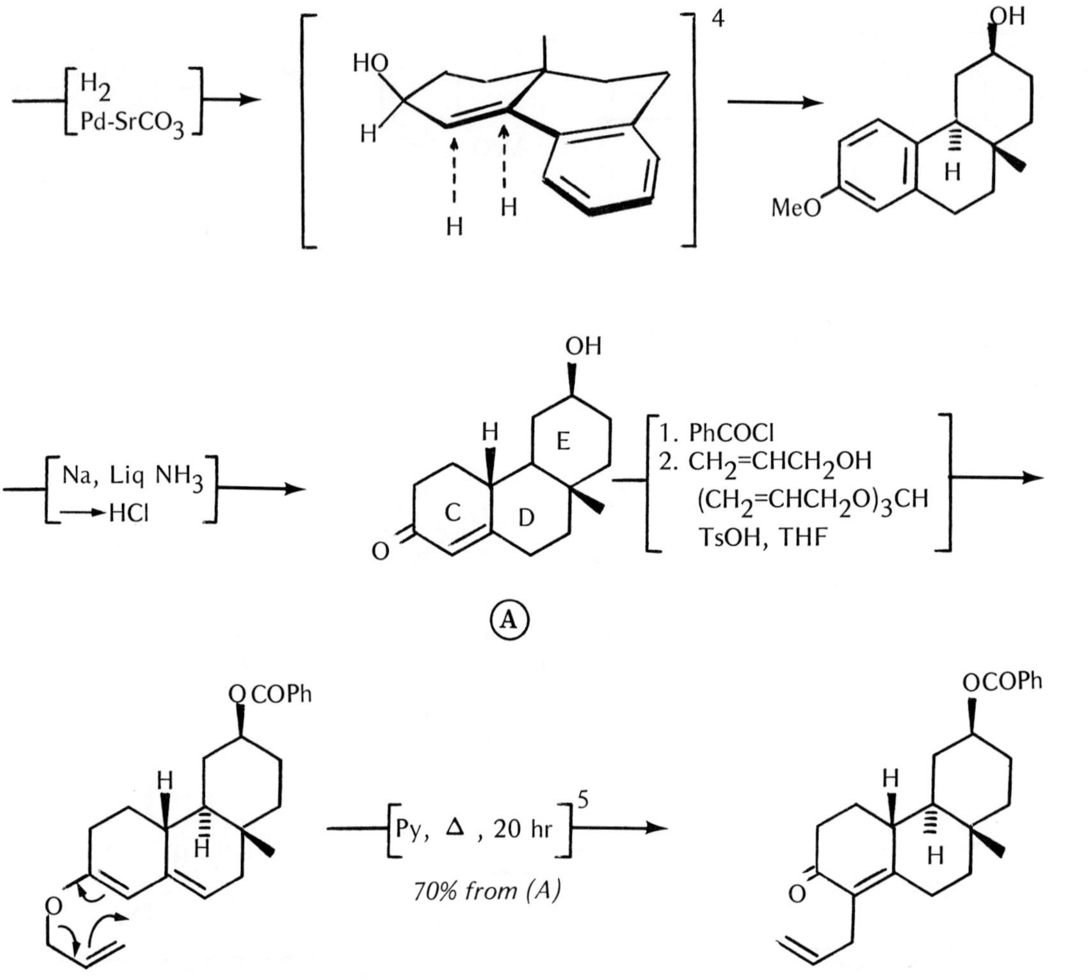
H2
Pd-SrCO3
4
HO
H
H
H
OH
H
MeO
Na, Liq NH3
HCl
OH
H
E
C
D
O
A
1. PhCOCl
2. CH2=CHCH2OH
(CH2=CHCH2O)3CH
TsOH, THF
OCOPh
H
H
O
Py, Δ, 20 hr
5
70% from (A)
OCOPh
H
H
O

4. Since the molecule is essentially flat and rigid, catalytic hydrogenation takes place *trans* to the axial methyl group, thus establishing the necessary stereochemistry of the erstwhile D/E junction.

5. Use of Claisen rearrangement for internal delivery of the alkyl substituent avoids the problem of polyalkylation which often attends alkylation of α,β-unsaturated ketones.

[Et_2AlCN] 90%

OCOPh, H, H, O, CN

[OH/OH, TsOH]

OCOPh, H, H, O, O, CN

[1. LAH, THF, Δ
2. aq AcOH, THF, MeOH
NaOAc]

OH, H, H, O, O, CHO

[PhCOCl]

OCOPh, H, H, O, O, CHO

[1. $NaBH_4$
2. MsCl]

OCOPh, H, H, O, O, CH_2OMs

[1. aq HCl, THF
2. B^-]
85%

OH, H, H, O, CH_2

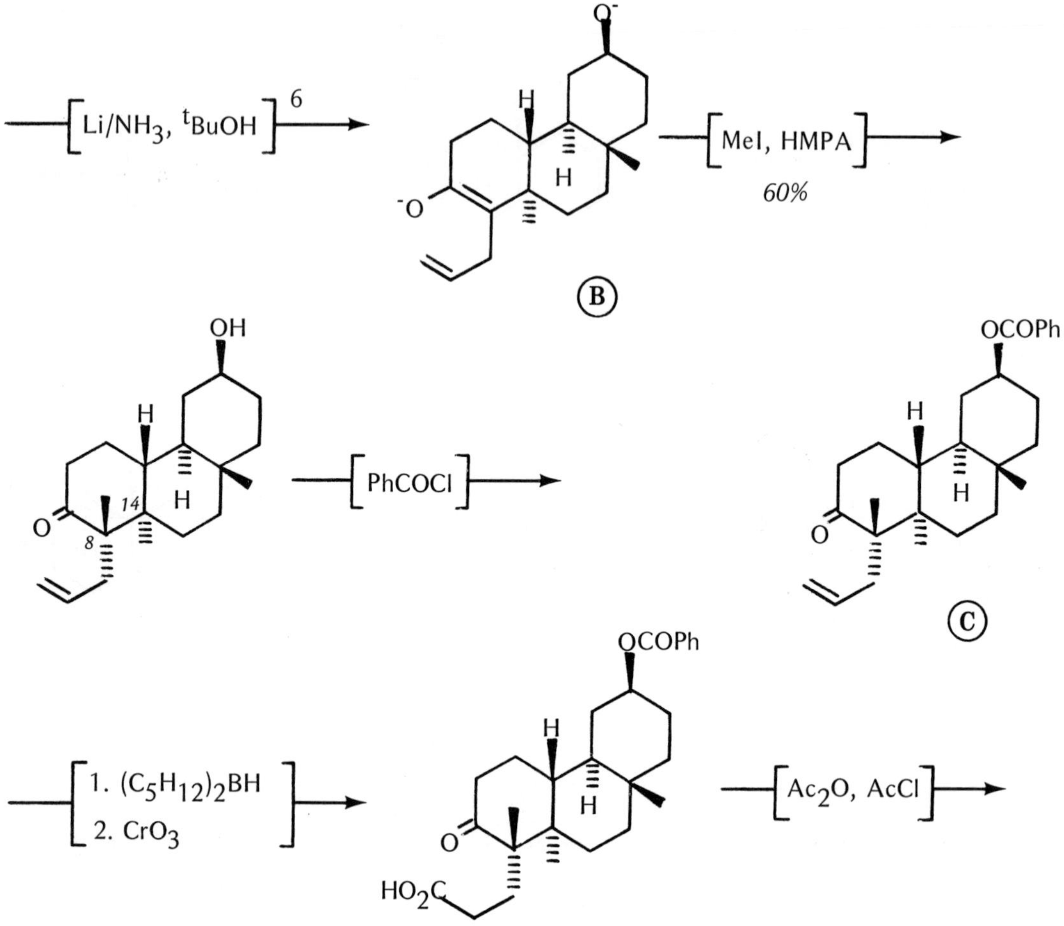

6. Methylation of the enolate (B) derived from a cyclopropyl ketone for stereospecific introduction of the two vicinal *trans* methyl groups in ring C further illustrates the concept of enolate trapping developed by Stork (ref. 2). In this instance, generation of the enolate has been accomplished by the reductive removal of an α-substitutent (i→ii); cf. W. G. Dauben and E. J. Deviny, J. Org. Chem., **31**, 3794 (1966).

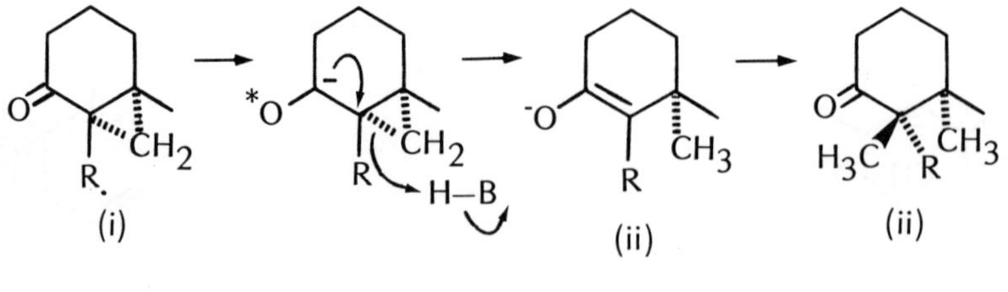

OCOPh

[1. EtMgBr 2. aq base][7]

OH

D

[Li, liq NH_3]

OH

[1. CH_2=$CHCH_2Br$ 2. PhCOCl][8]

50%

OCOPh

[1. Ketalize; 2. 9-BBN
3. CrO_3; 4. Ac_2O, AcCl
5. EtMgBr, Et_2O, THF, -30°
6. aq NaOH, MeOH]

---

7. *Cf.* R. B. Turner, J. Amer. Chem. Soc., **72**, 579 (1950); G. I. Fujimoto, *ibid.*, **73**,1856 (1951).

8. Regiospecific α-alkylation of the ketone occurs in a transition state involving the approach of the alkylating agent as depicted in (i). Approach of the alkylating agent from the opposite direction, however, is beset by unfavorable interaction between R and the axial substituents at C-ll and C-8 (ii).

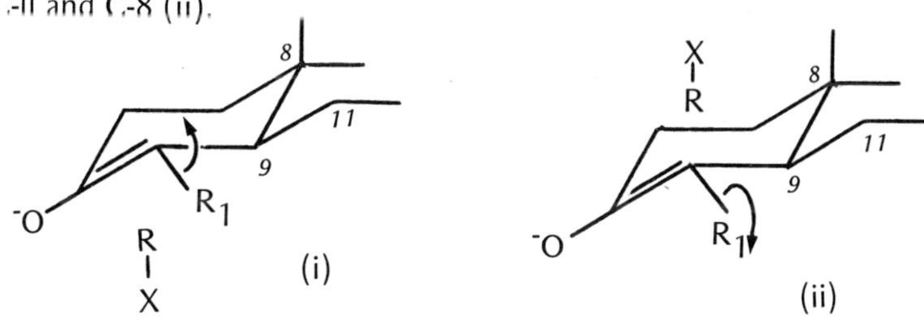

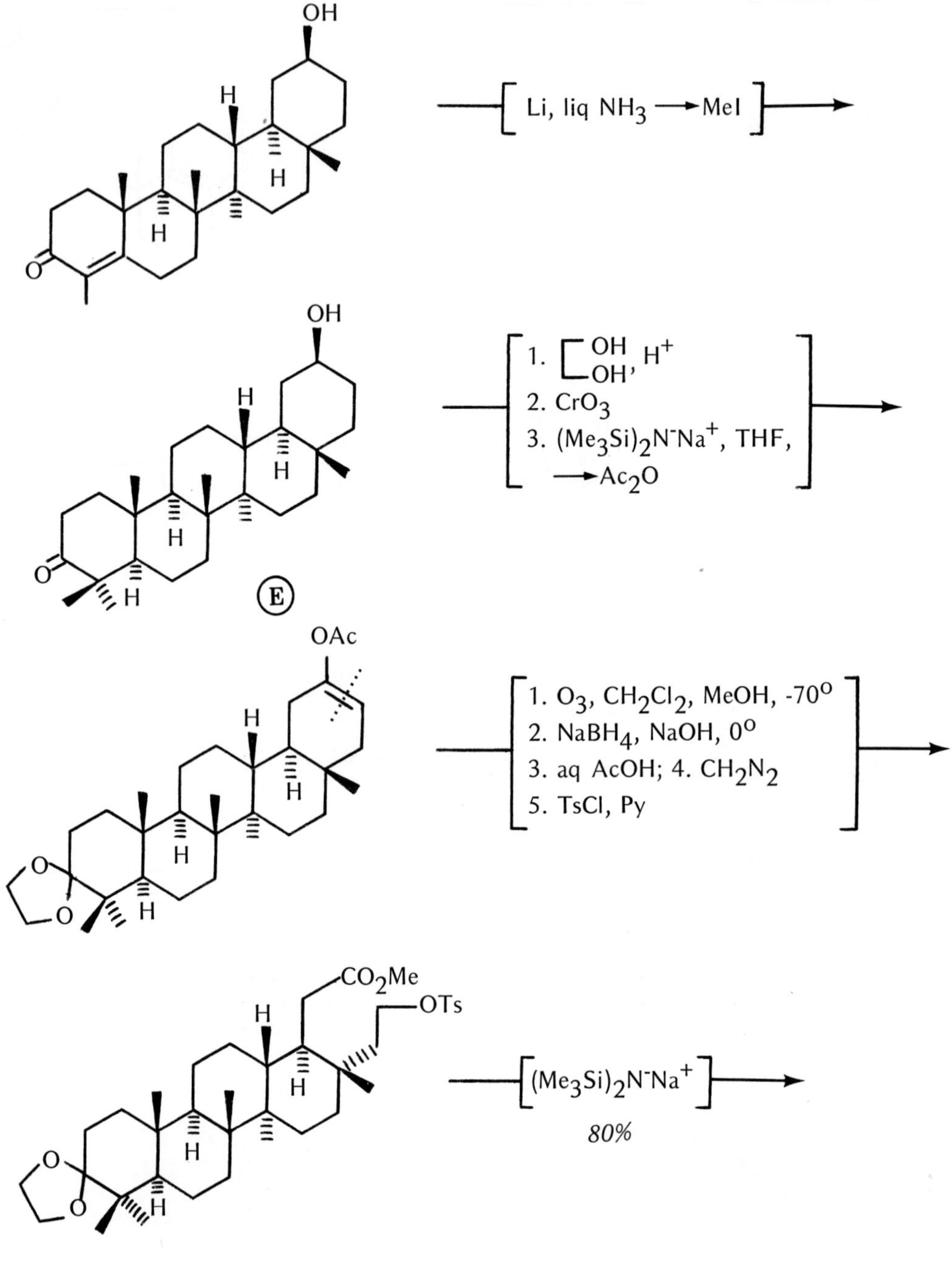

OH

H

H

H

O

[Li, liq NH_3 → MeI]

OH

H

H

H

O

H

(E)

[1. $\ulcorner$OH $\llcorner$OH, H^+
2. CrO_3
3. $(Me_3Si)_2N^-Na^+$, THF, → Ac_2O]

OAc

H

H

H

O

O

H

[1. O_3, CH_2Cl_2, MeOH, -70°
2. $NaBH_4$, NaOH, 0°
3. aq AcOH; 4. CH_2N_2
5. TsCl, Py]

CO_2Me

OTs

H

H

H

O

O

H

[$(Me_3Si)_2N^-Na^+$]
80%

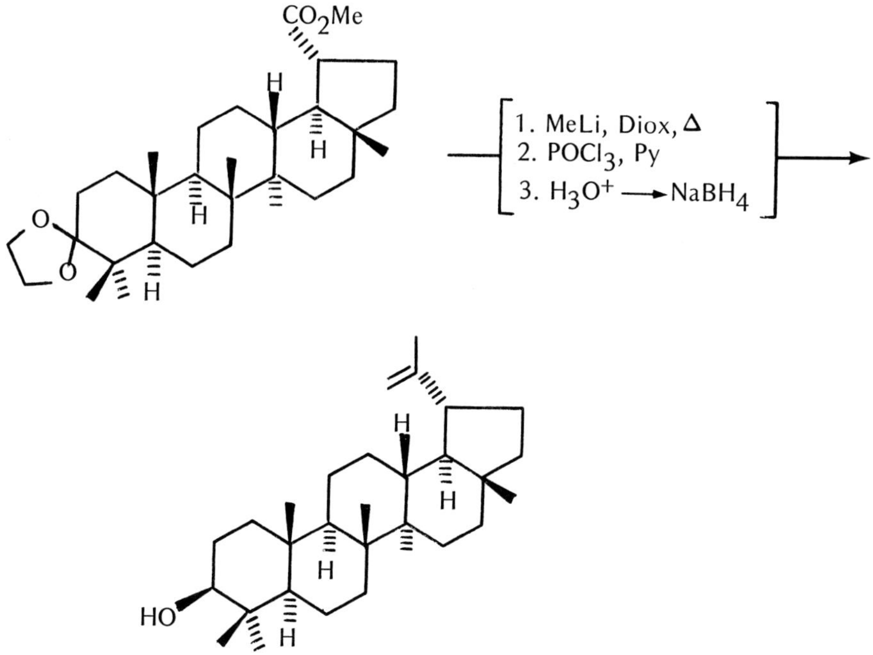

CO_2Me

1. MeLi, Diox, Δ
2. $POCl_3$, Py
3. $H_3O^+ \longrightarrow NaBH_4$

HO

Lupeol

LYSERGIC ACID

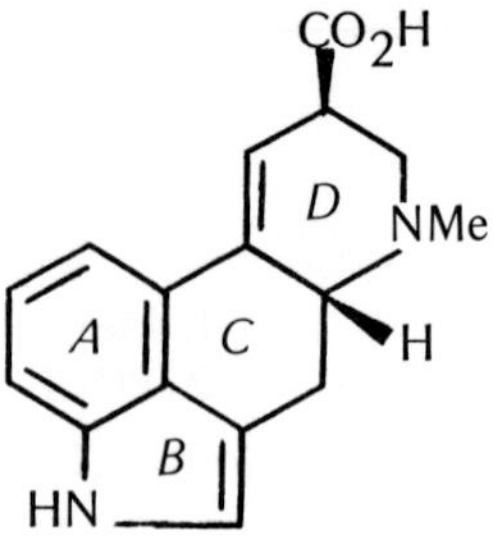

Julia's synthesis of lysergic acid[1,2] is based on internal nucleophilic trapping of a benzyne intermediate in the ring C-seco intermediate (A)[3]. As in an earlier approach, the synthesis proceeds *via* indoline derivatives and formation of the indole nucleus is deferred until the very end.[4]

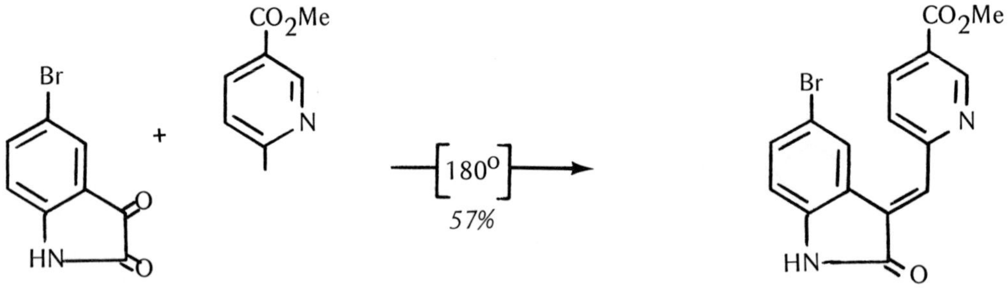

[180°]

57%

1. Review: A. Stoll and A. Hofman, The Alkaloids, **8**, 725 (1965).

2. M. Julia, F. LeGoffic, J. Igolen and M. Baillarge, Tetrahedron Lett., 1569 (1969).

3. For application of 'arynic substitution' in the synthesis of related hexahydrobenzo[f]quinolines, see M. Julia, J. Igolen and F. LeGroffic, Bull. Soc. Chim. France, 4436 (1968); H. Heaney [Chem. Rev., **62**, 81 (1962)] has reviewed the chemistry of benzynes.

4. This strategy for circumventing the rather marked propensity of benzindoles to isomerize to naphthalenoid structures was originally devised in the first synthesis of lysergic acid by E. C. Kornfeld, E. J. Fornefeld, G. B. Kline, M. J. Mann, D. E. Morrison, R. G. Jones and R. B. Woodward, J. Amer. Chem. Soc., **78**, 3087 (1956); Art Org. Synth., 243 (1970).

[Zn, AcOH, Δ] 95%

[B_2H_6, THF]

[Ac_2O]

[MeI, MeOH, Me_2CO 80° (sealed tube)]

[KBH_4, aq MeOH] 98%

[$NaNH_2$, NH_3-THF] 15%

Ⓐ

1. HCl
2. Na_2HAsO_4
Raney Ni, Xy, Δ [5]

CO_2H NMe H HN

Lysergic Acid [6]

---

5. Recently it has been found that active MnO_2 is a suitable reagent for dehydrogenation of indolines to indoles, A. B. A. Jansen, J. M. Johnson and J. R. Surtees, J. Chem. Soc., 5573 (1974).

6. For some recent attempts at the synthesis of lysergic acid, see R. E. Bowman, D. D. Evans, J. Guyett, J. Weak and A. C. White, J. Chem. Soc., Perkin I, 760 (1973).

MALABARICANEDIOL

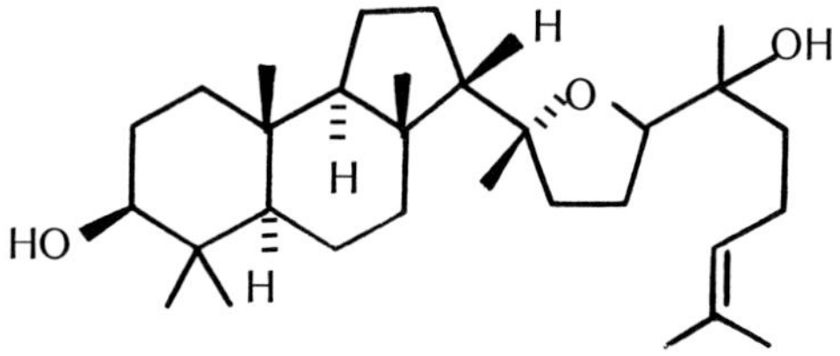

Malabaricanediol is the first squalene-derived natural product to be constructed by a nonenzymic biogenetic-type synthesis.[1] Its synthesis[2] involves conversion of the intact squalene molecule into a specific internal oxidation product, the epoxydiol (B),[3] which can be transformed to *dl*-malabricanediol by an acid-catalyzed cyclization.[4]

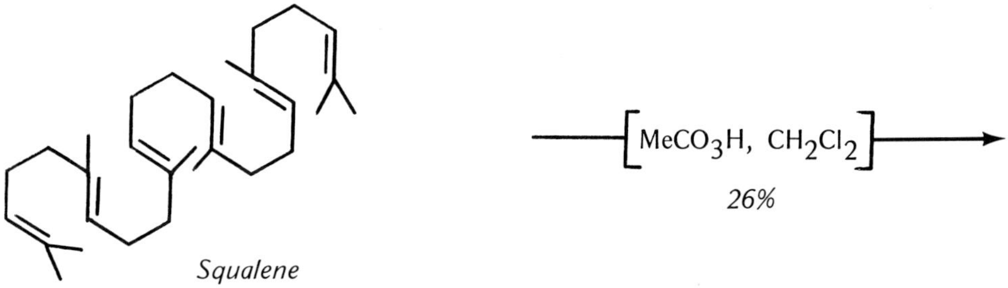

Squalene

[$MeCO_3H$, CH_2Cl_2]

26%

1. Although at the time of its isolation, the unusual perhydrocyclopenta[a]naphthalene skeleton of malabaricanediol was previously unknown among natural triterpenoids [A. Chawla and S. Dev., Tetrahedron Lett., 4837 (1967)], the ring system had been observed earlier as one of the non-enzymic cyclization products of squalene 2,3-oxide by E. E. van Tamelen, J. Willet, M. Schwartz and R. Nadeau, J. Amer. Chem. Soc., **88**, 5937 (1966).

2. K. B. Sharpless, J. Amer. Chem. Soc., **92**, 6999 (1970).

3. K. B. Sharpless, Chem. Commun., 1450, (1970).

4. For a recent review of biogenetic-type synthesis, see D. Goldsmith, Fortsch. Chem. Org. Naturst., **29**, 363 (1973).

Footnote 5

14 O 15 3 2 + 18 O 19 — [aq $HClO_4$, ^tBuOH][6] →

OH OH — [Ac_2O, Py] →

OAc OH — [NBS, ^tBuOH] →

---

5. Epoxidation of squalene afforded a mixture of squalene 2,3-oxide along with the two *trans* internal oxides shown above. Controlled acidic hydrolysis allowed selective opening of the terminal epoxide moiety to a diol from which the desired internal oxides were readily separated by thiourea clathrate formation.

6. The epoxide mixture was hydrolyzed to a mixture of the corresponding *erythro*-diols which were separated by chromatography on silver nitrate-impregnated silica gel. Only the desired diol arising from the 18,19-epoxide is shown above.

Br

OAc

(A)[7]

[H_2O, NBS ⟶ CO_3^{2-}] ⟶

Br

OAc

O

[1. Zn, AcOH (12 eq.)
2. KOH, MeOH][8] ⟶

OH

HO

(B)

H^+ O

[Picric Acid, $MeNO_2$, rt] ⟶

7%

7. Formation of the cyclic bromo-ether (A) serves to protect one terminal double bond during epoxidation of the other, and is regenerated during reduction by zinc dust.

8. The zinc dust reduction of the epoxy-bromo-ether requires only a calculated amount of acetic acid catalyst. If a large quantity of acetic acid is used, the 2,3-oxide function is reduced to the corresponding olefin.

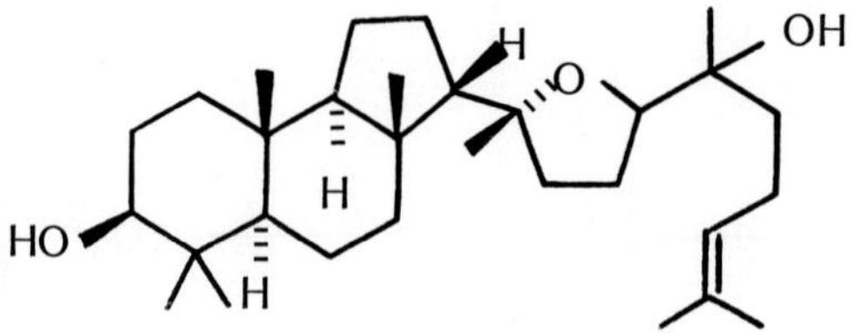

HO OH H H H O

Malabaricanediol

MARITIDINE

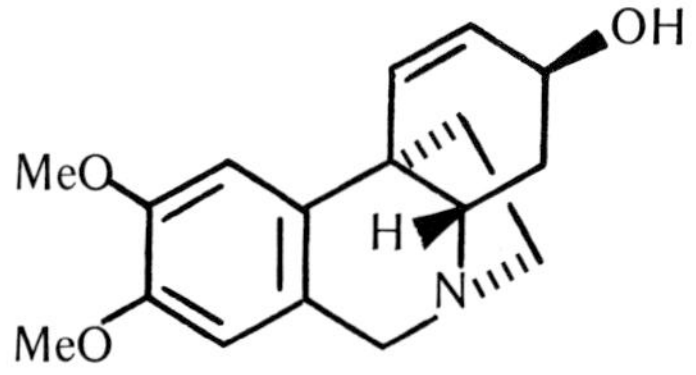
OH
MeO
H
N
MeO

Intramolecular oxidative phenol coupling has long been recognized as the key step in biosynthesis of *Amaryllidaceae* alkaloids.[1] Chemical simulation of this biogenetic process is expected to provide an exceedingly simple synthetic route to these alkaloids, and is illustrated in the synthesis of *dl*-maritidine[2] from O-methylnorbelladine utilizing an efficient new oxidative phenol coupling procedure.[3]

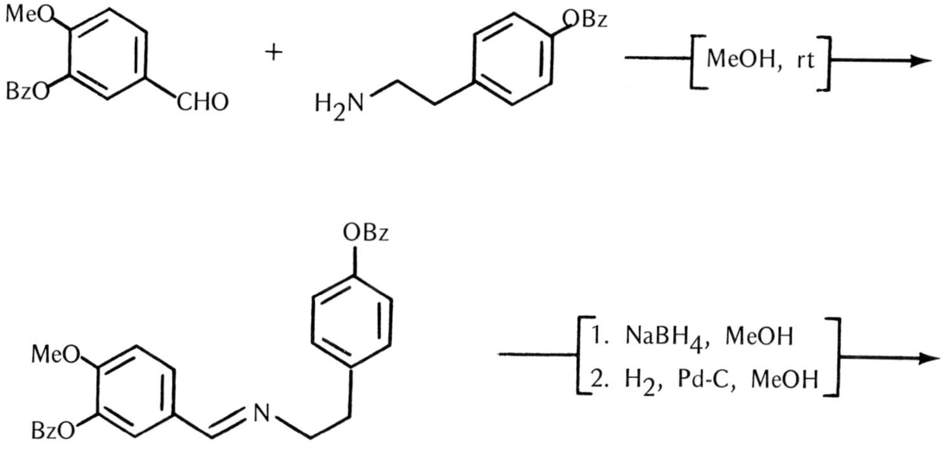
MeO
BzO
CHO
+
OBz
H2N
MeOH, rt
OBz
MeO
BzO
N
1. NaBH4, MeOH
2. H2, Pd-C, MeOH

1. D. H. R. Barton, Chem. Britain, **3**, 330 (1967); A. I. Scott, Quart. Rev., **19**, 1 (1965).

2. M. A. Schwartz and R. A. Holton, J. Amer. Chem. Soc., **92**, 1090 (1970).

3. In practice, intramolecular phenol coupling using external oxidizing agents (e.g., potassium ferricyanide) often results in rather extensive polymerization. However, utilizing vanadium oxytrichloride which reacts with phenols to give a phenoxyvanadium(V) compound, the oxidizing agent is incorporated into the diphenol prior to actual oxidation. The electron-transfer step can then be profitably carried out in an inert solvent and in the absence of excess oxidizing agent [M. A. Schwartz, R. A. Holton and S. W. Scott, J. Amer. Chem. Soc., **91**, 2800 (1969)]. For oxidative coupling utilizing thallium(III) trifluoroacetate, see M. A. Schwartz, B. F. Rose and B. Vishnuvajjala, J. Amer. Chem. Soc., **95**, 612 (1973). For analogous oxidative couplings see also ref. 4.

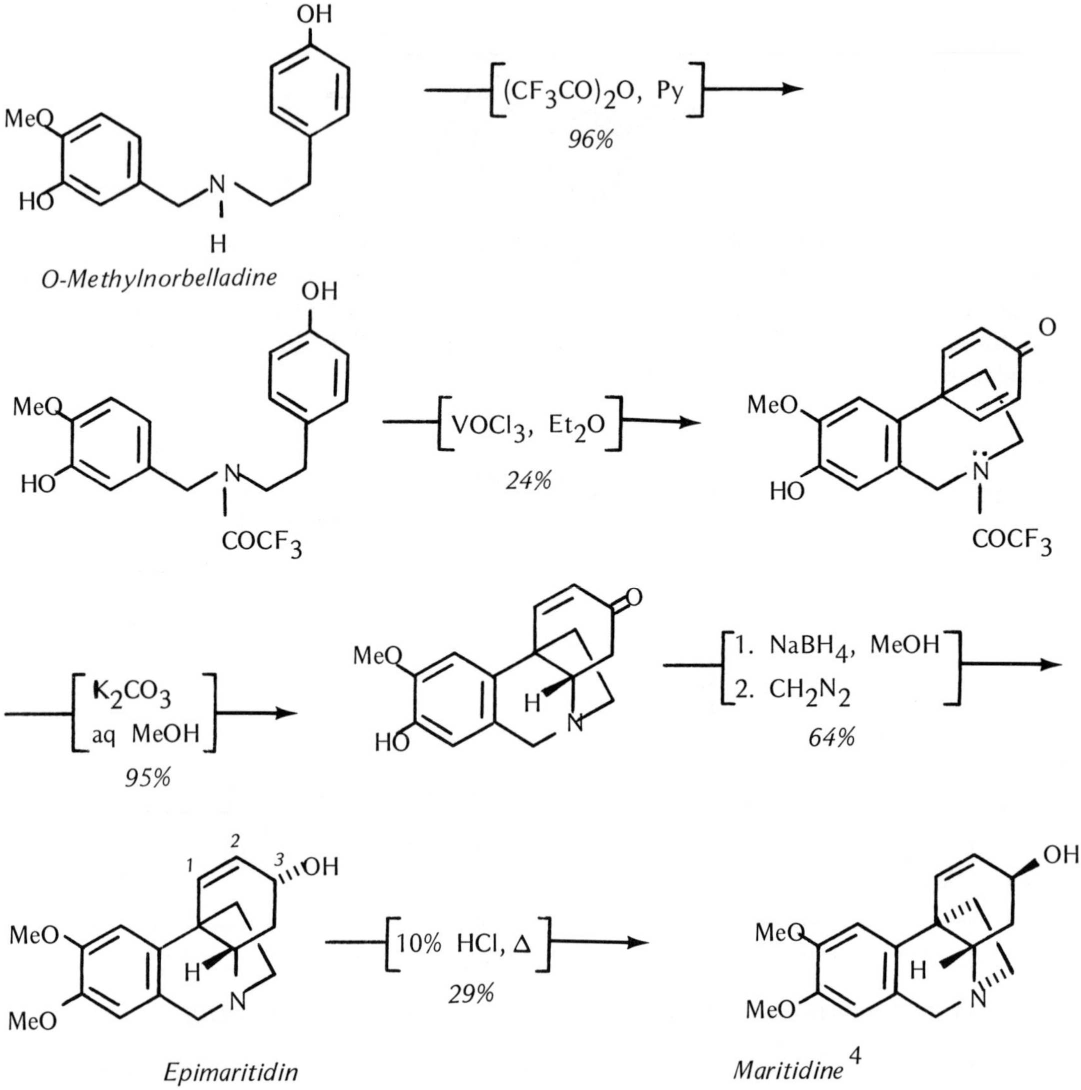

OH

MeO

HO

N

H

O-Methylnorbelladine

[$(CF_3CO)_2O$, Py] 96%

OH

MeO

HO

N

$COCF_3$

[$VOCl_3$, Et_2O] 24%

MeO

HO

N

$COCF_3$

O

[K_2CO_3 aq MeOH] 95%

MeO

HO

H

N

O

[1. $NaBH_4$, MeOH 2. CH_2N_2] 64%

1 2 3 OH

MeO

MeO

H

N

Epimaritidin

[10% HCl, Δ] 29%

OH

MeO

MeO

H

N

Maritidine[4]

4. A formal total synthesis of *dl*-maritidine, utilizing the photochemical cyclization of a bromoaromatic compound (i, x=Br) in one step to the crinan ring system (ii) has been reported by T. Kametani, T. Kohno, S. Shibuya and K. Fukumoto, Chem. Commun., 774 (1971). Oxomaritidine is also obtained in good yield via anodic oxidation of the

OMe OH X O O N H (i) hν aq NaOH EtOH OMe O O O N (ii)

trifluoroacetyl derivative of (i, x=H) [E. Kotani, N. Takeuchi and S. Tobinaga, Chem. Commun., 550 (1973)], or utilization of iron-DMF complex $[Fe(DMF)_3Cl_2][FeCl_4]$ for the oxidative coupling step [E. Kotani, N. Takeuchi and S. Tobinaga, Tetrahedron Lett., 2735 (1973)].

5. For a synthesis of the crinan ring system by photocyclization of enamides, see I. Ninomiya, T. Naito and T. Kiguchi, J. Chem. Soc. Perkin I, 2261 (1973).

6. For a review of synthetic approaches to crinine, see Art Org. Synth., 138 (1970).

METAPHANINE

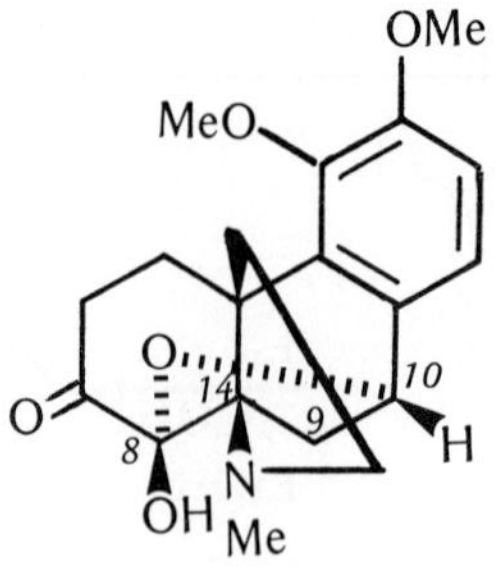

The interesting aza-propellane skeleton of metaphanine, a complex hasubanan alkaloid,[1] has been effectively constructed by Ibuka *et al.*[2-4] using a modified Robinson annelation reaction.[3,5] The synthesis is completed by stepwise oxidative introduction of oxygen functions at C-8 and C-10 which eventually furnishes the unique hemiketal ring in the molecule.

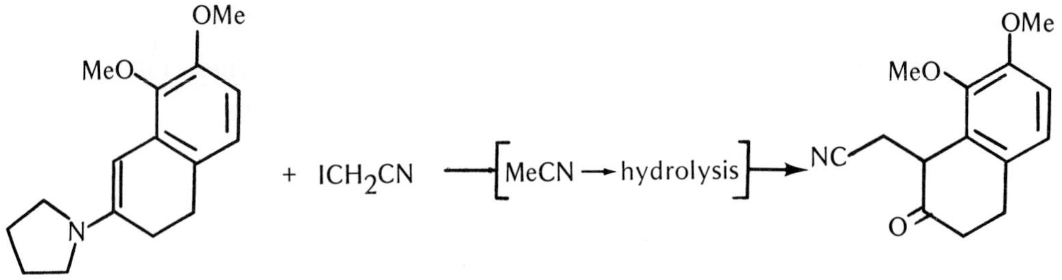

+ ICH_2CN —[MeCN → hydrolysis]→

1. The hasubanan alkaloids, a small group of alkaloids isolated from *Stephania* species of Menispermaceae, are structurally related to morphine. The hasubanan skeleton, however, is different from morphinan in that it contains a five-membered heterocyclic ring formed by linkage of the ethanamine nitrogen with C-14 instead of C-9.

2. T. Ibuka, K. Tanaka and Y. Inubushi, Tetrahedron Lett., 1393 (1972); Chem. Pharm. Bull., **22**, 907 (1974).

3. Details of construction of the hasubanan skeletal ring system and total synthesis of *dl*-cepharamine, the functionally least complicated of hasubanan alkaloids have been reported earlier by T. Ibuka, K. Tanaka and Y. Inubushi, Tetrahedron Lett., 1611 (1969); Y. Inubushi, M. Kitano and T. Ibuka, Chem. Pharm. Bull. Japan, **19**, 1820 (1971).

4. For the total synthesis of *dl*-hasubanonine, see T. Ibuka, K. Tanaka and Y. Inubushi, Tetrahedron Lett., 4811 (1970); Chem. Pharm. Bull., **22**, 782 (1974).

5. For an alternative synthesis of the hasubanan ring system, see M. Tomita, M. Kitano and T. Ibuka, Tetrahedron Letter, 3391 (1968).

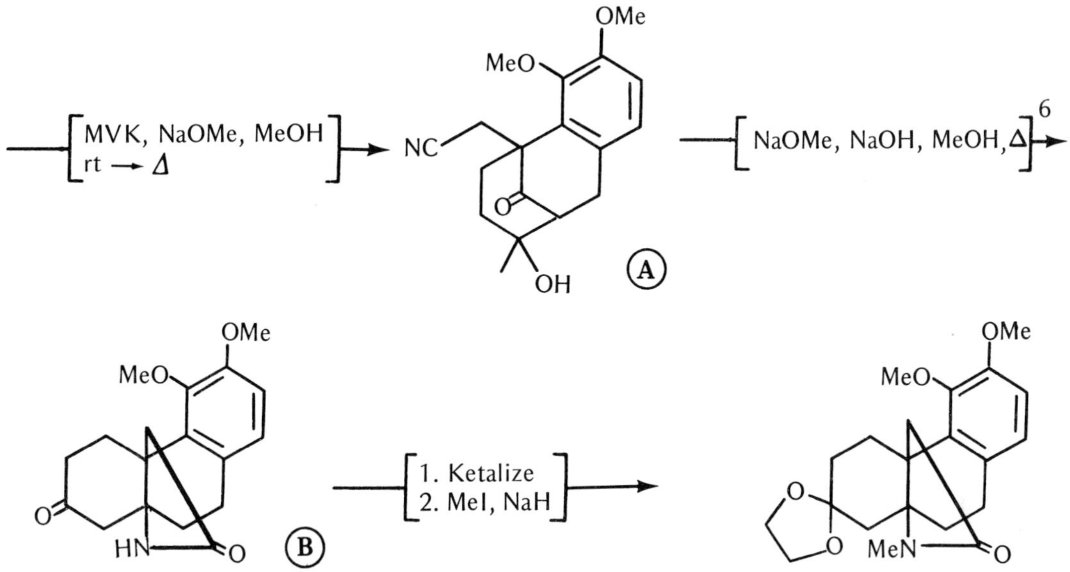

6. The annelation presumably proceeds by the pathway outlined below:

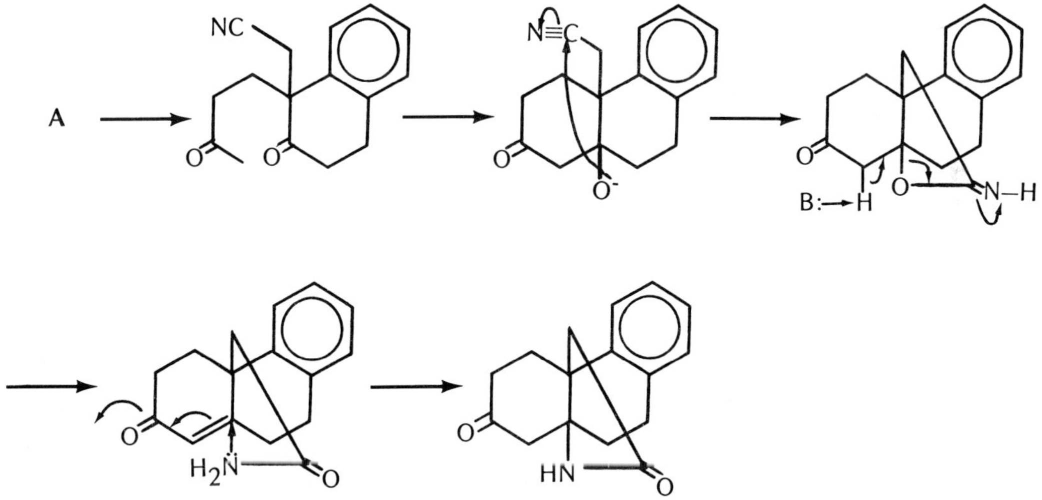

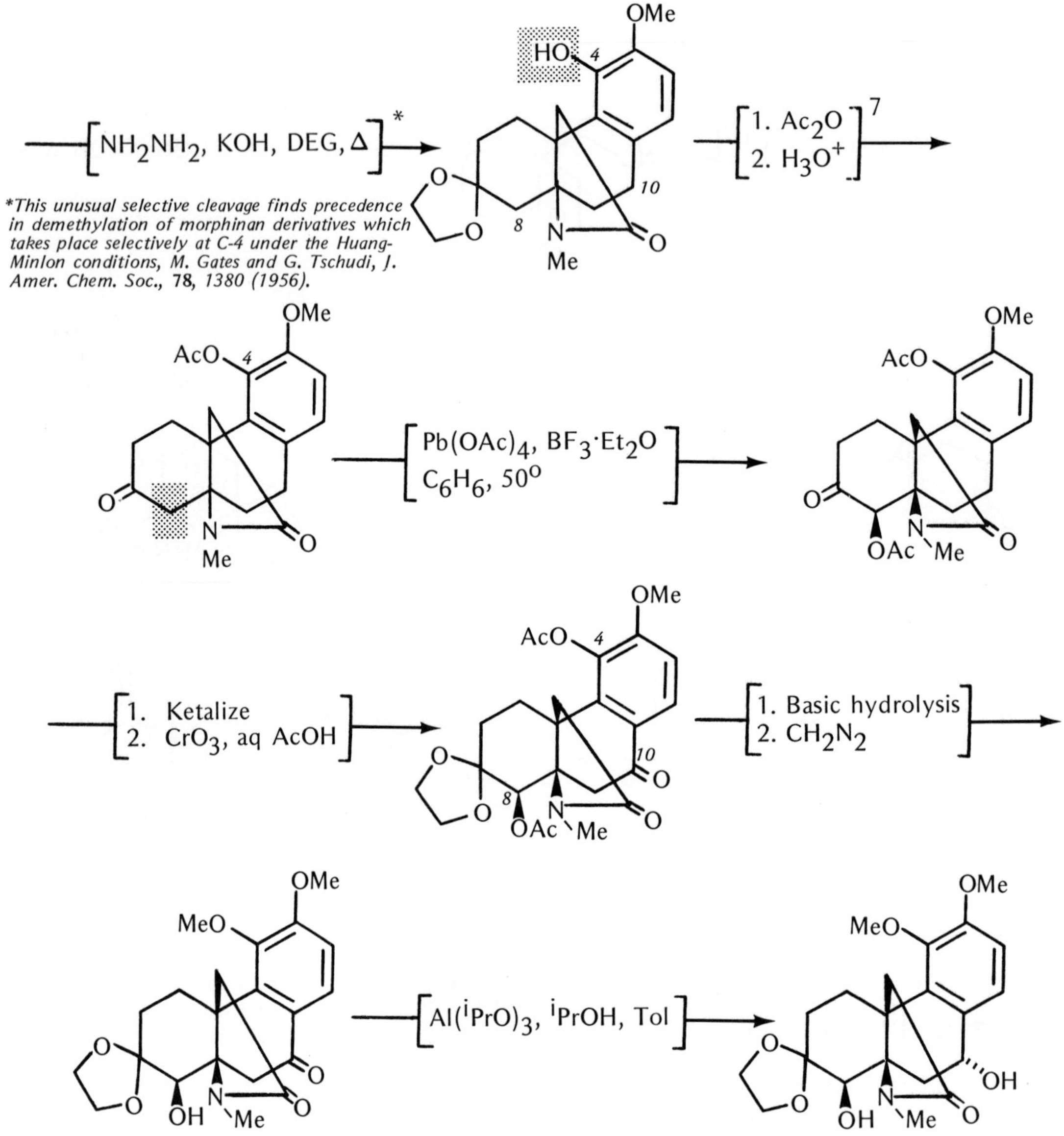

[NH_2NH_2, KOH, DEG, Δ]*

[1. Ac_2O 2. H_3O^+][7]

This unusual selective cleavage finds precedence in demethylation of morphinan derivatives which takes place selectively at C-4 under the Huang-Minlon conditions, M. Gates and G. Tschudi, J. Amer. Chem. Soc., 78, 1380 (1956).

[$Pb(OAc)_4$, $BF_3 \cdot Et_2O$ C_6H_6, 50°]

[1. Ketalize 2. CrO_3, aq AcOH]

[1. Basic hydrolysis 2. CH_2N_2]

[$Al(^iPrO)_3$, ^iPrOH, Tol]

7. Attempts at introduction of the acetoxy group at C-8 and oxidation of the benzylic methylene at C-10 in the dimethoxy keto lactam derived from (B) were fruitless. However, satisfactory results were obtained when the C-4 methoxy group was replaced by an acetoxy group!

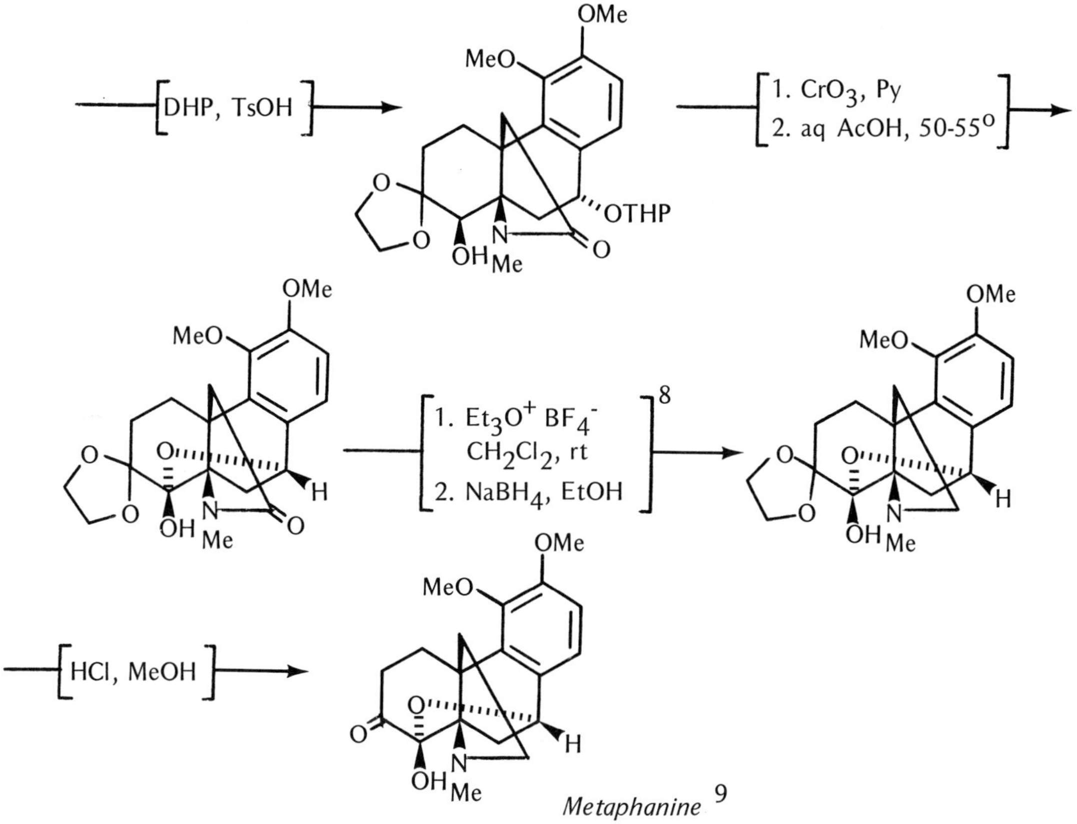

[DHP, TsOH] →

[1. CrO_3, Py
2. aq AcOH, 50-55°] →

[1. $Et_3O^+ BF_4^-$
CH_2Cl_2, rt
2. $NaBH_4$, EtOH][8] →

[HCl, MeOH] →

Metaphanine[9]

---

8. R. F. Borch, Tetrahedron Lett., 61 (1968).

9. Two independent but related approaches to the hasubanan carbocyclic system, based on the endocyclic enamine annelation technique (i→ii), have been devised by D. A. Evans, Tetrahedron Lett., 1573 (1969); D. A. Evans, C. A. Bryan and G. M. Wahl, J. Org. Chem., **35**, 4122 (1970); and S. L. Keely, A. J. Martinez and F. C. Tahk, Tetrahedron Lett., 2763 (1969); Tetrahedron, **26**, 6729 (1970).

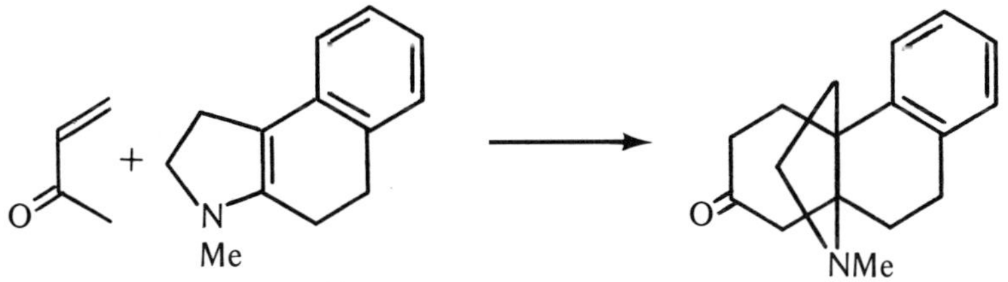

(i) (ii)

MINOVINE

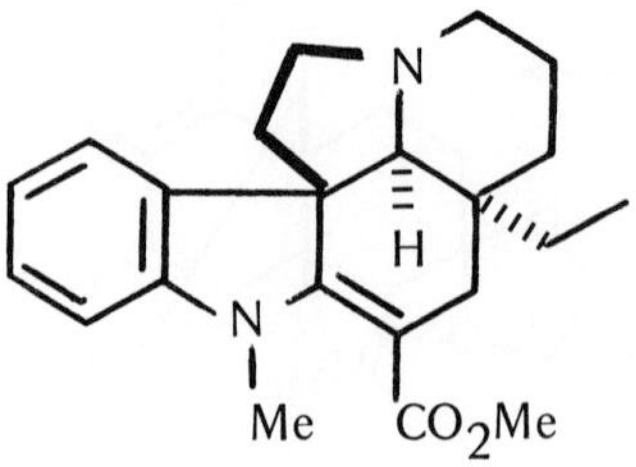

The synthesis of minovine is modeled after Wenkert's speculations about the biosynthesis of indole alkaloids.[1] Based on the consideration that the sequence *Corynanthe→Aspidosperma→Iboga* proceeds through an indolyl acrylic ester (A)[2], Ziegler and Spitzner[3] elaborated the two characteristic reactive functionalities in (A), *viz.*, the endocyclic enamine (C) and the 2,2'-indole acrylic ester moiety (B), in separate substrates and brought these together in a Mannich reaction, effecting closure of ring C. Incorporation of the two-carbon tryptamine residue has been effected by an initial alkylation of the secondary amine (N_b) followed by β-indole alkylation using an appropriate 1,2-disubstituted electrophilic ethane.

Preparation of indole component:

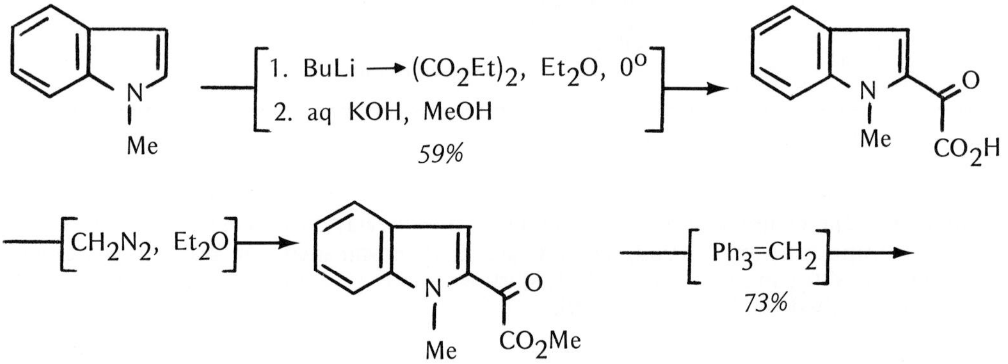

N
Me
1. BuLi → (CO2Et)2, Et2O, 0°
2. aq KOH, MeOH
59%
N
Me
O
CO2H
CH2N2, Et2O
N
Me
O
CO2Me
Ph3=CH2
73%

1. E. Wenkert, J. Amer. Chem. Soc., **84**, 98 (1962).

2. A. I. Scott, Accounts Chem. Res., **3**, 151 (1970).

3. F. E. Ziegler and E. B. Spitzner, J. Amer. Chem. Soc., **92**, 3492 (1970); **95**, 7146 (1973).

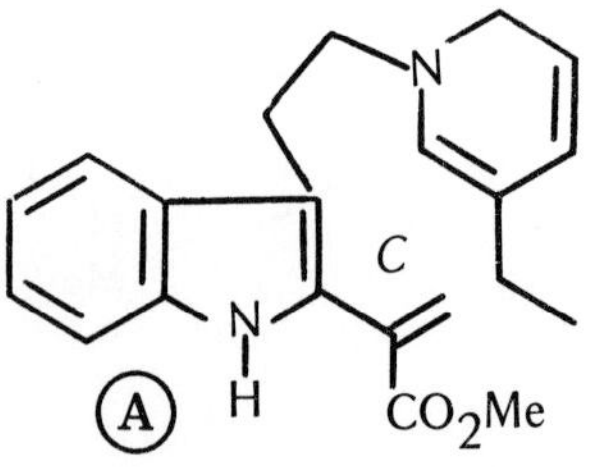

N
C
N
H
A
CO2Me

Me CO_2Me (B)

Preparation of piperidine part (C):

N + CN → → CHO CN —[$(CH_2OH)_2$, TsOH]→

CN —[1. LAH, Et_2O 2. C_6H_5CHO → H_2, Pd-C]→ O O NHBz

—[1. Conc HCl 2. Na_2CO_3]→ Bz N (C)

Preparation of minovine:

N Me C CO_2Me (B) + Bz N D (C) —[MeOH, Δ, 13 hr][4]→ 55%

Bz N H N Me CO_2Me —[H_2, Pd-C / MeOH, HCl]→ HN_b H N Me CO_2Me

—[Br, Br', Na_2CO_3, DMF, Δ]→ N H N Me CO_2Me

20%

Minovine[5]

4. The stereochemical outcome of the annelation sequence is dictated by the transition state (i) which maintains complete orbital overlap during closure of the incipient ring C. On the other hand, the transition state (ii) leading to the undesired C/D-*trans* ring junction is destabilized with respect to (i) since it necessitates ring D to be in a half-boat conformation to obtain overlap, resulting in unfavourable steric interaction between the indole β-H and the ethyl group.

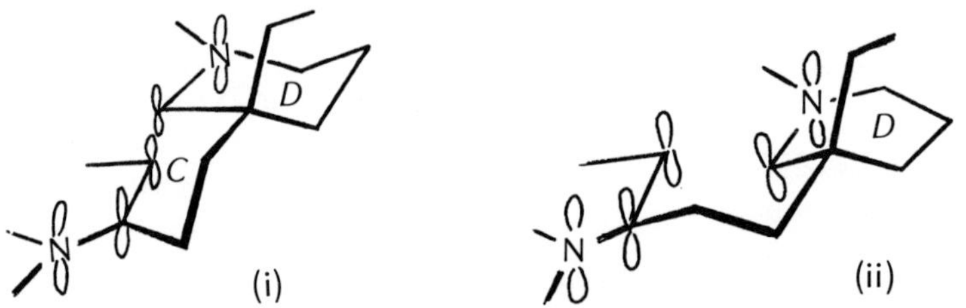

5. For an alternative total synthesis of *dl*-minovine based on the transannular cyclization reaction, see J. P. Kutney, K. K. Chan, A. Failli, J. M. Fromson, G. Gletsos and V. R. Nelson, J. Amer. Chem. Soc., **90**, 3891 (1968). This approach has been illustrated earlier in the elegant synthesis of quebrachamine and aspidospermidine by J. P. Kutney, N. Abdurahman, P. Le Quesne, E. Piers and I. Vlattas, J. Amer. Chem. Soc., **88**, 3656 (1966), Art Org. Synth., 33 (1970).

MUSCONE

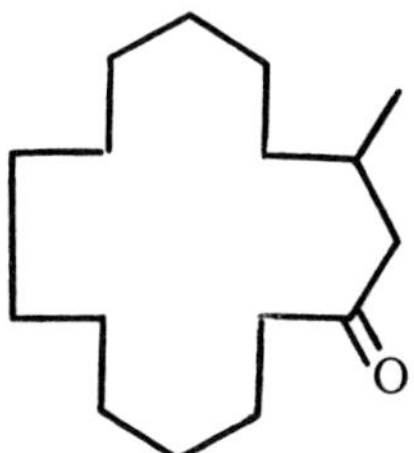

Cyclododecanone, which became the most accessible of medium-ring ketones following Wilke's[1] discovery of the butadiene-cyclotrimerization process, is an important starting point for manufacture of the valuable musk ketones.[2] Two syntheses of muscone, commencing from cyclododecanone are outlined below. The first synthesis[3], which emanated jointly from Eshenmoser's group in Zurich and laboratories of the Swiss perfume firm, Firmenich & Cie., illustrates a three-carbon ring expansion reaction based on the widely applicable tosylhydrazone version of the oxidoketone→alkynone fragmentation.[4]

---

1. G. Wilke, German patent 1,050,333 (1959).

2. For other syntheses of muscone, see B. D. Mookherjee, R. Trenkle and R. R. Patel, J. Org. Chem., **36**, 3266 (1971), and references cited therein.

3. A. Eschenmoser, D. Felix and G. Ohloff, Helv. Chim. Acta, **50**, 708 (1967); D. Felix, J. Schreiber, G. Ohloff and A. Eschenmoser, Helv. Chim. Acta, **54**, 2896 (1971).

4. This elegant and versatile synthesis of acetylenic ketones and aldehydes through fragmentation of α,β-epoxy-ketones with p-toluenesulfonyl hydrazine was developed by J. Shreiber, D. Felix, A. Eschenmoser, M. Winter, F. Gautschi, K. H. Shulte-Elte, E. Sundt, G. Ohloff, J. Kalvoda, H. Kaufmann, P. Wieland and G. Anner, Helv. Chim. Acta, **50**, 2101 (1967).

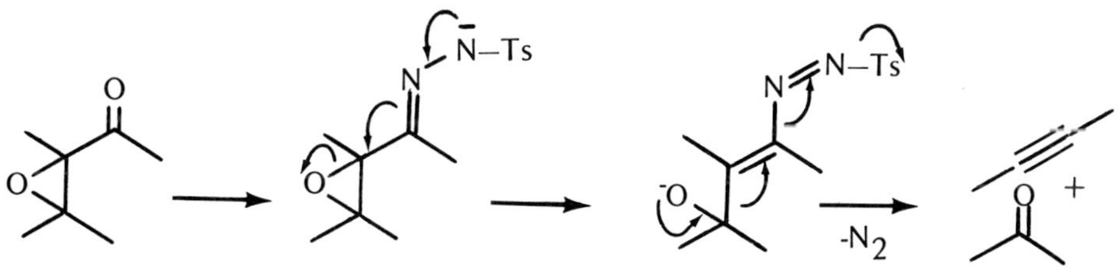

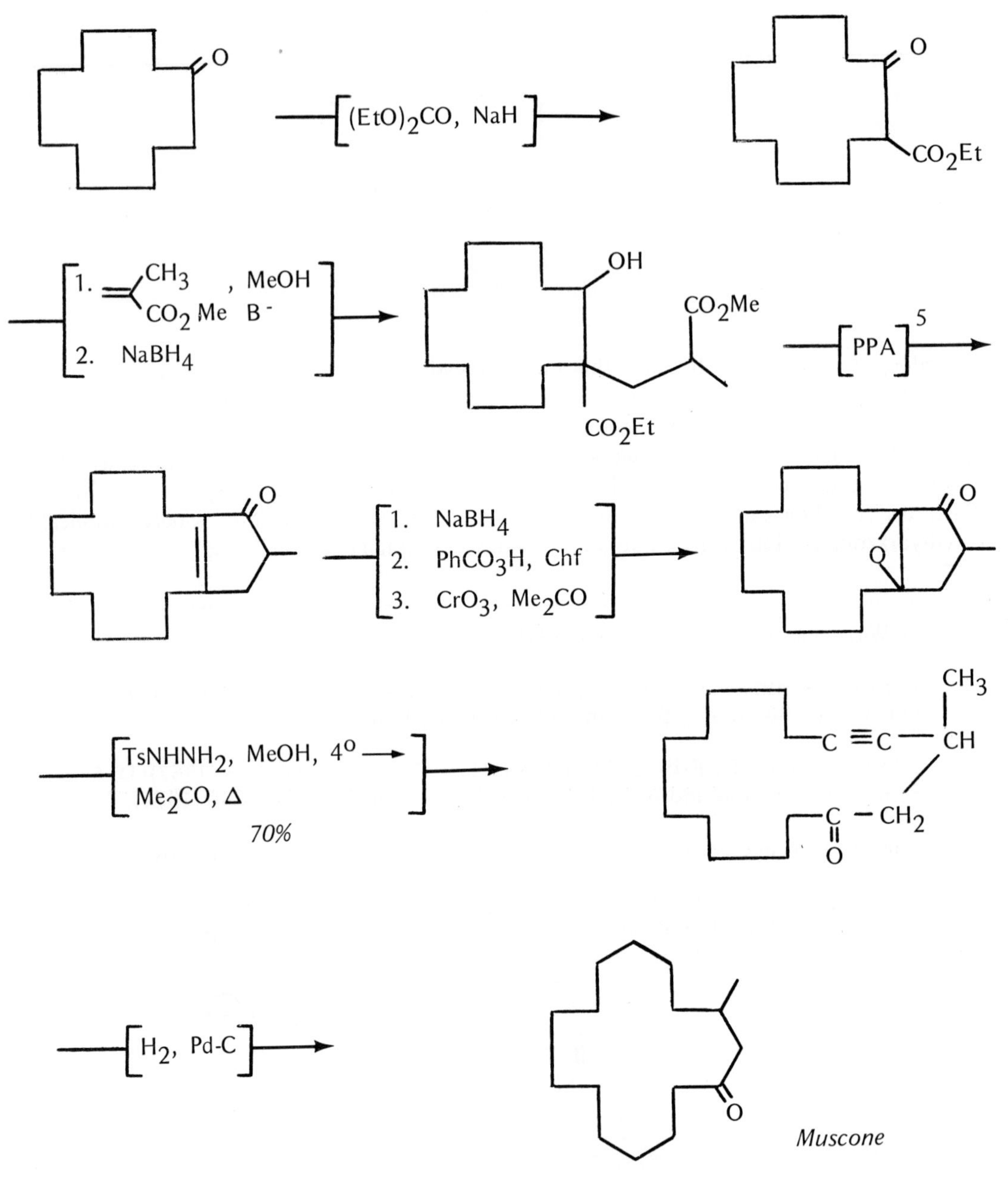

5. Cf. K. Biemann, G. Büchi and B. H. Walker, J. Amer. Chem. Soc., **79**, 5558 (1957).

Another synthesis[6] of *dl*-muscone, commencing from cyclododecanone, is based on application of Gutsche's photolytic method[7] for the cleavage of key bicycloketone (A) to methyl 3-methylcyclopentadecanecarboxylate (B).

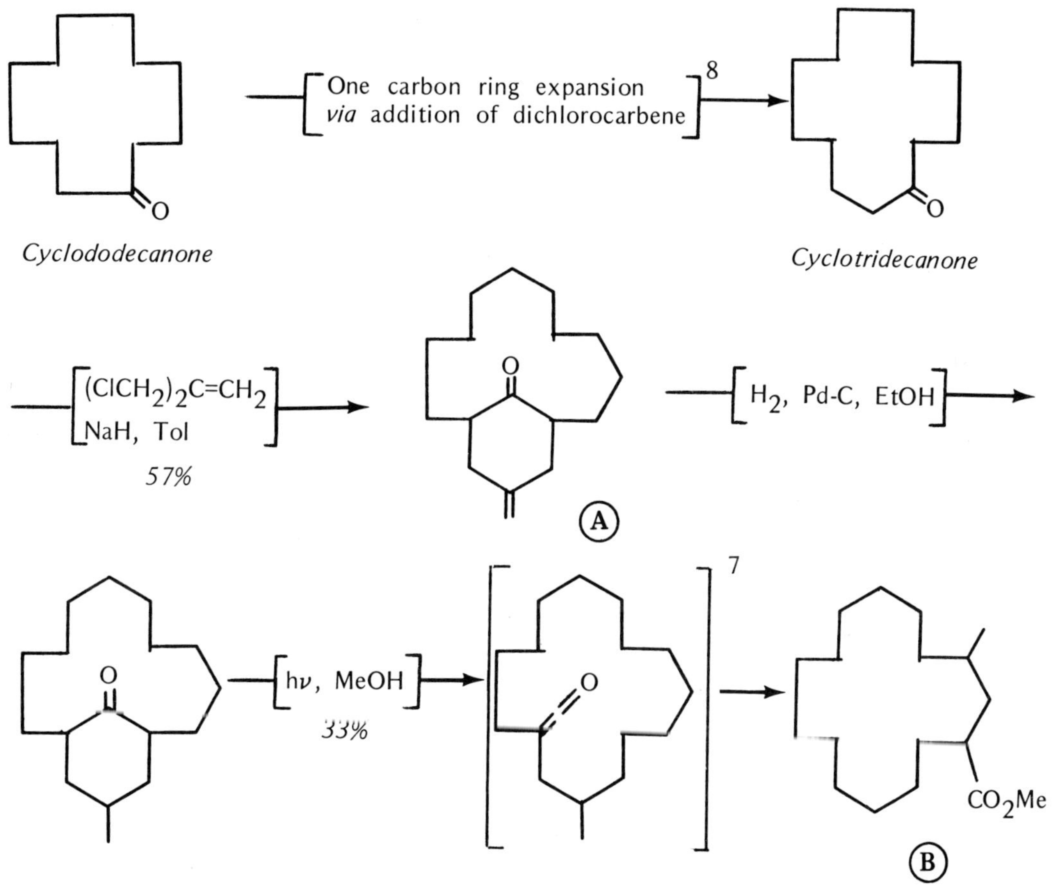

6. H. Nozaki, H. Yamamoto and T. Mori, Canad. J. Chem., **47**, 1107 (1969).

7. C. D. Gutsche and J. W. Baum, J. Amer. Chem. Soc., **90**, 5862 (1968).

8. W. E. Parham and R. J. Sperley, J. Org. Chem., **32**, 926 (1967).

[KOH, EtOH] →

CO_2H → [HgOAc, C_6H_6, Δ, hν → I_2, C_6H_6 → HgOAc; C_6H_6, Δ][9] →

82%

OAc → [1. B^-; 2. $Na_2Cr_2O_7$, H_2SO_4, Et_2O] → *Muscone*

The reaction of dodecatrienylnickel with allene results in the formation of a number of cyclic products including the unsaturated ketone (B), which furnishes muscone on hydrogenation.[10] The sequence outlined below is believed to proceed through the bis-π-allyl intermediate (A) which, under carbonylation conditions, either undergoes ring closure to cyclic hydrocarbons or carbonyl insertion to yield the muscone precursor (B).

Ni → [$CH_2{=}C{=}CH_2$, Et_2O, -20°] → Ni (A)

---

9. D. H. R. Barton, H. P. Faro, E. P. Serebryakov and N. F. Woolsey, J. Chem. Soc., 2438 (1965).

10. R. Baker, B. N. Blackett and R. C. Cookson, Chem. Commun., 802 (1972).

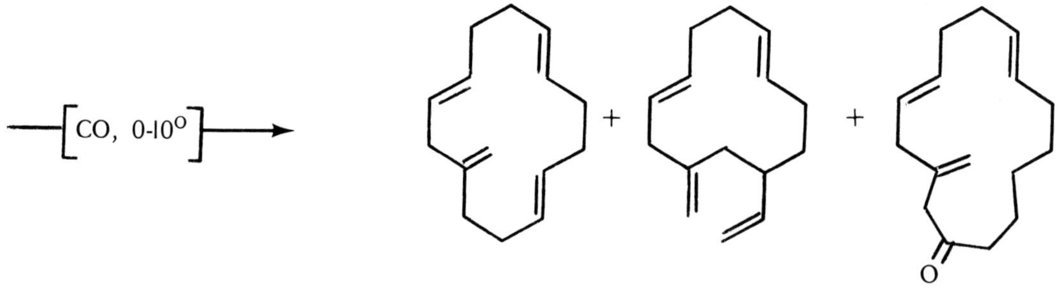

—[CO, 0-10°]→ + +

4-5% (B)

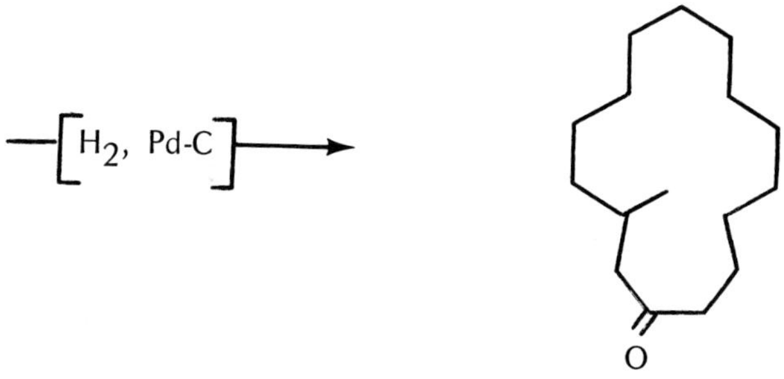

—[H_2, Pd-C]→

Muscone[11]

11. For a review of methods for the synthesis of macrocyclic compounds, see P. R. Story and P. Busch, Adv. Org. Chem., **8**, 67 (1972).

NOOTKATONE

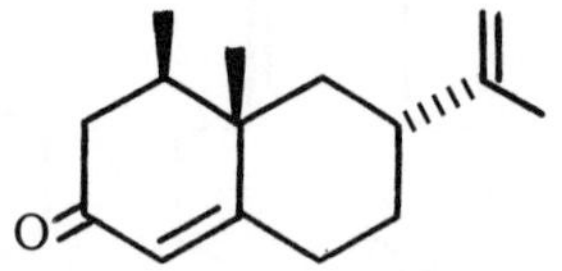

Several of the earlier syntheses of eremophilane sesquiterpenes have utilized the Robinson annelation reaction as a key step[1-6]. However, the synthesis of nootkatone outlined below[7] is based on acid catalyzed rearrangement[9] of the Diels-Alder adduct (B), directly available from a 1-methoxycyclohexa-1,4-diene (A), to a 4-substituted cyclohexenone[10] which undergoes ring closure in formic acid to give the desired eremophilane skeleton.

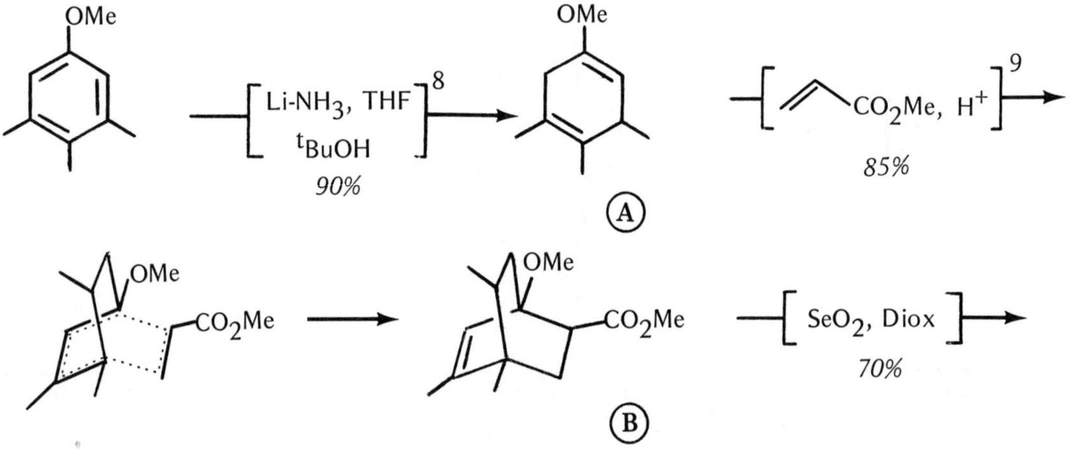

OMe

[Li-NH_3, THF ; ${}^tBuOH$][8] 90%

OMe

(A)

[CO_2Me, H^+][9] 85%

OMe CO_2Me → OMe CO_2Me

(B)

[SeO_2, Diox] 70%

1. M. Pesaro, G. Bozzata and P. Schudel, Chem. Commun., 1152 (1968); H.C. Odom and A. R. Pinder. Chem. Commun., 26 (1969).

2. J. A. Marshall and R. A. Ruden, Tetrahedron Lett., 1239 (1970).

3. A. van der Gen, L. M. van der Linde, J. G. Witeveen and H. Boelens, Recueil, **90**, 1034 (1971).

4. J. A. Marshall, H. Faubl and T. M. Warne, Chem. Commun., 753 (1967).

5. R. M. Coates and J. E. Shaw, Chem. Commun., 47 (1968).

6. E. Piers and R. J. Keziere, Tetrahedron Lett., 583 (1968).

7. K. P. Dastur, J. Amer. Chem. Soc., **95**, 6509 (1973); **96**, 2605 (1974).

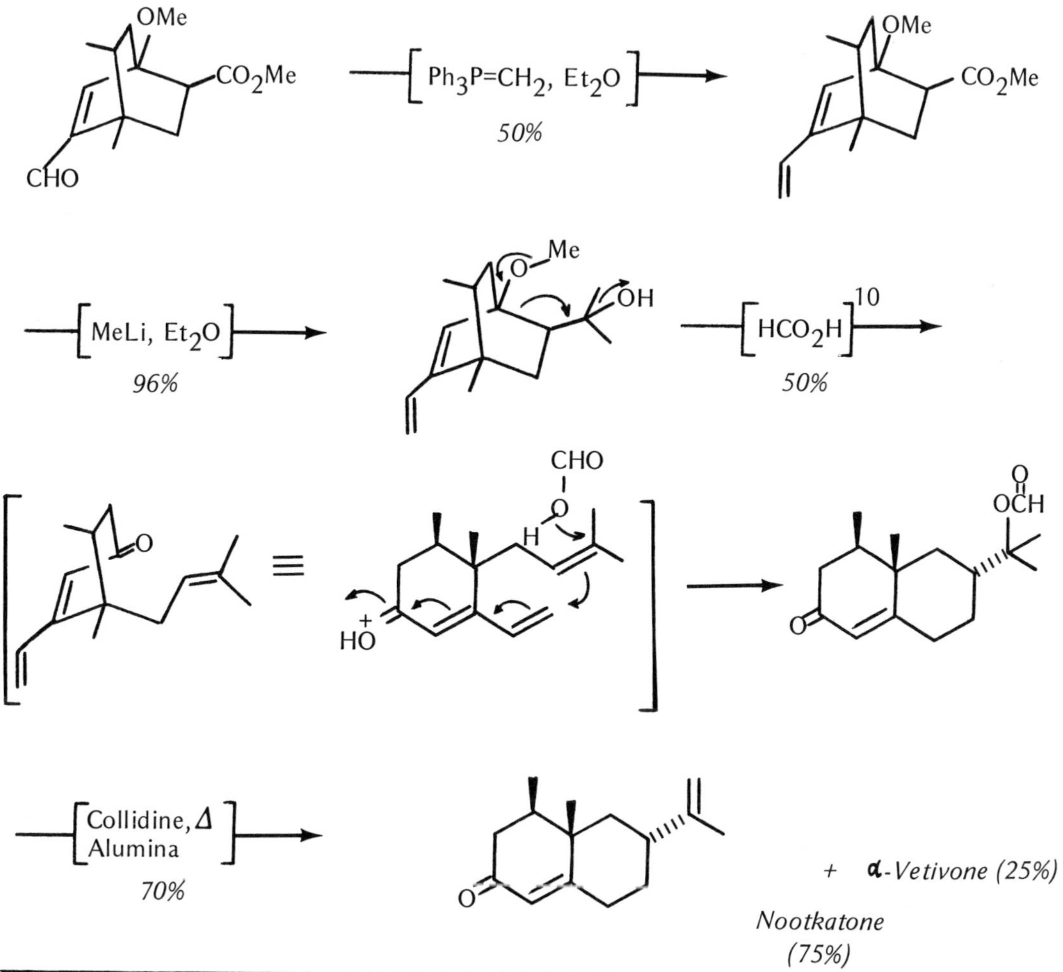

$Ph_3P{=}CH_2$, Et_2O 50%

MeLi, Et_2O 96%

HCO_2H[10] 50%

Collidine, Δ Alumina 70%

+ α-Vetivone (25%)

Nootkatone (75%)

8. For a definitive review on Birch reduction, see A. J. Birch and G. Subba Rao, Adv. Org. Chem., **8**, 1 (1972).

9. The direct reaction of dienophiles with 1,4-dienes to give Diels-Alder adducts of the 1,3-dienes has been interpreted in terms of equilibration through change-transfer complexes, A. J. Birch and K. P. Dastur, Tetrahedron Lett., 4195 (1972); A. J. Birch, P. L. Macdonald and V. H. Powell, J. Chem. Soc. (C), 1469 (1970).

10. A. J. Birch and J. S. Hill, J. Chem. Soc. (C), 419 (1966).

OCCIDENTALOL

Occidentalol is a eudesmane-type sesquiterpene isolated from the wood of Eastern white cedar (*Thuja occidentalis Linn.*), and is characterized by the presence of a *cis*-fused decalin system and a homoannular 1,3-diene unit in the molecule, an intriguing combination of functional groups rarely encountered in this class of natural products. Three related syntheses of occidentalol, all of which are based on hexahydronaphthalen-2-one intermediates have been reported, and utilize the carbonyl group for introduction of the diene chromophore in ring A.[1-3,5] The synthesis outlined below,[2,3] utilizing a *cis*-fused decalone derivative (A) illustrates several of the key features of these syntheses.

HgOAc, H_2O-THF → $NaBH_4$, NaOH

93%

Dihydrocarvone

, KOH, EtOH

31%

HCl, THF

74%

1. M. Ando, K. Nanaumi, T. Nakagawa, T. Asao and K. Takase, Tetrahedron Lett., 3891 (1970).

2. Y. Amano and C. H. Heathcock, Can. J. Chem., **50**, 340 (1972).

3. M. Sergent, M. Mongrain and P. Deslongchamps, Can. J. Chem., **50**, 336 (1972).

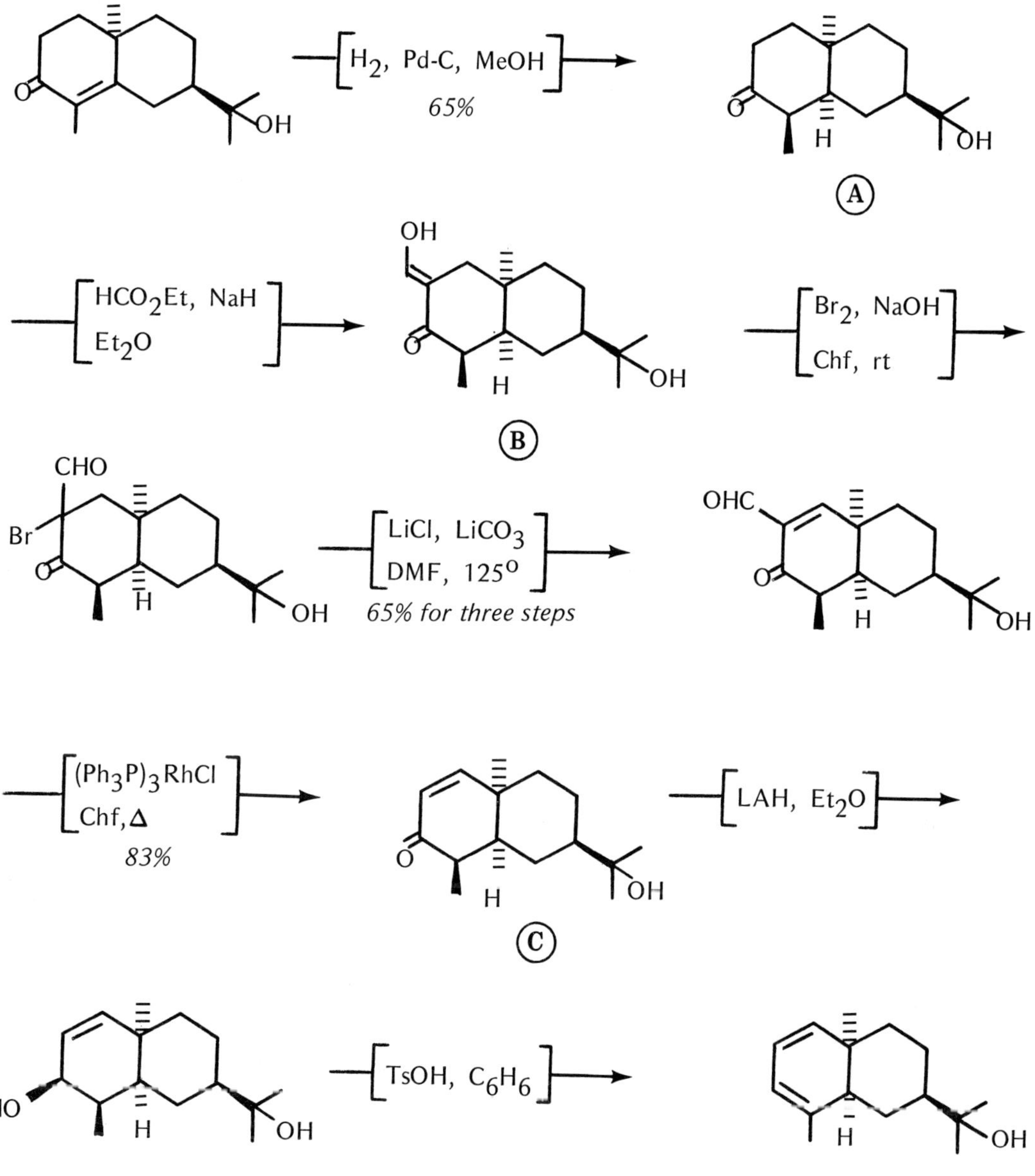

[H_2, Pd-C, MeOH] 65% → (A)

[HCO_2Et, NaH, Et_2O] → (B)

[Br_2, NaOH, Chf, rt] →

[LiCl, $LiCO_3$, DMF, 125°] 65% for three steps →

[$(Ph_3P)_3RhCl$, Chf, Δ] 83% → (C)

[LAH, Et_2O] →

[TsOH, C_6H_6] →

Occidentalol

4. Bromination and deformylation of (B) could also be carried out in a single operation to give directly a bromoketone, which was dehydrobrominated to the desired enone (C).

An interesting feature of the synthesis of occidentalol by Hortmann *et al.*[5] is conversion of a *trans* to a *cis*-fused bicyclic diene by a photolysis-recyclization sequence.[6] The key steps of this synthesis are outlined below.

CO_2Me H → HO CO_2Me H

[Al_2O_3, Py, 220°] 25% → CO_2Me H ⇌ (hν, −78°) CO_2Me

[−20°][6] → CO_2Me H ⇌ (K^tBuO, ^tBuOH, Δ) CO_2Me H

[MeLi, Et_2O] → H OH *Occidentalol*

---

5. A. G. Hortmann, D. S. Daniel and J. E. Martinelli, J. Org. Chem., **38**, 728 (1973).

6. This represents an interesting route to polyfunctionally substituted *cis* decalins from the more readily available and predictably substituted *trans* decalins. The key elements of this approach have previously been portrayed by photolytic cleavage of the *trans*-eudesm-1,3-dienolide (i) to a cyclodecatriene (ii) which undergoes disrotatory thermal cyclization to the *cis*-fused diene (iii), E. J. Corey and A. G. Hortmann, J. Amer. Chem. Soc., **87**, 5736 (1965).

(i) H O O ⇌ (hν) (ii) O O → (Δ) (iii) H O O

In a direct approach for the synthesis of occidentalol, Corey and Watt[7] used the Diels-Alder reaction with an α-pyrone[8] for establishing the *cis*-fused decalin system, followed by elimination of the lactone bridge in the adduct for generation of the homoannular 1,3-diene unit.

CO_2Me + [150°, 24 hr][9] → 25-40% → CO_2Me

→ [$(CH_2OH)_2$, TsOH, C_6H_6, Δ] → H CO_2Me

H CO_2Me [LAH, Et_2O] → H CH_2OH

7. D. S. Watt and E. J. Corey, Tetrahedron Lett., 4651 (1972).

8. Prepared from 1,1,3,3-tetramethoxypropane and dimethyl malonate, cf. T. B. Windholz, L. H. Peterson and G. J. Kent, J. Org. Chem., **28**, 1443 (1963).

9. 4-Methyl-3-cyclohexenone was prepared by Birch reduction of *p*-methylanisole followed by hydrolysis of the resulting enol ether, cf. A. J. Birch, J. Chem. Soc., 596 (1946).

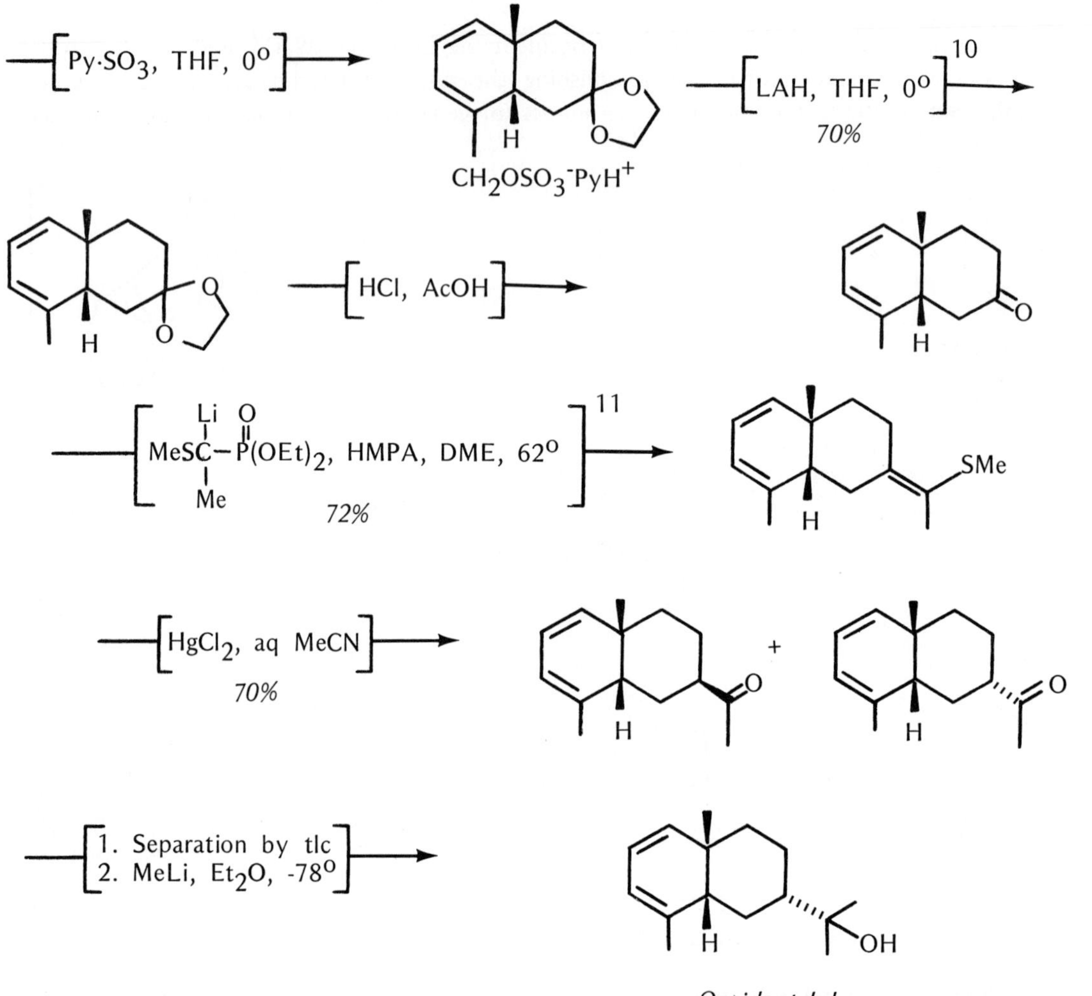

—[Py·SO_3, THF, 0°]→ $CH_2OSO_3^-PyH^+$ —[LAH, THF, 0°][10]→ 70%

H H

—[HCl, AcOH]→

H H

—[MeSC(Li)(Me)–P(O)(OEt)$_2$, HMPA, DME, 62°][11]→ SMe
72%

H

—[$HgCl_2$, aq MeCN]→ + O O
70%

H H

—[1. Separation by tlc 2. MeLi, Et_2O, -78°]→ OH

H

Occidentalol

---

10. This efficient procedure for deoxygenation of allylic alcohols involving the reduction of an intermediate sulfate monoester was developed earlier by E. J. Corey and K. Achiwa, J. Org. Chem., **34**, 3667 (1969).

11. For the scope and development of this synthesis of vinyl sulfides with lithio (1-methylthio)alkylphosphonate esters (i) *via* the Emmons-Horner phosphonate reaction, see E. J. Corey and J. I. Shulman, J. Org. Chem., **35**, 777 (1970). This synthesis utilizes the ability of electron-withdrawing substituents, such as a thioalkyl moiety, on the carbon α to the phosphorus function to promote Wittig elimination (i→ii) of β-hydroxy phosphonates which are generally unreactive. The resulting vinyl sulfides readily undergo mercury(II)-promoted hydrolysis to ketones.

$$CH_3SCH_2\overset{O}{\overset{\|}{P}}(OEt)_2 \longrightarrow CH_3S\underset{R}{\underset{|}{C}}H\overset{O}{\overset{\|}{P}}(OEt)_2 \xrightarrow{n\text{-BuLi}}$$

$$CH_3S\overset{Li}{\overset{|}{\underset{R}{\underset{|}{C}}}}-\overset{O}{\overset{\|}{P}}(OEt)_2 \xrightarrow{R_1R_2C=O} \underset{\text{(i)}}{R_1R_2\overset{Li^+\ O^-}{\overset{|}{C}}-\overset{P(=O)(OEt)_2}{\overset{|}{\underset{SCH_3}{\underset{|}{C}}}}-R}$$

$$\xrightarrow{\Delta} (EtO)_2\overset{O}{\overset{\|}{P}}-O^-\ \ Li^+ + \underset{\text{(ii)}}{R_1R_2C=C\begin{matrix} R \\ SCH_3 \end{matrix}}$$

PATCHOULI ALCOHOL

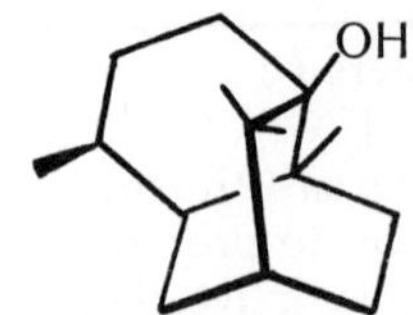

The first synthesis of patchouli alcohol was actually directed towards an α-patchoulane skeleton (A), the structure then believed to represent the natural product.[1] It was only by a fortuitous rearrangement during the final stages of the synthesis that Büchi arrived at patchouli

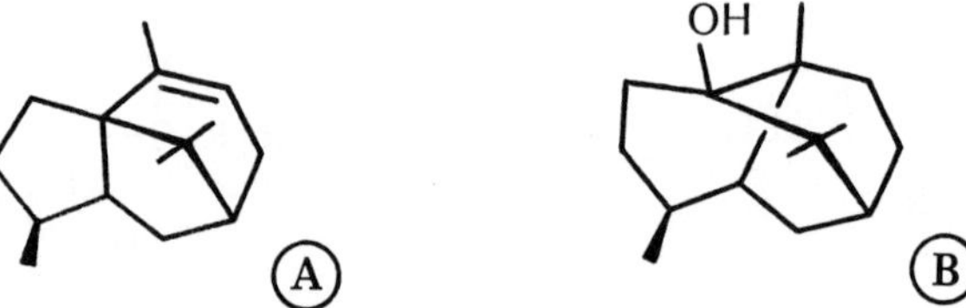

alcohol (B), the true structure of which was eventually revealed during the course of an X-ray study.[2] With the revised structure now firmly established, Danishefsky and Dumas[3] executed a short and direct synthesis of patchouli alcohol which is outlined below. A key feature of this synthesis is the efficient utilization of an intramolecular reductive cyclization process[4] for building the three-carbon bridge in the bicyclo[2·2·2]octane intermediate (C).

O^-Li^+ —[MeI, 150°]→ O —[ ]→ O 93%

---

1. G. Büchi, W. D. McCleod and J. Padilla, O., J. Amer. Chem. Soc., **86**, 4438 (1964); Art. Org. Synth., 271 (1970).

2. M. Dobler, J. D. Dunitz, B. Gubler, H. P. Weber, G. Büchi and J. Padilla O., Proc. Chem. Soc., 383 (1962).

3. S. Danishefsky and D. Dumas, Chem. Commun., 1287 (1968).

4. For early studies dealing with this type of reductive cyclization (i→ii), see H. O. House, J. J. Riehl and C. G. Pitt, J. Org. Chem., **30**, 650 (1965).

O, X (i) $+ 2e^-$ → O^- $+ X^-$ (ii)

[H_2, Pd-C]

NaOMe, MeOH

[Li, THF]

HO

[HCl, Chf]

Cl

H

[aq Diox]
85%

OH

H

[H_2, Pd-C / EtOAc, $NaNO_2$][5]

OH

*

[PBr_3]
85%

Separated from corresponding epimer by chromatography.

Br

[Na, THF][4]
27%

OH

Patchouli Alcohol[6,7]

Ⓒ

---

5. Addition of sodium nitrite inhibits hydrogenolysis of the allylic alcohol, H. Dart and H. Henbest, J. Chem. Soc., 3053 (1960).

6. For a related approach to patchouli alcohol, see R. N. Mirrington and K. J. Schmalzl, J. Org. Chem., **37**, 2871 (1972).

7. An elegant general approach to the synthesis of α-patchoulane sesquiterpenes has been recently disclosed by W. F. Erman and L. C. Stone [J. Amer. Chem. Soc., **93**, 2821 (1971)]. The key to this approach, illustrated by a synthesis of patchoulenone and epipatchoulenone outlined below, is the intramolecular Lewis acid catalyzed cyclization of the diazo ketone (i) to give a carbethoxy cyclopropyl ketone (ii). The carbethoxy group serves as an electron-withdrawing substituent to direct ring opening and to provide a handle for annelation of the fused five-membered ring (→iii).

EtO_2C + CO_2H → EtO_2C / CO_2H —2 steps→

EtO_2C / $COCHN_2$ (i) —$BF_3.Et_2O$, *30%*→ H / EtO_2C / O →

EtO_2C / O (ii) —1. LAH 2. Oxidation→ OHC / O

—1. $(EtO)_2\overset{O}{\overset{\|}{P}}\bar{C}HCO_2Et$ 2. Hydrogenation→ CO_2Et / O —1. MeLi 2. Oxidation→

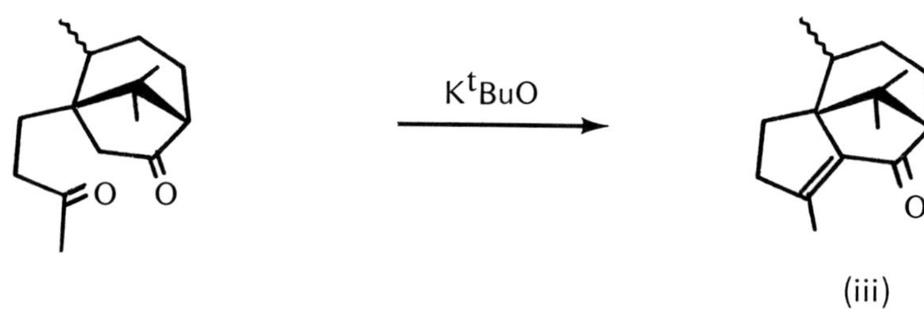

Patchoulenone and epipatchoulenone

PORANTHERINE

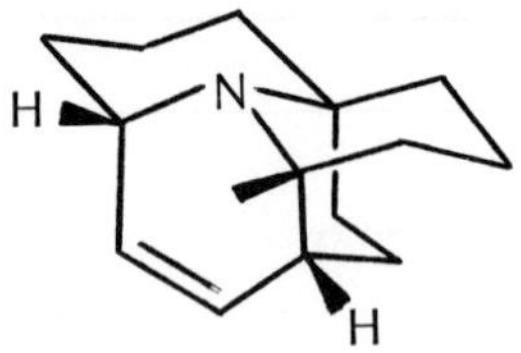

The synthesis of porantherine[1,2] vividly illustrates the application of antithetic (retrosynthetic) analysis in developing synthetic routes to complex molecules.[3] Systematic bond disconnection suggests structures A-F (or their equivalents) as key intermediates and also provides the overall plan of synthesis.

Antithetic Analysis of Porantherine:[3]

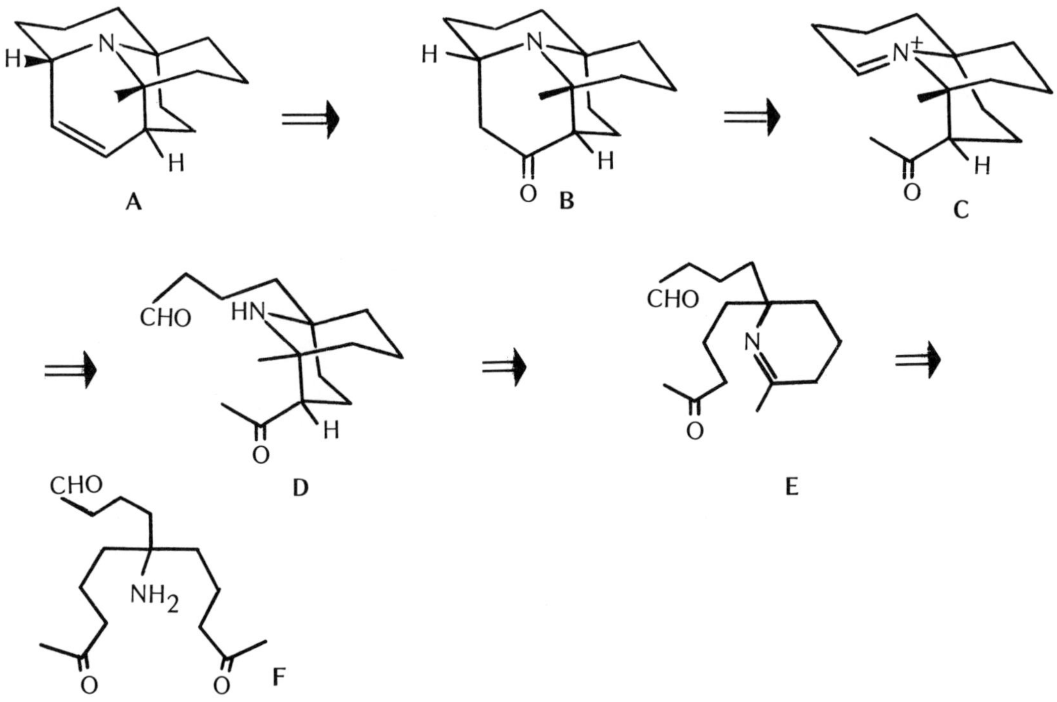

A ⇒ B ⇒ C ⇒ D ⇒ E ⇒ F

---

1. E. J. Corey and R. D. Balanson, J. Amer. Chem. Soc., **96**, 6516 (1974).

2. Porantherine, characterized by a bridged 9b-aza-phenalene network, is the major alkaloid of the low, woody shrub, *Poranthera corymbosa* growing in northern New South Wales.

3. Corey has described the process of generating an antithetic tree and its application to synthetic planning and problem simplification. For two very lucid and definitive reviews

Synthesis of Porantherine:

2 [ketal]MgCl + HCO_2Et —[THF, Δ]→ 95%

—[Collins' Reagent, CH_2Cl_2, rt]→ 93%

—[$MeNH_2$, Tol, 110°, Molecular Sieves (Sealed tube)]→ 100%

NMe

—[Li, C_6H_6, rt]→ 65%

NHMe

(A)

—[10% HCl→base]→ 90%

Me

N

O

(B)

—[AcO, TsOH, C_6H_6, Δ][4]→ 45%

---

on various aspects of the theory of synthetic analysis, see E. J. Corey, Pure Appl. Chem., **14**, 19 (1967), Quart. Rev., **25**, 455 (1971).

4. This reaction may be regarded as a Mannich type process proceeding *via* enol acetate of the conjugate acid (iminium ion) of (B).

Me-N

O H

(C)

[Collins' reagent, 72 hr][5]

80%

OHC-N

O H

[OsO_4, $NaIO_4$]

OHC

OHC-N

O H

[1. $(CH_2OH)_2$, TsOH
2. 3N KOH, EtOH, 110°
(Sealed tube)]

HN

O O

H

O O

[10% HCl → base]

85%

N

O H

[TsOH, Tol, Δ]

45%

N

H

O H

---

5. A. Cave, C. Kan-Fan, P. Potier, J. LeMen and M. -M. Janot, Tetrahedron, **23**, 4691 (1967).

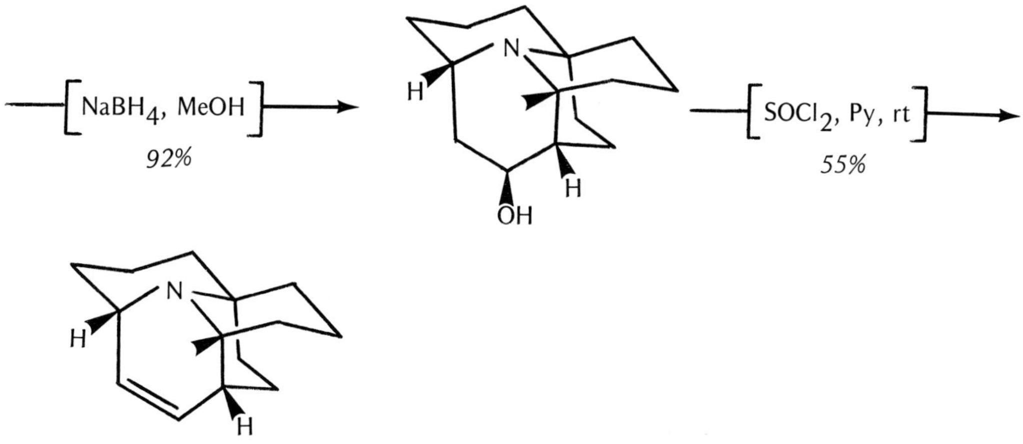

[$NaBH_4$, MeOH] 92%

H N H OH

[$SOCl_2$, Py, rt] 55%

H N H

Porantherine

PRECALCIFEROL$_3$

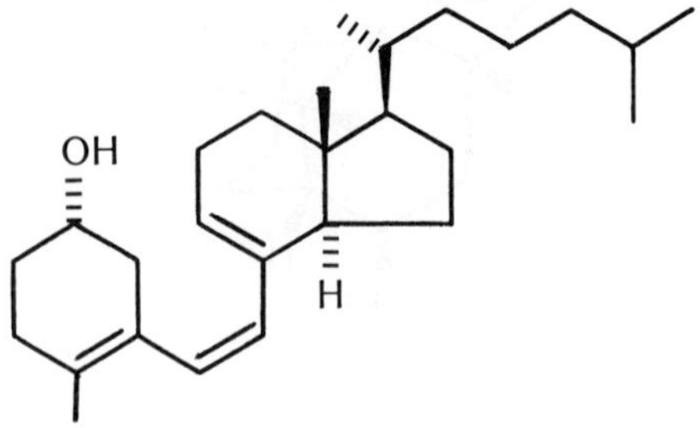

From the extensive program on the chemical synthesis of calciferol and related compounds[1,2] by Lythgoe *et al.* at the University of Leeds, the total synthesis of precalciferol$_3$[3] has been selected to portray their approach to 9,10-seco-steroids by non-photochemical means. The central *cis*-double bond linking rings A and C of precalciferol$_3$ has been generated by partial hydrogenation of an acetylenic intermediate (B). The key en-yne (A) has been utilized as the source of both ring A and the acetylene link, thus securing the location of the ring A double bond unambiguously. Similarly, proper location of the ring C double bond has been established by introducing it, not by a dehydration reaction, but by elimination from a *trans*-diaxial bromohydrin.[10]

---

1. Precalciferols are the first isolable products of irradiation of the pro-vitamins. Precalciferol$_3$, the immediate product of irradiation of 7-dehydrocholesterol is converted by further irradiation into the corresponding 6,7-*trans*-isomer, tachysterol$_3$, and by thermal equilibration into vitamin D_3. A photochemical route to vitamin D_3 was disclosed by Inhoffen in 1958 [reviewed by H. H. Inhoffen, Angew. Chem., **72**, 875 (1960); Art Org. Synth., 364 (1970)].

2. For the synthesis of tachysterol$_3$, see R. S. Davidson, S. M. Waddington-Feather, D. H. Williams and B. Lythgoe, J. Chem. Soc. C, 2534 (1967).

3. J. Dixon, P. S. Littlewood, B. Lythgoe and A. K. Saksena, Chem. Commun., 993 (1970); T. M. Dawson, J. Dixon, P. S. Littlewood, B. Lythgoe and A. K. Saksena, J. Chem. Soc. C, 2960 (1971).

Construction of C/D Ring Components – Ring C:[4]

+ CO_2H + CO_2Me —[115°, 20 hr → aq. NaOH]→ CO_2H CO_2H + CO_2H CO_2H

—[Ac_2O, Δ]→ O O O —[H_2O, Δ]→

CO_2H CO_2H —[m-$ClC_6H_4CO_3H$, Et_2O]→ CO_2H CO_2H O

—[5N H_2SO_4, Diox]→ O HO O CO_2H —[-CO_2H, $(CF_3CO)_2O$]→

O O Mes CO_2H —[2N NaOH]→ CO_2H CO_2H O Mes OH

Mes=mesitoate

4. I. J. Bolton, R. G. Harrison, B. Lythgoe and R. S. Manwaring, J. Chem. Soc. C, 2944 (1971).

1. $PhCHN_2$, Et_2O
2. Ac_2O

CO_2Bz CO_2Bz O Mes OAc

H_2, Pd-C

CO_2H CO_2H O Mes OAc

$Pb(OAc)_4$, Py, 70°*

*Cf. C. M. Cimarusti and J. Wolinsky, J. Amer. Chem. Soc., **90**, 113 (1968).

O Mes OAc

1. NaOMe, MeOH aq
2. CH_2=CH-OEt, TsOH

O Mes O O

1. LAH, Et_2O
2. PhCOCl, Py
3. AcOH, aq Diox 60°

O COPh OH

– *Isooctyl Chain:*[5]

H CN

H_2, Pd-C, EtOH

H CN

Citronellonitrile M

5. I. J. Bolton, R. G. Harrison and B. Lythgoe, J. Chem. Soc. C, 2950 (1971).

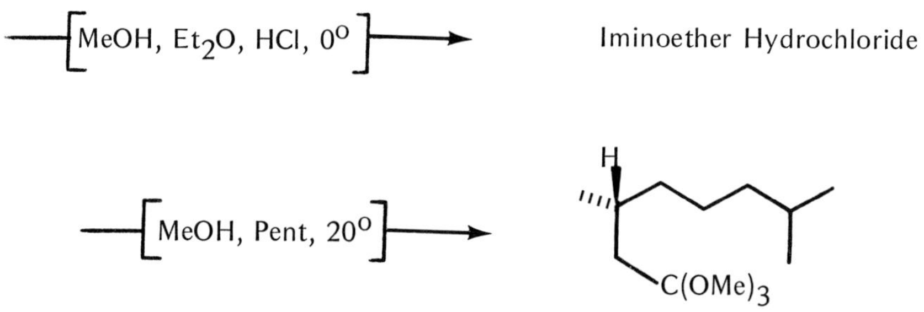

[MeOH, Et_2O, HCl, 0°] → Iminoether Hydrochloride

[MeOH, Pent, 20°] → $C(OMe)_3$

– *Indane Nucleus:*[5,6]

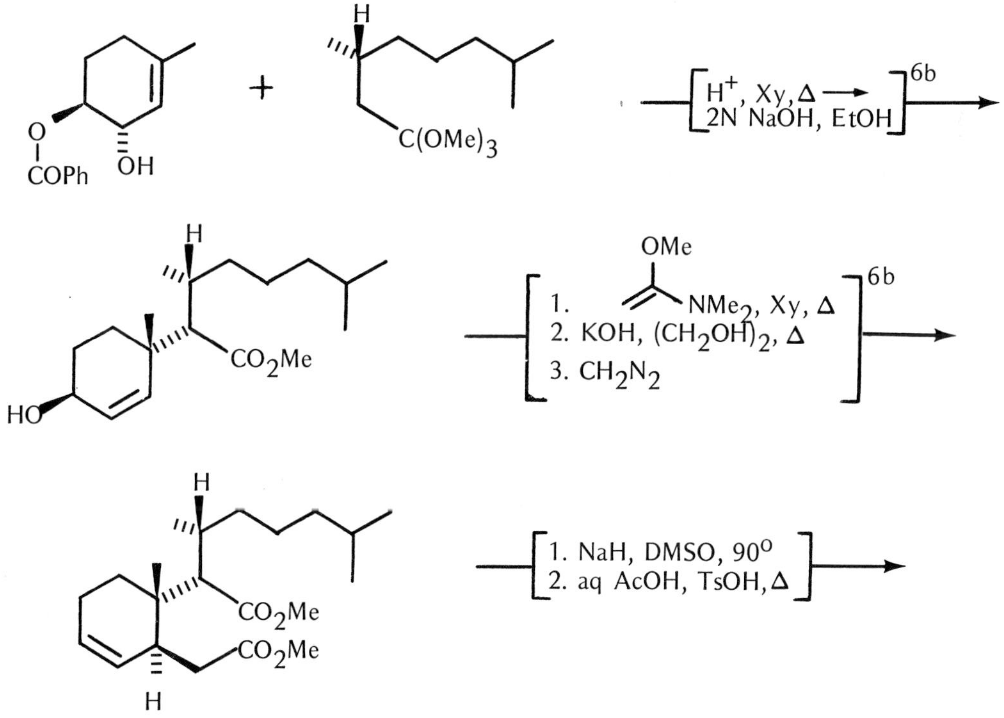

O COPh OH + $C(OMe)_3$ [H^+, Xy, Δ → 2N NaOH, EtOH][6b] →

HO CO_2Me

[1. OMe NMe$_2$, Xy, Δ
2. KOH, $(CH_2OH)_2$, Δ
3. CH_2N_2][6b] →

CO_2Me CO_2Me H

[1. NaH, DMSO, 90°
2. aq AcOH, TsOH, Δ] →

---

6. P. S. Littlewood, B. Lythgoe and A. K. Saksena, J. Chem. Soc. C, 2955 (1971).

6b. Note the application of two successive Claisen-type rearrangements to generate the C_{13}-C_{17} link and to deliver the remaining elements of ring D at C_{14}, Cf. A. E. Wick, D. Felix and A. Eschenmoser, Helv., **47**, 2425 (1964).

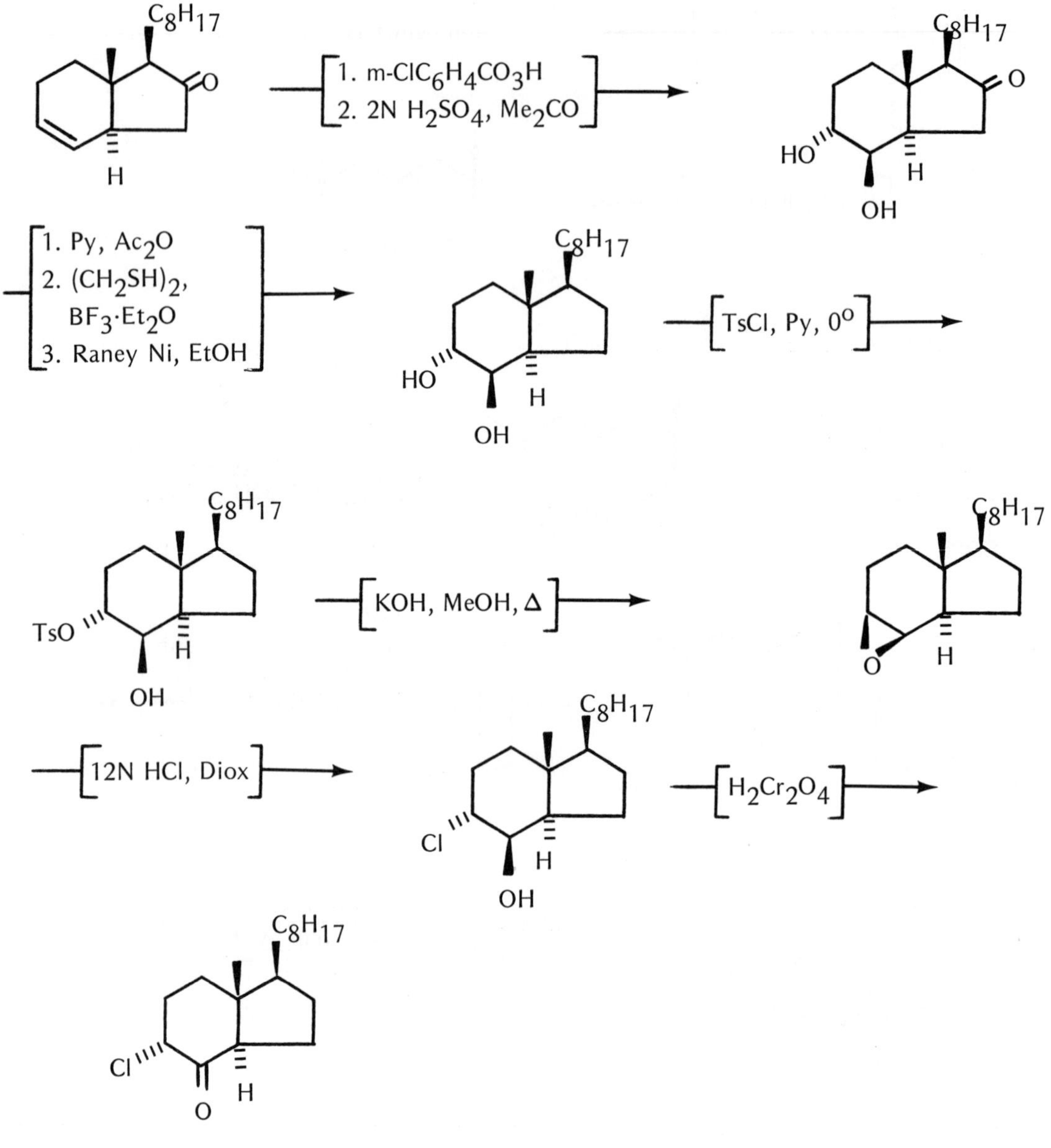

C_8H_{17}, O, H

→ [1. m-$ClC_6H_4CO_3H$; 2. 2N H_2SO_4, Me_2CO] →

C_8H_{17}, O, HO, H, OH

→ [1. Py, Ac_2O; 2. $(CH_2SH)_2$, $BF_3 \cdot Et_2O$; 3. Raney Ni, EtOH] →

C_8H_{17}, HO, H, OH

→ [TsCl, Py, 0°] →

C_8H_{17}, TsO, H, OH

→ [KOH, MeOH, Δ] →

C_8H_{17}, O, H

→ [12N HCl, Diox] →

C_8H_{17}, Cl, H, OH

→ [$H_2Cr_2O_4$] →

C_8H_{17}, Cl, O, H

Construction of Ring A:[7-9]

MeO CO$_2$H — [Li-liq NH$_3$, -40°] → MeO CO$_2$H →

O CO$_2$H — [NaBH$_4$, 2N Na$_2$CO$_3$ → 2N HCl, 100°] → O O

— [NaOMe, MeOH] → HO CO$_2$Me — [LAH, Et$_2$O, -25°] →

HO CH$_2$OH — [MnO$_2$, Me$_2$CO] → HO CHO

— [Cl$^-$ Ph$_3$P$^+$CH$_2$Cl BuLi, Et$_2$O, C$_6$H$_6$] → HO Cl — [Na, liq NH$_3$] →

7. J. Dixon, B. Lythgoe, I. A. Siddiqui and J. Tideswell, J. Chem. Soc. C, 1301 (1971).

8. P. R. Bruck, R. D. Clark, R. S. Davidson, W. H. H. Gunther, P. S. Littlewood, and B. Lythgoe, J. Chem. Soc. C, 2529 (1967).

9. T. M. Dawson, J. Dixon, P. S. Littlewood and B. Lythgoe, J. Chem. Soc. C, 2352 (1971).

HO CH [1. TMSCl, Py, Et_2O 2. BuLi, Et_2O, 0°] TMSO C^- Li^+ (A)

Preparation of Precalciferol:[3]

C^- Li^+ OTMS + C_8H_{17} Cl O H [Et_2O, 0°] 65%

C_8H_{17} Cl HO H OH [Bis(ethylene diamine) Chromium (II), DMF][10] 65% C_8H_{17} H OH (B)

10. Elimination of chlorohydrin was accomplished using the method of J. K. Kochi, D. Singleton and L. Andrews, Tetrahedron, **24**, 3505 (1968); J. K. Kochi and D. Singleton, J. Amer. Chem. Soc., **90**, 1582 (1968).

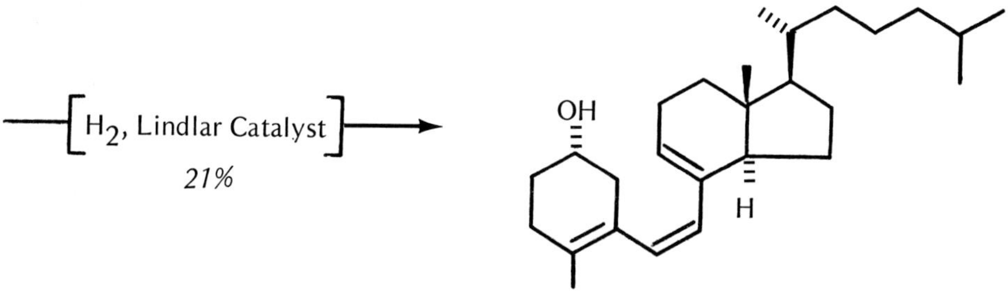

[H_2, Lindlar Catalyst]
21%

OH

H

Precalciferol$_3$[11]

11. Product was purified *via* 3,5-dinitrobenzoate.

PROGESTERONE

The application of nonenzymic biogenetic-like olefinic cyclizations[1] for the stereospecific construction of polycyclic systems, developed by W. S. Johnson and his school, represents an exciting new approach to steroid total synthesis. The synthesis of progesterone[2], outlined below, is based on the finding that an appositely placed acetylenic bond participates in polyolefinic cyclizations directly to produce the C/D *trans*-fused hydrindane system (B→C).[3]

Preparation of Acetylenic Aldehyde (A):

CHO + BrMg → HO — [$(EtO)_3CCH_3$, $EtCO_2H$, 138°][4] →

EtO, O → CO_2Et — [LAH, Et_2O, 0°] 90% →

53% overall yield

---

1. Review: W. S. Johnson, Accts. Chem. Res., **1**, 1 (1968).

2. W. S. Johnson, M. B. Gravestock and B. E. McCarry, J. Amer. Chem. Soc., **93**, 4333 (1971).

3. W. S. Johnson, M. B. Gravestock, R. J. Parry, R. F. Myers, T. A. Bryson and D. H. Miles, J. Amer. Chem. Soc., **93**, 4330 (1971).

4. Ortho acetate Claisen reaction: W. S. Johnson, L. Werthemann, W. R. Bartlett, T. J. Brocksom, T. Li, D. J. Faulkner and M. R. Petersen, J. Amer. Chem. Soc., **92**, 741 (1970).

CH_2OH —[CrO_3, Py, CH_2Cl_2]→ CHO
86%

(A)

Preparation of Progesterone:

PPh_3 + CHO (A) —Footnote 5→

—[0.1N HCl, MeOH]→ O O

—[aq NaOH, EtOH, Δ]→ O —[MeLi, Et_2O]→

40% based on (A)

5. Preparation of the ylid is already described under synthesis of estrone.

"The crude carbinol in 1,2-dichloroethane containing ethylene carbonate (to trap the vinyl cation) was treated with TFA for 3 hr at 0° under nitrogen."

71%

OH (B) (C)

[1. O_3, MeOH, CH_2Cl_2, -70°
2. Zn, AcOH, -15 to -23°]

[5% KOH]

Progesterone

Recently, Johnson and coworkers reported the direct construction of a complete steroid nucleus, in optically active form, by nonenzymic biogenetic-like cyclization.[6,7] This was accomplished by incorporating, along with the terminal acetylenic residue into the key

6. R. L. Markezich, W. E. Willy, B. E. McCarry and W. S. Johnson, J. Amer. Chem. Soc., **95**, 4414 (1973).

7. B. E. McCarry, R. L. Markezich and W. S. Johnson, J. Amer. Chem. Soc., **95**, 4416 (1973).

cyclization substrate, trienyol (E), a cyclohexenol moiety previously shown to be capable of initiating stereoselective cyclization to A/B *cis* decalin structures.[8] Commencing with a substrate bearing a chiral C-5 the cyclization proceeds stereospecifically to yield an optically active tetracyclic product.

Preparation of Ring A/B Component:

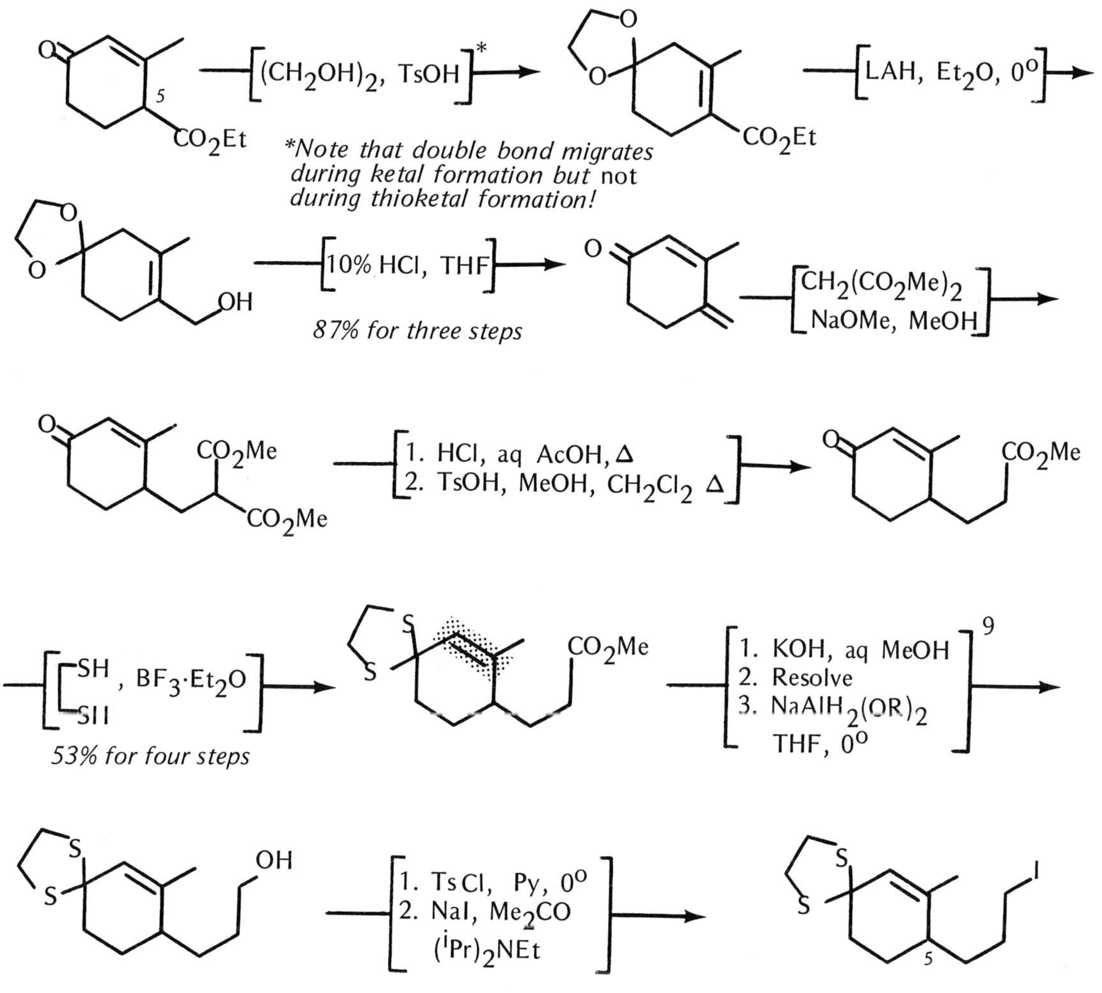

$[(CH_2OH)_2$, TsOH]* → ; [LAH, Et_2O, 0°] →

CO_2Et ; CO_2Et

**Note that double bond migrates during ketal formation but* not *during thioketal formation!*

[10% HCl, THF] →

87% for three steps

[$CH_2(CO_2Me)_2$, NaOMe, MeOH] →

CO_2Me ; CO_2Me

[1. HCl, aq AcOH, Δ
2. TsOH, MeOH, CH_2Cl_2 Δ] →

CO_2Me

[HS–CH₂CH₂–SH, $BF_3{\cdot}Et_2O$] →

53% for four steps

CO_2Me

[1. KOH, aq MeOH
2. Resolve
3. $NaAlH_2(OR)_2$, THF, 0°][9] →

[1. TsCl, Py, 0°
2. NaI, Me_2CO, $(^iPr)_2NEt$] →

8. W. S. Johnson, P. J. Neustaedter and K. K. Schmiegel, J. Amer. Chem. Soc., **87**, 5148 (1965).

9. Resolution may be effected at this stage *via* α-methyl benzylamine salt.

—[Ph_3P, MeCN, $(^iPr)_2NEt$, 50°]* →

I^- P^+Ph_3 S S 5 (D)

Diisopropylethylamine prevented migration of double bond to β,γ-position.

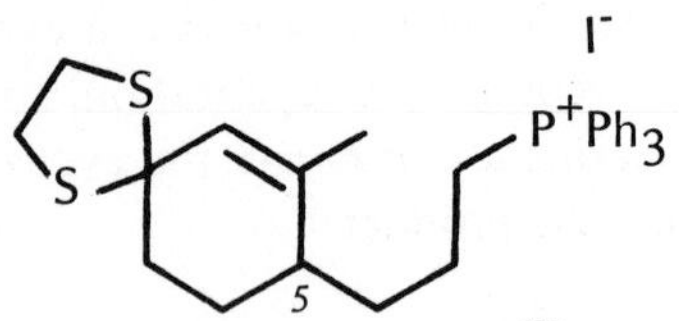

Preparation of progesterone:

D + **A** —[Ph Li, Et_2O, -70°]→ *65-72%*

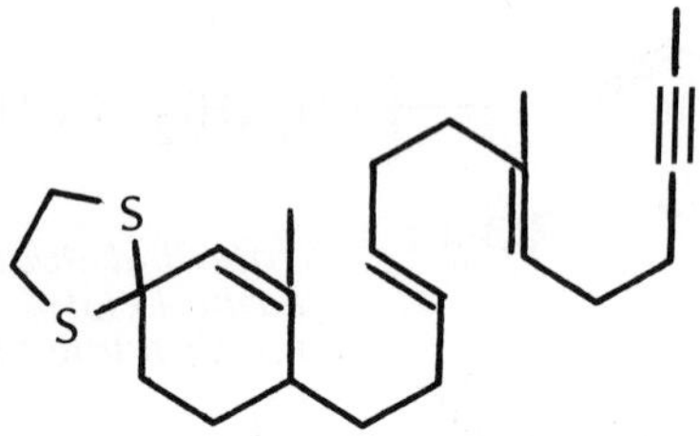

S S

—[MeI, aq DMF, $CaCO_3$]→ *40-50%*

O

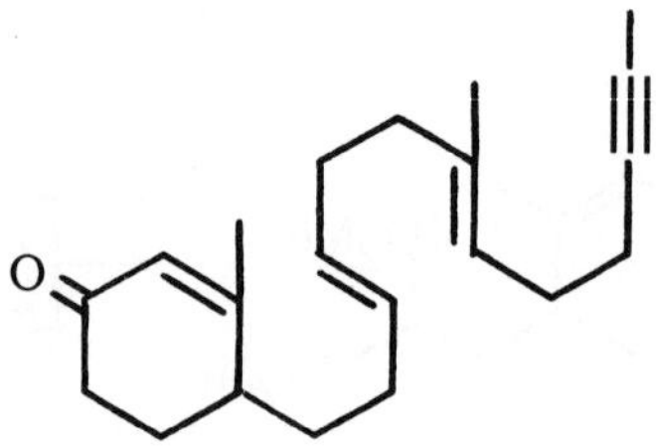

—[$NaAlH(OCH_2CH_2OMe)_2$, THF, 0°]→ *100%*

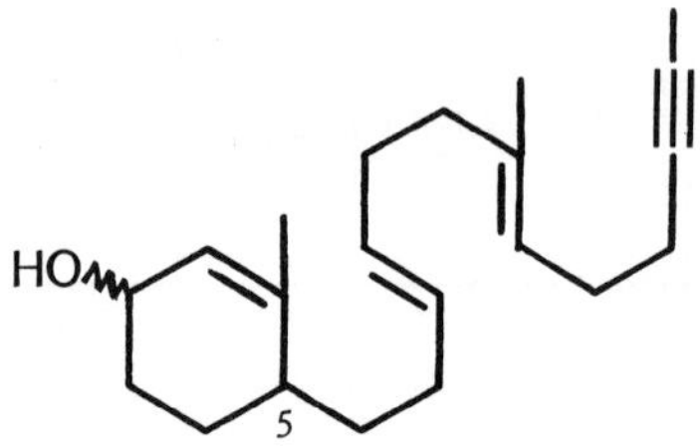

HO 5 (E)

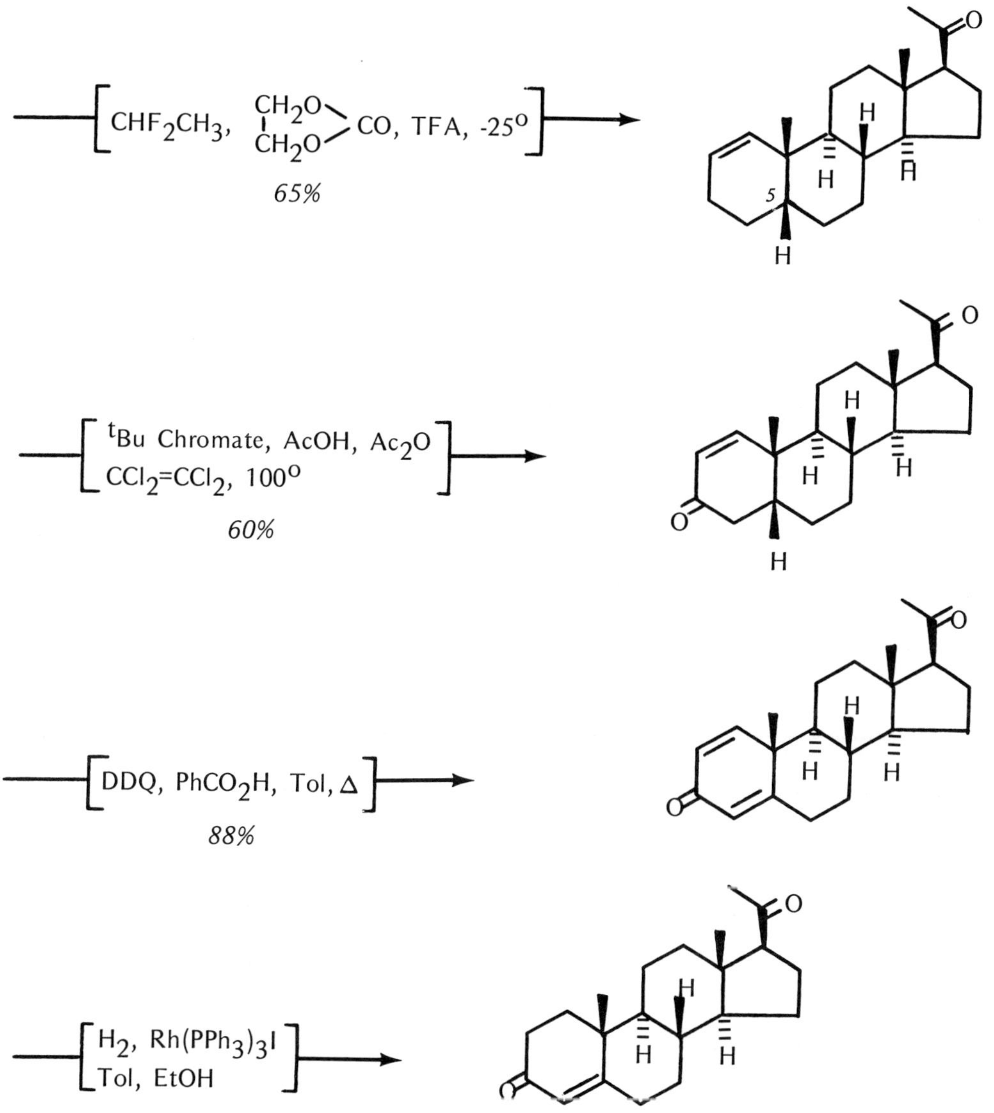

[CHF_2CH_3, CH_2O / CH_2O > CO, TFA, -25°] →
65%

H
H
H
5
H

[${}^tBu$ Chromate, AcOH, Ac_2O
CCl_2=CCl_2, 100°] →
60%

O
H
H
H
O
H

[DDQ, $PhCO_2H$, Tol, Δ] →
88%

O
H
H
H
O

[H_2, $Rh(PPh_3)_3I$
Tol, EtOH] →

O
H
H
H
O

Progesterone

PROSTAGLANDINS

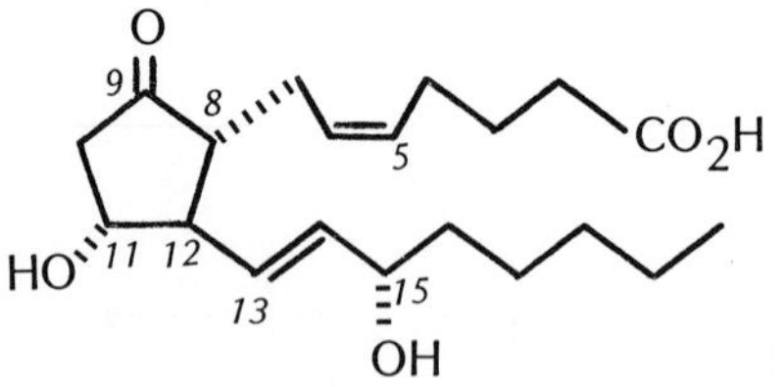

$$\text{[structure: prostaglandin E}_2\text{ with positions 5, 8, 9, 11, 12, 13, 15; HO, OH, }CO_2H]$$

The elegant work of Corey and his associates at Harvard, who have successfully executed the synthesis of all the major naturally occurring prostaglandins in pure optically active form, is an outstanding example of the application of organic synthesis to solve the problem of a limited supply of a natural product.[1-3]

The practical approach to prostaglandins outlined here was developed by Corey with the objective of securing the syntheses of all the primary prostaglandins from a single resolved precursor. This approach permits optical resolution at an early stage and appears suited for large-scale production. A noteworthy feature of this synthesis is the use of Diels-Alder reaction for construction of the key bicyclic intermediate in which four of the ring appendages and stereocentres of the basic prostaglandin nucleus are established efficiently and with complete stereospecificity. Generation of the 9- and 11-hydroxyl groups is accomplished by Baeyer-Villiger oxidation of the ketone to a lactone, followed by saponification and iodolactonization. Sequential introduction of the C-12 and C-8 side chains containing the *cis*-Δ^5 and *trans*-Δ^{13} linkages by modified Wittig reactions completes this versatile synthesis.[3-9] Minor modifications lead to the synthesis of other primary prostaglandins[10].

---

1. Reviews: J. S. Bindra and R. Bindra, Progress in Drug Res., **17**, 410 (1973); J. E. Pike, Scient. Amer., **225** (5), 84 (1971).

2. For a review of the earlier synthetic approaches to prostaglandins, see Art Org. Synth., 291 (1970). Subsequent synthetic work has been reviewed each year in Annual Reports in Medicinal Chemistry, Academic Press: J. F. Bagli, Ann. Repts. Med. Chem. **5**, 170 (1970); G. L. Bundy, *ibid.*, **6**, 137 (1971); **7**, 157 (1972); R. A. Mueller, *ibid.*, **8**, 172 (1973); **9**, 162 (1974). See also, J. E. Pike, Fortschr. Chem. Org. Naturstoffe, **28**, 313 (1970).

3. For an excellent introduction to the strategy underlying the program of total synthesis of prostaglandins at Harvard, see E. J. Corey, Ann. N. Y. Acad. Sci., **180**, 24 (1971).

4. E. J. Corey, N. M. Weinshenker, T. K. Schaaf and W. Huber, J. Amer. Chem. Soc. **91**, 5675 (1969).

5. E. J. Corey, T. K. Schaaf, W. Huber, U. Koelliker and N. M. Weinshenker, J. Amer. Chem. Soc. **92**, 397 (1970).

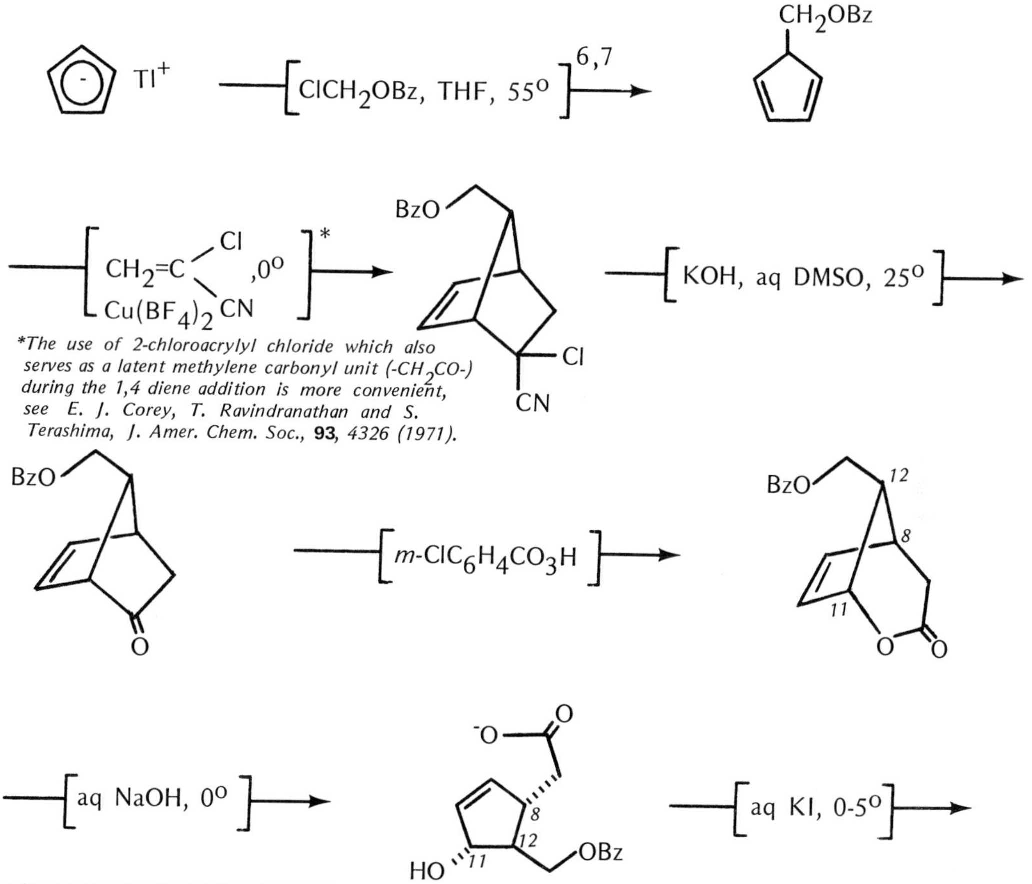

Tl^+ —[$ClCH_2OBz$, THF, 55°][6,7]→

CH_2OBz

—[CH_2=C(Cl)(CN), $Cu(BF_4)_2$, 0°]*→ BzO … Cl … CN —[KOH, aq DMSO, 25°]→

*The use of 2-chloroacrylyl chloride which also serves as a latent methylene carbonyl unit ($-CH_2CO-$) during the 1,4 diene addition is more convenient, see E. J. Corey, T. Ravindranathan and S. Terashima, J. Amer. Chem. Soc., **93**, 4326 (1971).

BzO … O —[m-$ClC_6H_4CO_3H$]→ BzO 12 8 11 O O

—[aq NaOH, 0°]→ $^-$O O 8 12 HO 11 OBz —[aq KI, 0-5°]→

6. The use of thallous cyclopentadienide offers several advantages over the commonly used alkali metal salts, E. J. Corey, U. Koelliker and J. Neuffer, J. Amer. Chem. Soc., **93**, 1489 (1971).

7. a) The use of a benzyl ether instead of a methyl ether offers an advantage since its eventual removal is conveniently effected by catalytic hydrogenation. b) The use of p-phenylbenzoyl group gave crystalline, easily separable reduction products. Reduction with hydride reagent (i) gave a better ratio of $15\alpha/15\beta$ reduction products: E. J. Corey, S. M. Albonico, U. Koelliker, T. K. Schaaf and R. K. Varma, J. Amer. Chem. Soc., **93**, 1491 (1971).

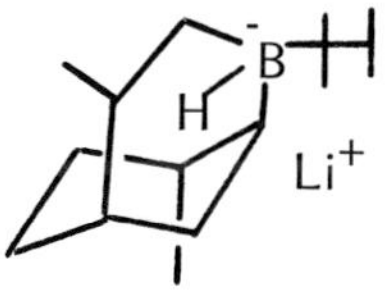

B^- H H Li^+

(i)

[PBPCOCl*, Py, 25°]

*PBP= biphenyl

[(^{n}Bu)$_3$SnH]

[H_2, Pd-C][7]

[CrO_3, Py]

(A)

[$(MeO)_2\overset{O}{P}CH_2COC_5H_{11}$ BuLi, DME][8]

[$Zn(BH_4)_2$, DME][7,9]

8. For application of the β-oxido phosphonium ylid (ii) for direct introduction of 15(S)-configuration, see E. J. Corey, H. Shirahama, H. Yamamoto, S. Terashima, A. Venkateswarlu and T. K. Schaaf, J. Amer. Chem. Soc., **93**, 1490 (1971).

$Ph_3P{=}CH$ (ii)

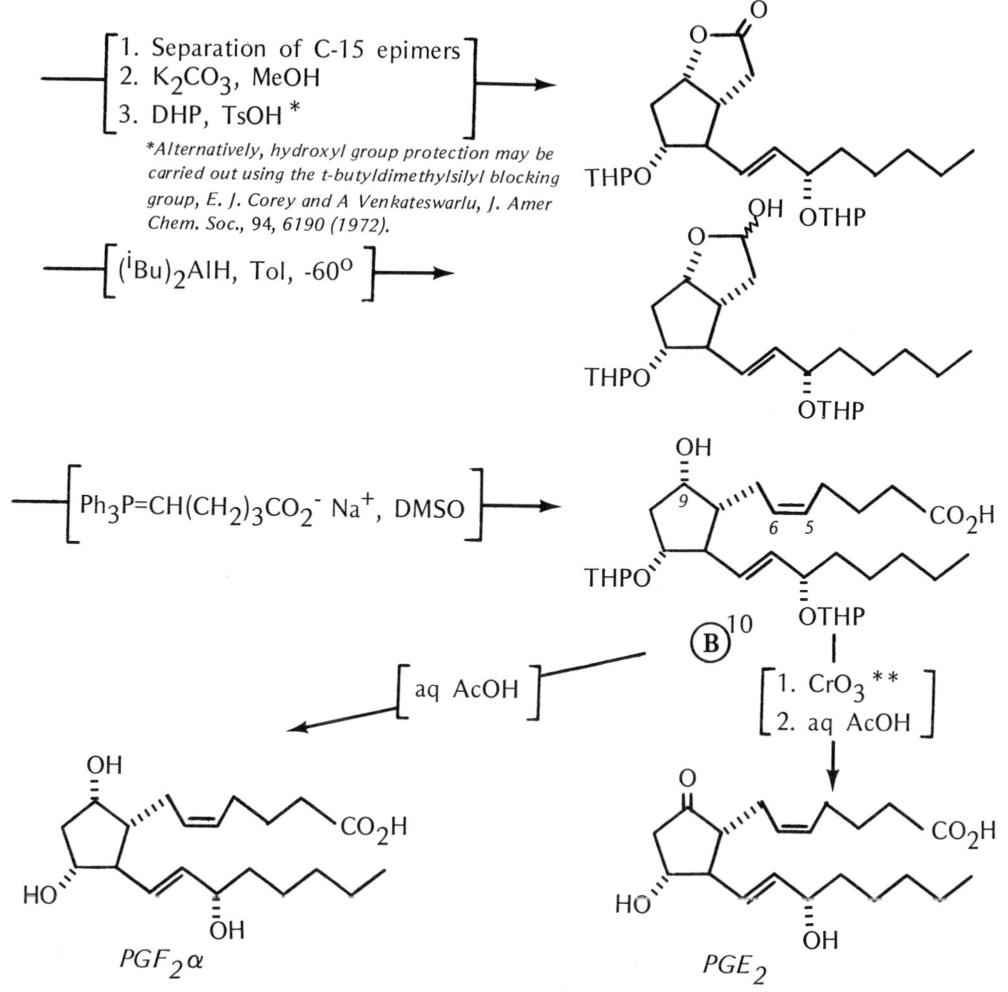

9. For alternative methods of stereoselective reduction to the 15(S) hydroxyl, see E. J. Corey, K. B. Becker and R. K. Varma, J. Amer. Chem. Soc., **94**, 8616 (1972).

10. The bis-tetrahydropyranyl ether (B) can be selectively reduced to the corresponding 5,6-dihydro compound, which can be transformed either to PGE_1 or $PGF_1\alpha$: E. J. Corey, R. Noyori and T. K. Schaaf, J. Amer. Chem. Soc., **92**, 2586 (1970).

***For oxidation of the 9-hydroxy group using N-chlorosuccinimide-dimethyl sulfide complex, see E. J. Corey and C. U. Kim, J. Org. Chem.,* **38**, *1233 (1973).*

An alternative route to the Corey aldehyde (A), developed at Pfizer,[11] commences with an unusual Prins reaction on norbornadiene. Oxidation and acid catalyzed opening of the cyclopropane ring in (C), followed by Baeyer-Villiger oxidation gave the key chloro-lactone (D) which is cleanly converted to the desired δ-lactone upon treatment with base.

[HCHO, HCO_2H, H_2SO_4]

[CrO_3]

(C)

[1. Resolve 2. aq HCl, Δ]

[m-$ClC_6H_4CO_3H$]

[EtOCCl, TEA]

[$NaBH_4$, THF, 0°]

11. J. S. Bindra, A. Grodski, T. K. Schaaf and E. J. Corey, J. Amer. Chem. Soc., **95**, 7522 (1973).

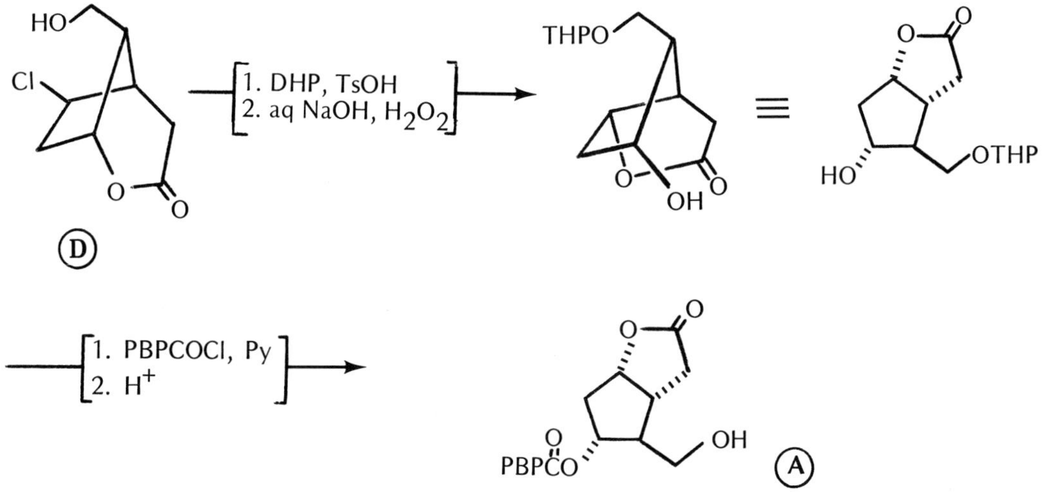

HO Cl O O (D) [1. DHP, TsOH 2. aq NaOH, H_2O_2] → THPO O OH O ≡ O O HO OTHP

[1. PBPCOCl, Py 2. H^+] → O O O PBPCO OH (A)

The stereospecific synthesis of PGF$_2\alpha$ by Woodward and his colleagues at Basel illustrates the ingenious use of functional groups as internal protecting agents and their application to exert stereochemical control during synthetic operations. The synthesis commences with cyclohexane-1,3,5-triol, leading to a cyclohexyl amino diol (E) which undergoes ring contraction to establish the prostanoid cyclopentane nucleus (F).[12]

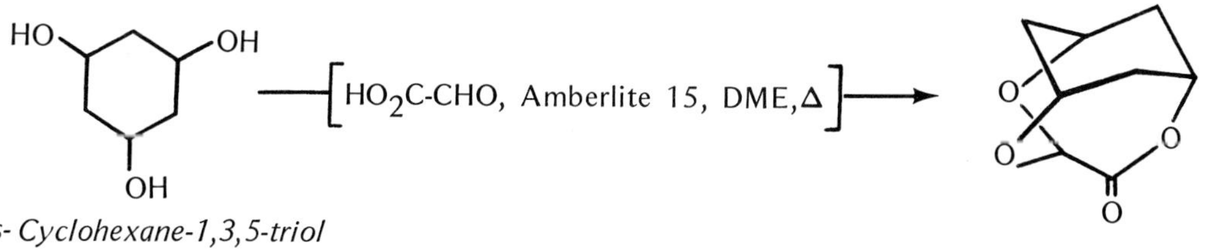

HO OH OH [HO_2C-CHO, Amberlite 15, DME,Δ] → O O O O

cis-Cyclohexane-1,3,5-triol

12. R. B. Woodward, J. Gosteli, I. Ernest, R. J. Friary, G. Nestler, H. Raman, R. Sitrin, Ch. Suter, J. K. Whitesell, J. Amer. Chem. Soc., **95**, 6853 (1973).

[$NaBH_4$, EtOH] →

OH OH

[MsCl, Py, -20°] →

H MsO OMs

[EtOH, KOH, Δ] →

OMs

[K_2CO_3, aq DME, Δ] →

HO

[MsCl, TEA, MeOH, 0°] →

MsO

[^{i}PrOH, KOH, Δ] →

[H_2O_2, PhCN, K_2CO_3, MeOH] →

[aq NH_3, 100° (sealed tube)] →

MeOH, HCl

E

1. $NaNO_2$, AcOH, NaOAc, 0-5°
2. Na_2CO_3-CH_2Cl_2

12b

F

$(MeO)_2\overset{O}{\overset{\|}{P}}CH_2COC_5H_{11}$
BuLi, DME

$PGF_{2}\alpha$

12b. Tiffeneau-Demjanov rearrangement: see D. Redmore and C. D. Gutsche, Adv. Alicyclic Chem., **3**, 2 (1971).

Recently Sih and his associates at the University of Wisconsin reported a short efficient synthesis of prostaglandins. The problem of introducing the asymmetric centre at C-15 in this synthesis is solved by introduction of the lower side chain, already bearing the C-15 asymmetric centre, by a 1,4-addition reaction.[13, 14]

Li+

+

Br

CO_2Et

CO_2Et

H_2O_2, NaOH *

*Generates singlet oxygen.

H O O H

OH

CO_2Et

O

+

O

CO_2Et

HO

2 steps

1. DHP, TsOH

2. Li O O , $(^nBu)_3PCuI$ 0°, Et_2O

O

CO_2Et

THPO

15

O O

1. AcOH, H_2O, THF

2. Baker's yeast

PGE_1

13. C. J. Sih, R. G. Salomon, P. Price, G. Peruzzotti and R. Sood, Chem. Commun., 240 (1972).

14. C. J. Sih, P. Price, R. Sood, R. G. Salomon, G. Peruzzotti and M. Casey, J. Amer. Chem. Soc., **94**, 3643 (1972).

PYRENOPHORIN

Methods for the construction of large ring compounds and their application to macrolide synthesis continue to attract attention. The synthesis of pyrenophorin, a macrolide antibiotic, illustrates the effective use of protecting groups and functional group activation for construction of the sixteen-membered bis-lactone ring.[1]

$NaAlH_4$

$HS(CH_2)_3SH$, $BF_3.Et_2O$ [2]

1. DHP, TsOH
2. $^nBuLi \rightarrow HCO_2Et$

OH

OTHP

CHO

(A)

Ph_3P ... $SO_2C_6H_4Me$

OTHP

$SO_2C_6H_4Me$

---

1. E. W. Colvin, T. A. Purcell and R. A. Raphael, Chem. Commun., 1031 (1972).
2. E. J. Corey and D. Seebach, Angew. Chem. Int. Ed., 4, 1077 (1965).

[1. H_3O^+ 2. $BrCOCH_2Br$]

$OCOCH_2Br$ $SO_2C_6H_4Me$

[1. Ph_3P 2. aq NaOH]

$OCOCH_2{=}PPh_3$ $SO_2C_6H_4Me$

[(A), C_6H_6, Δ]

OTHP $SO_2C_6H_4Me$

[1. H_3O^+ 2. DBN, C_6H_6][3]

OH CO_2H

---

3. The p-toluenesulfonylethyl protecting group [A. W. Miller and C. J. M. Stirling, J. Chem. Soc. C, 2612 (1968)] is selectively removed by treatment with DBN. Use of stronger bases resulted in extensive cleavage of the central ester linkage.

4. The reagent reacts with carboxylic acids to form imidazolides which give esters with alcohols under basic catalysis: H. A. Staab and A. Mannschreck, Chem. Ber., **95**, 1284 (1962).

5. Cf. E. J. Corey and D. Crouse, J. Org. Chem., **33**, 298 (1968).

QUININE

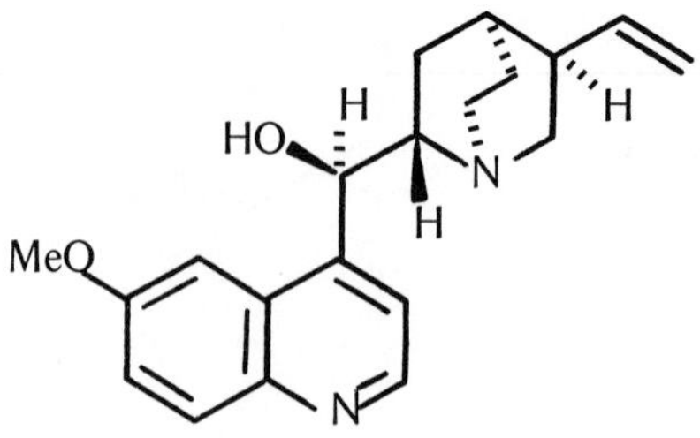

Following the classic synthesis of quinine by Woodward and Doering in 1944,[1] no major synthetic achievements were forthcoming in this area for nearly twenty five years. Recently, however, there has been a remarkable resurgence of interest in the synthesis of Cinchona alkaloids, culminating in several ingenious approaches to quinine.[2,3] An elegant stereoselective approach to quinine developed at Hoffmann-La Roche[4] is outlined below. In this synthesis the desired *cis* vinyl and acetic acid side-chains of the key intermediate N-benzoyl meroquinene (C) have been fashioned from the carbocyclic ring of a hexahydroquinolone (A) by Schmidt rearrangement and pyrolytic fragmentation of an N-nitrosolactam (B). Construction of the quinuclidine nucleus has been accomplished by an intramolecular nucleophilic addition of the piperidine nitrogen to the β-carbon of the substituted 4-vinylquinoline (F).

---

1. R. B. Woodward and W. E. Doering, J. Amer. Chem. Soc., **67**, 860 (1945); Art. Org. Synth., p. 296 (1970).

2. For a review of the classic work on synthesis of Cinchona alkaloids, see R. B. Turner and R. B. Woodward, The Alkaloids, **3**, 1 (1953). The more recent developments in this field have been reviewed by M. R. Uskoković and G. Grethe, The Alkaloids, **14**, 181 (1973).

3. A common feature of all these syntheses is that they all proceed through meroquinene derived intermediates bearing a properly positioned functional group which facilitates the quinuclidine ring forming reaction (i→ii).

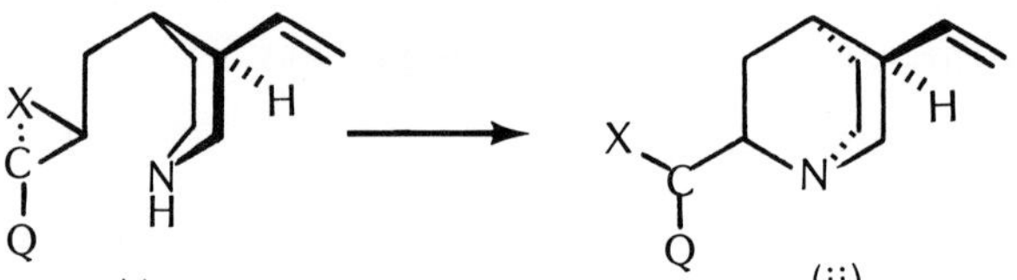

(i) (ii)

4. M. Uskoković, J. Gutzwiller and T. Henderson, J. Amer. Chem. Soc., **92**, 203, 204 (1970).

Preparation of N-Benzoyl Meroquinene:

[Pyrrolidine, C_6H_6, Δ]

O, N, Ph

[NaN$_3$, PPA, 60°]
67%

(A)[5]

O, NH, N, Ph

[H_2, Rh-Al_2O_3]

[N_2O_4]
100%

H, N, Ph

[125°][6]

(B)

O, N, H, 11, 10, Ph

5. R. L. Augustine, J. Org. Chem., **23**, 1853 (1958).

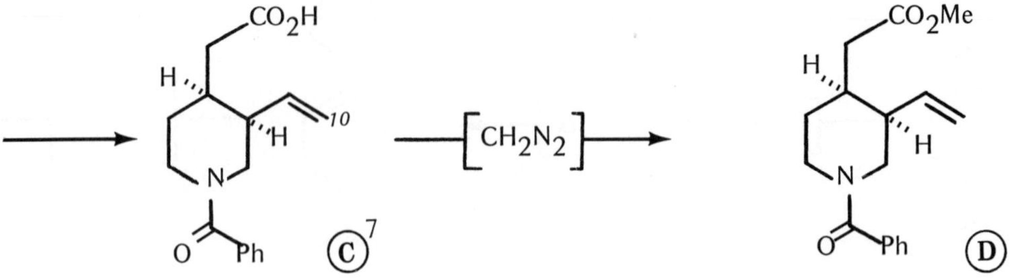

(C)[7] —[CH_2N_2]→ (D)

N-Benzoyl Meroquinene Methyl Ester

Preparation of Quinine:

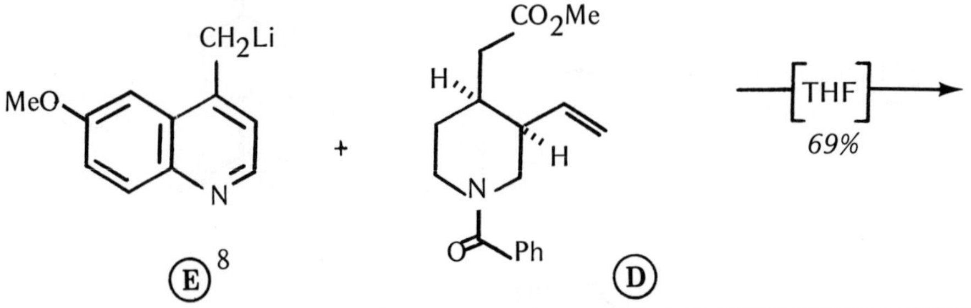

(E)[8] + (D) —[THF]→ *69%*

6. Cf. R. Huisgen, J. Reinertshofer, Ann., **574**, 174 (1952); **575**, 197 (1952).

7. N-Benzoylmeroquinene (C) was accompanied by the seven-membered lactone (i). The two compounds were formed in 50 and 30% yield respectively. The latter is apparently formed by nucleophilic attack of one of the ester oxygens on the C-10 carbon. Lactone (i) could be converted to (D) by two different routes.

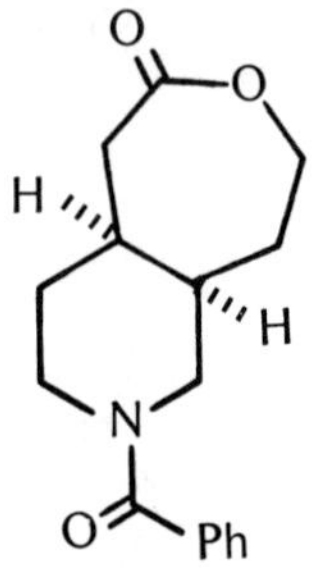

(i)

8. Prepared from 6-methoxylepidine and lithium diisopropylamide.

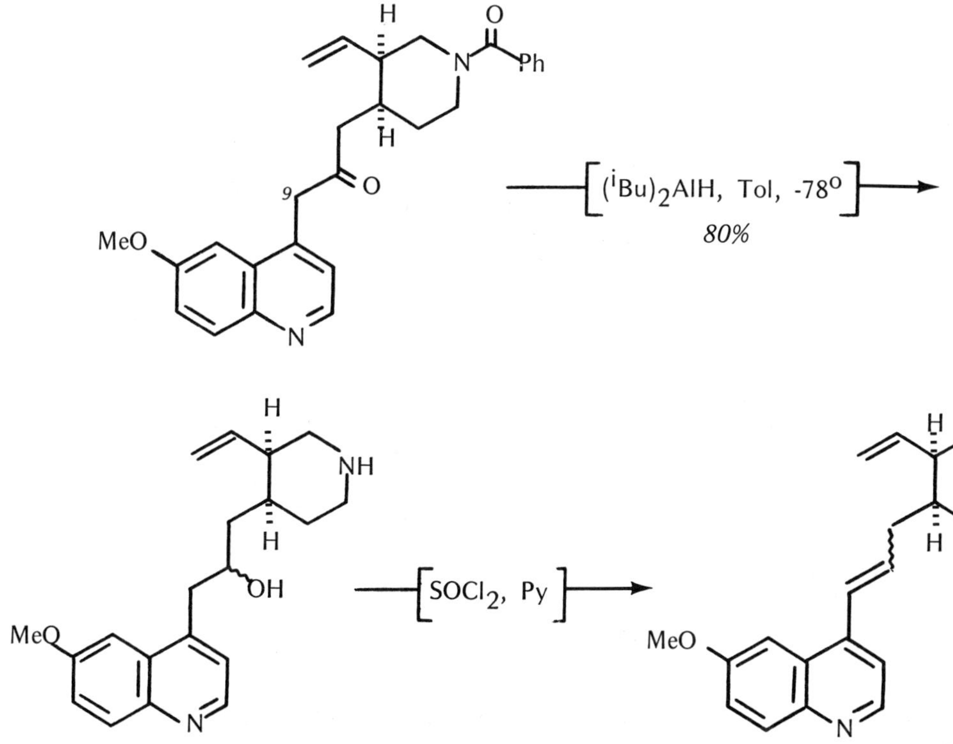

H
O
N
Ph
H
9
O
MeO
N

[$(^iBu)_2AlH$, Tol, -78°]
80%

H
NH
H
OH
MeO
N
Footnote 9

[$SOCl_2$, Py]

H
NH
H
MeO
N
(F)[10]

9. Resolved *via* dibenzoyl-(+)-tartrate.

10. An alternative non-stereoselective route from olefin (F) to quinine proceeds *via* the epoxide (ii).

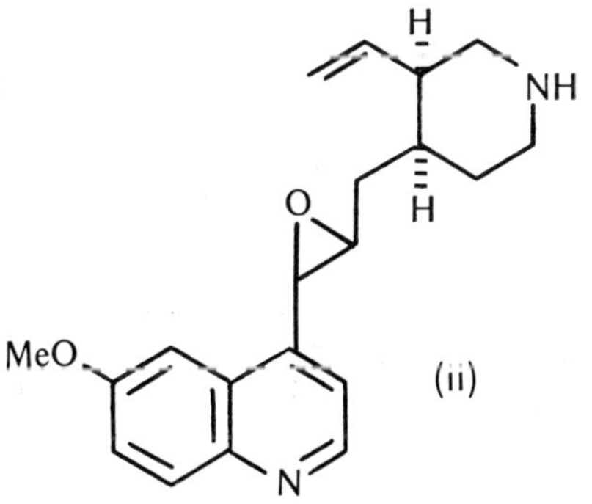

H
NH
O
H
MeO
(ii)
N

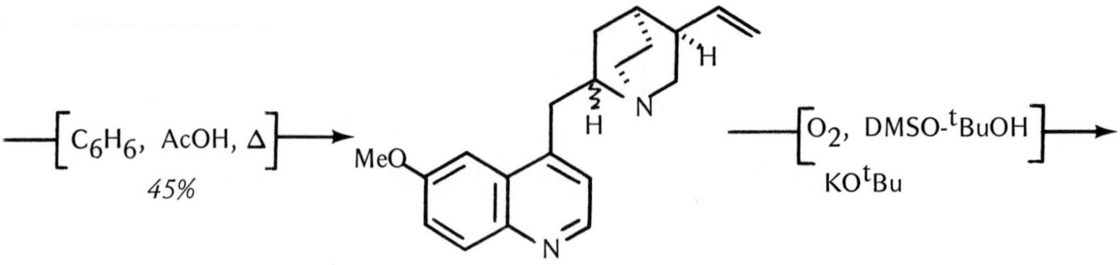

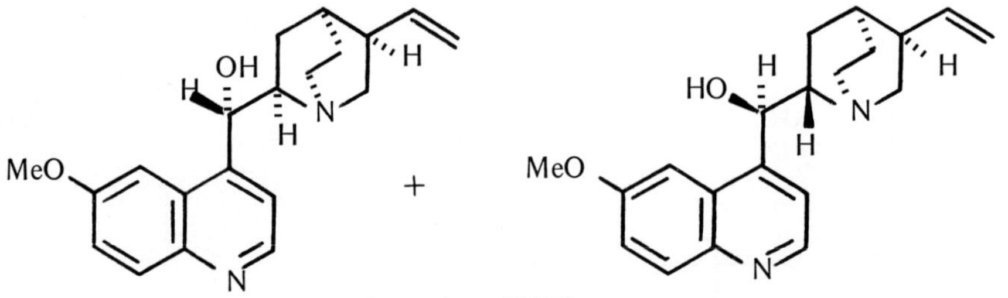

[C_6H_6, AcOH, Δ] 45%

[O_2, DMSO-^tBuOH, KO^tBu]

MeO

Quinidine (41%)

Quinine (32%)

An alternative synthesis[11] of meroquinene (D) is outlined below. A key feature of this synthesis is application of the photolytic Löffler-Freytag reaction for generating the vinyl side-chain from an ethyl group in the corresponding saturated precursor cincholoipon methyl ester.

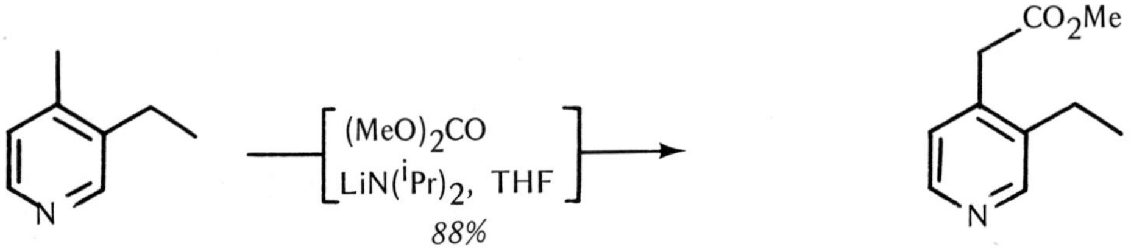

[$(MeO)_2CO$, $LiN(^iPr)_2$, THF] 88%

CO_2Me

11. M. Uskoković, C. Reese, H. L. Lee, G. Grethe and J. Gutzwiller, J. Amer. Chem. Soc., **93**, 5902 (1971); G. Grethe, H. L. Lee, T. Mitt and M. Uskoković, Helv., **56**, 1485 (1973).

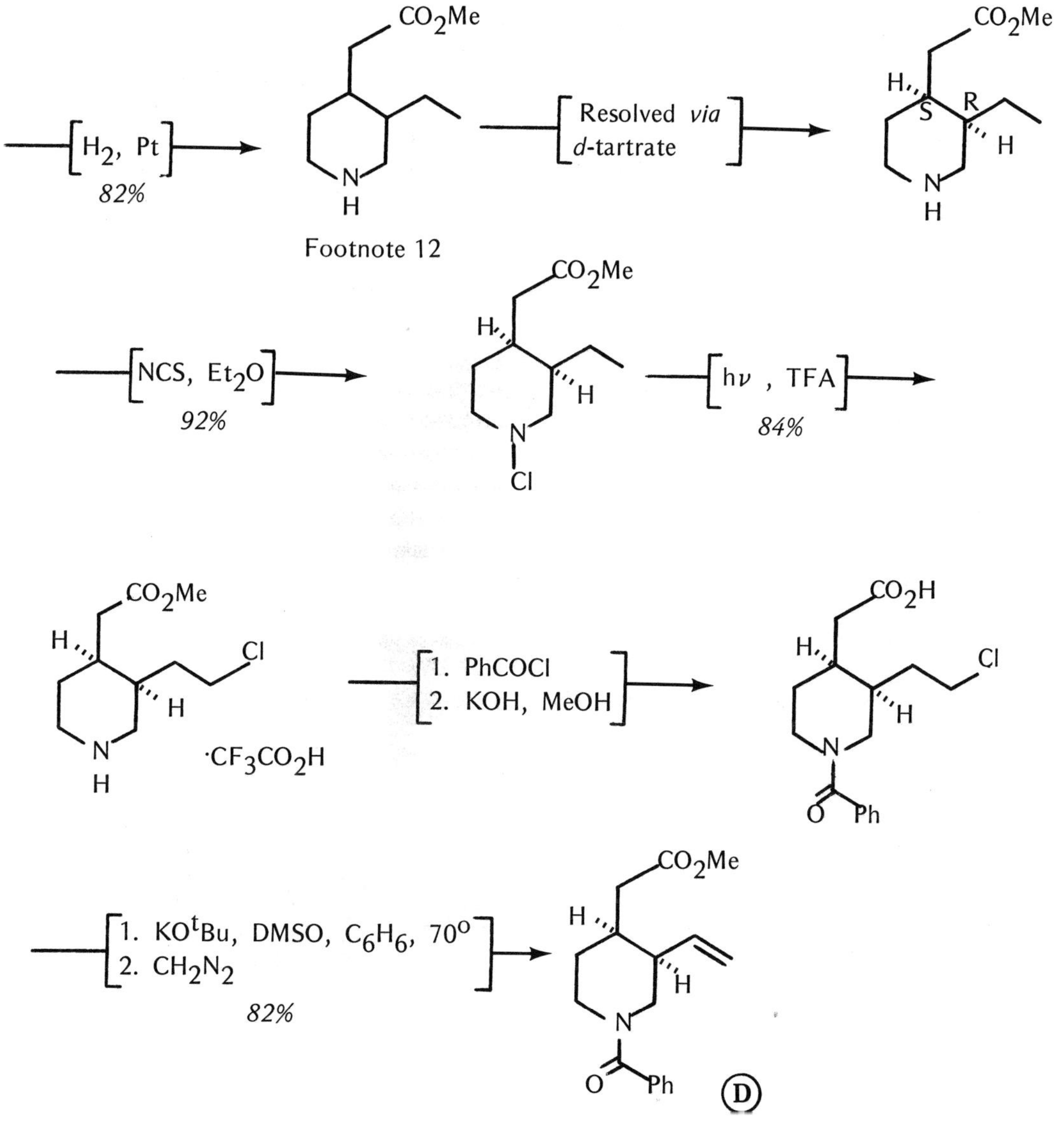

[H_2, Pt] 82%

[Resolved *via* *d*-tartrate]

CO_2Me

N H

Footnote 12

H, S, R, H

[NCS, Et_2O] 92%

N Cl

[hν, TFA] 84%

H Cl

$\cdot CF_3CO_2H$

[1. PhCOCl 2. KOH, MeOH]

CO_2H

O Ph

[1. KO^tBu, DMSO, C_6H_6, 70° 2. CH_2N_2] 82%

(D)

---

12. The corresponding ethyl ester has been previously synthesized by G. Stork and S. M. McElvain, J. Amer. Chem. Soc., **68**, 1053 (1946).

13. In the synthesis developed by Gates *et al.*, the key vinylquinoline (i) was obtained by the Wittig reaction of quininaldehyde with an ylid prepared from meroquinene alcohol, M. Gates, B. Sugavanam and W. L. Schreiber, J. Amer. Chem. Soc., **92**, 205 (1970). On the other hand, Taylor and Martin synthesized the vinylquinoline (i) from meroquinene aldehyde and a quinoline Wittig reagent obtained directly by reaction of 4-chloro-6-methoxyquinoline and methylenetriphenylphosphorane, E. C. Taylor and S. F. Martin, J. Amer. Chem. Soc., **94**, 6218 (1972).

Gates' Synthesis:

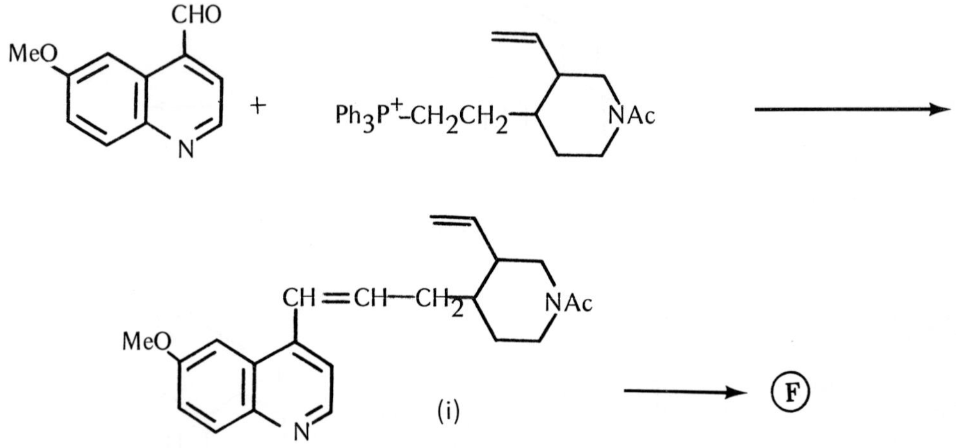

CHO, MeO, N + $Ph_3P^+-CH_2CH_2-$ NAc ⟶

MeO, N, $CH{=}CH-CH_2$, NAc (i) ⟶ (F)

Taylor's Synthesis:

Cl, MeO, N + $2CH_2{=}PPh_3$ ⟶ MeO, N, $CH{=}PPh_3$

+ $OHC-CH_2$ NAc ⟶ (i)

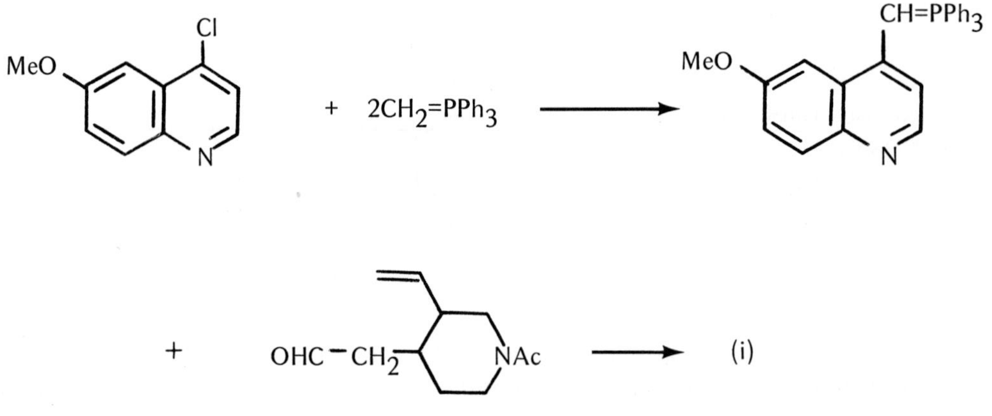

RHOEADINES

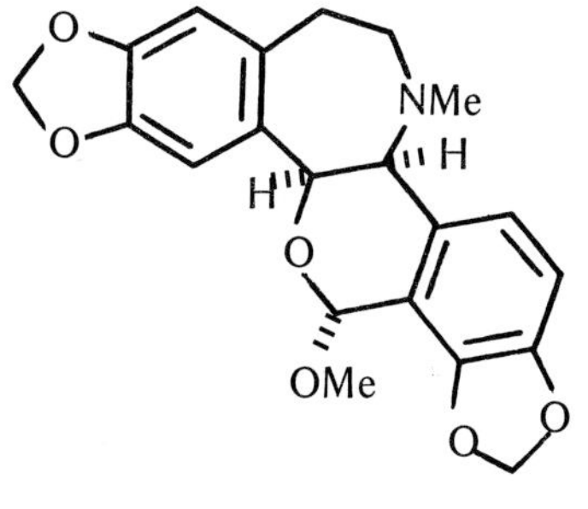
O
O
NMe
H
H
O
OMe
O
O

The rhoeadines are a group of alkaloids isolated from plants of the genus *Papaver* (Papaveraceae), and are characterized by a benzazepine structure and a *cis*-fused six-membered acetal ring. In the synthesis of rhoeadine outlined below, the requisite 3-benzazepine skeleton has been constructed from a benzylic spiroisoquinoline intermediate (B) by a skeletal rearrangement.[2]

Preparation of Indandione:[3]

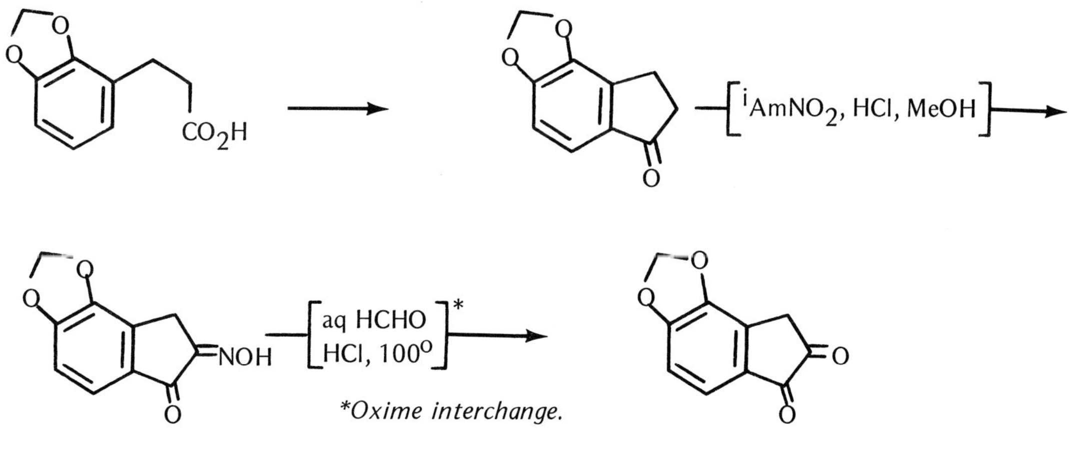
CO_2H
[$^iAmNO_2$, HCl, MeOH]
=NOH
[aq HCHO
HCl, 100°]*
*Oxime interchange.

---

1. Review: M. Shamma, "The Isoquinoline Alkaloids," Academic Press, New York (1972) p. 399.

2. H. Irie, S. Tani and H. Yamane, Chem. Commun., 1713 (1970); J. Chem. Soc. Perkin I, 2986 (1972).

3. Initial steps leading to the spiroisoquinoline (A) were developed earlier in the synthesis of ochotensine by H. Irie, T. Kishimoto and Uyeo, J. Chem. Soc. (C), 3051 (1968).

Preparation of Rhoeadine:

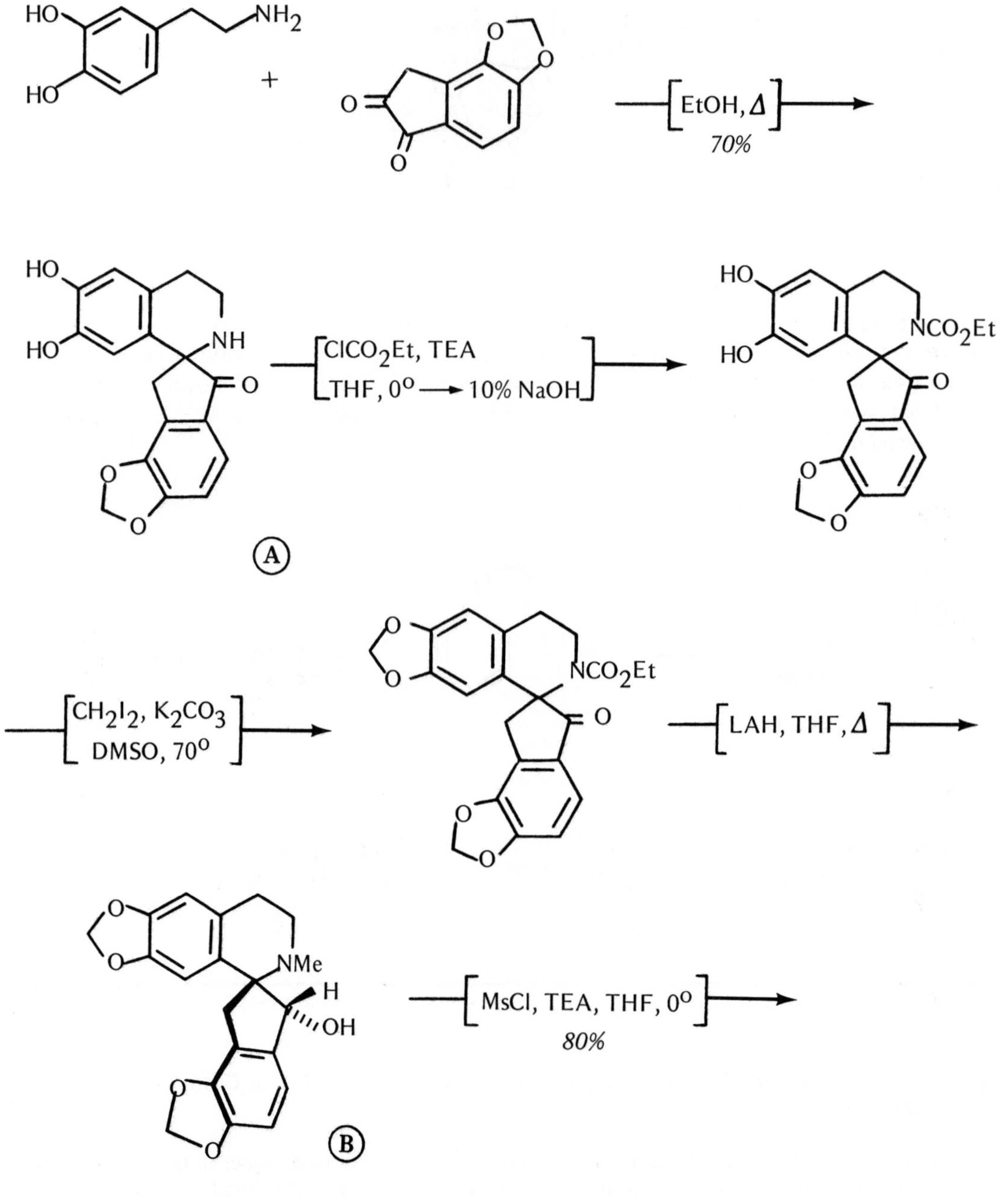

HO, HO, NH_2 + O, O, O, O — [EtOH, Δ] 70% →

HO, HO, NH, O, O, O (A) — [$ClCO_2Et$, TEA, THF, 0° → 10% NaOH] → HO, HO, NCO_2Et, O, O, O

— [CH_2I_2, K_2CO_3, DMSO, 70°] → O, O, NCO_2Et, O, O, O — [LAH, THF, Δ] →

O, O, NMe, H, OH, O, O (B) — [MsCl, TEA, THF, 0°] 80% →

NMe

NMe

[OsO_4, Py]

HO

HO

NMe

[1. $NaIO_4$, HCl, 0°
2. $NaBH_4$]

NMe

H

H

HO

HO

(C)

Rhoeagenine diol

[1. MnO_2, Chf
2. $HC(OMe)_3$] 4

NMe

H

H

OMe

Rhoeadine

---

4. The final stages of the synthesis, *viz.*, conversion of rhoeagenine diol (C) to rhoeadine had already been accomplished during work on constitution of rhoeadine by F. Santavy, J. L. Kaul, L. Hruben, L. Dolejs, V. Hanus, K. Blaha and A. D. Cross, Coll. Czech. Chem. Commun., **30**, 3479 (1965).

The synthesis of alpinine[5] illustrates a different approach to rhoeadine alkaloids, and is based on the photosensitized oxidation of the enaminoketone (A) and rearrangement of the resulting dioxetan to a spiro keto lactone (B). The rest of the synthesis closely follows the pathway laid down earlier at Hoffman-La Roche for the synthesis of rhoeadine.[6]

[NaH→MeI, THF–DMF]

[HCHO, HCl]

[NaCN, DMSO]

[1. 2,3-dimethoxybenzaldehyde; 2. Na–Hg, EtOH]

[1. EtOH, HCl; 2. 5% MeOH, KOH]

[1. $SOCl_2$; 2. $AlCl_3$, $PhNO_2$]

5. K. Orito, R. H. Manske and R. Rodrigo, J. Amer. Chem. Soc., **96**, 1944 (1974).

MeO, MeO, N–Me, Ac, OMe, OMe, O

[10% KOH, aq EtOH, Δ]
62%

MeO, MeO, N–Me, MeO, OMe

[Triton-B, Py, O_2]
79%

MeO, MeO, N–Me, O, MeO, OMe (A)

[air, EtOH, 10 days (Rose Bengal)]
37%

MeO, MeO, O–O, N–Me, O, MeO, OMe

MeO, MeO, N^+–Me, O, O, ^-O, MeO, OMe

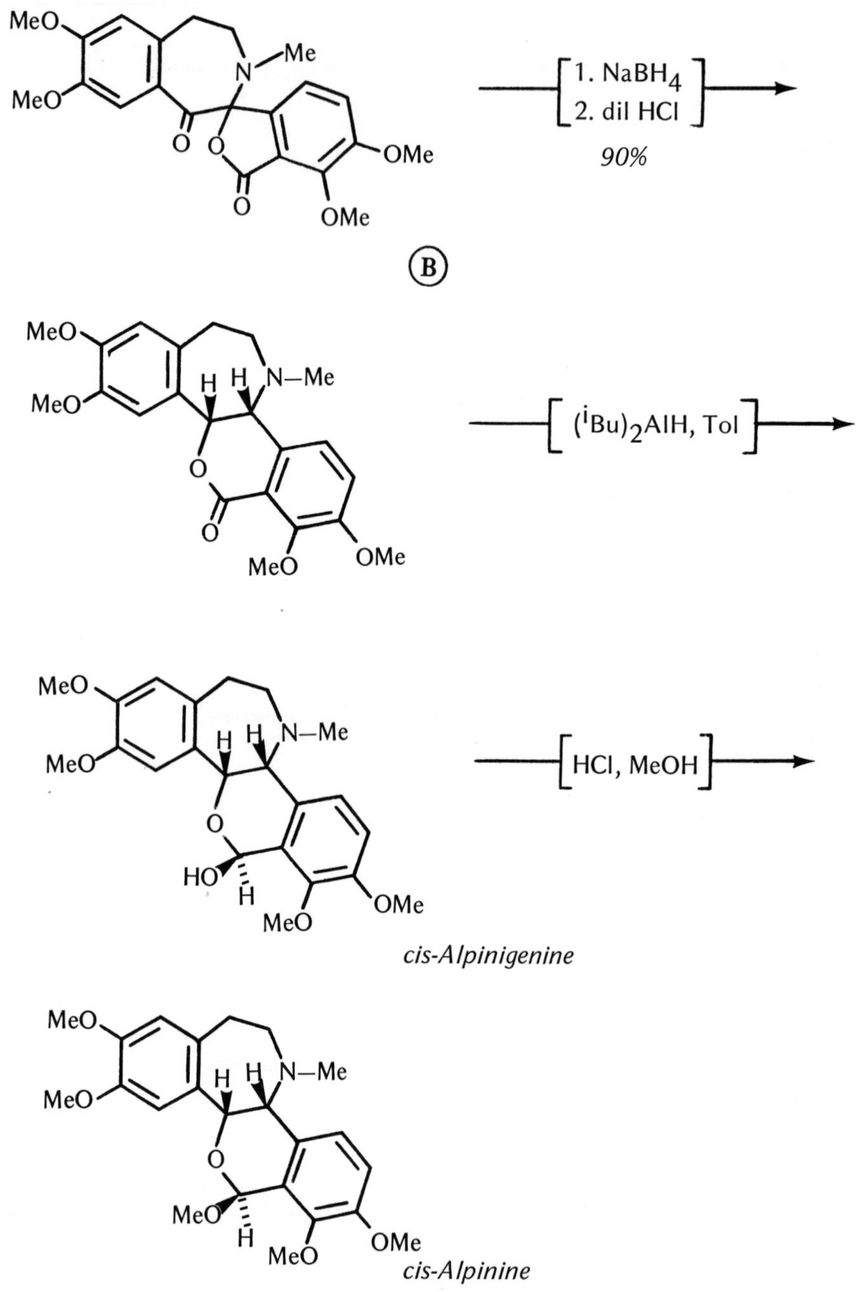

MeO MeO N—Me O O O OMe OMe

$\left[\begin{array}{l}1.\ NaBH_4\\2.\ \text{dil HCl}\end{array}\right]$

90%

(B)

MeO MeO H H N—Me O O MeO OMe

$[(^iBu)_2AlH,\ Tol]$

MeO MeO H H N—Me O HO H MeO OMe

cis-Alpinigenine

$[HCl,\ MeOH]$

MeO MeO H H N—Me O MeO H MeO OMe

cis-Alpinine

6. For an alternative total synthesis of rhoeadine based on conversion of phthalideisoquinolines into benzazepines, see W. Klotzer, S. Teitel and A. Brossi, Helv., **54**, 2057 (1971); **55**, 2228 (1972); N. Klotzer, S. Teitel, J. F. Blount and A. Brossi, J. Amer. Chem. Soc., **93**, 4321 (1971).

7. Recently Shamma and Toke have reported a potentially useful route to *trans* B/D-rhoeadanes based on a new method of ring expansion for the formation of benzazepines [M. Shamma and L. Toke, Chem. Commun., 740 (1973)]. Elements of this approach are outlined below.

MeO, MeO, Br^-, N^+, EtO_2C → PhCOCl / aq NaOH → MeO, MeO, N^+–COPh, O, H, EtO_2C

→ MeO, MeO, CHO, N—COPh, EtO_2C → KO^tBu, DMSO →

MeO, MeO, A, B, N-COPh, H, H, D, O, C, O

RHYNCHOPHYLLINE

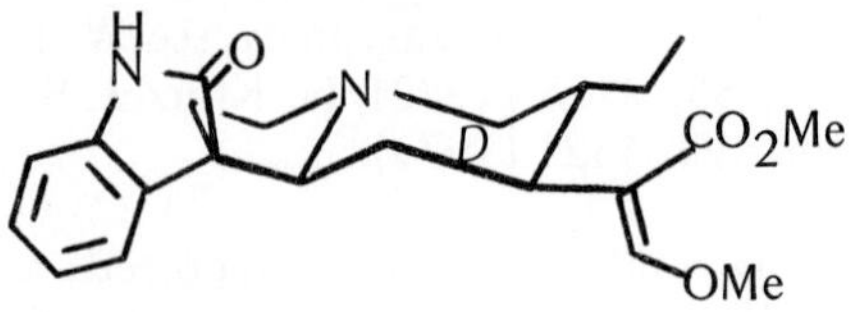
H
N
O
N
D
CO2Me
OMe

Although a partial synthesis of rhynchophylline[1] from dihydrocorynantheine was accomplished in 1961,[1,2] the total synthesis of this oxindole alkaloid remained incomplete for a long time.[3] In the synthesis finally reported by Ban *et al.*,[4] the spiro oxindole nucleus has been established at an early stage, followed by stepwise construction of the functionalized piperidine ring D.

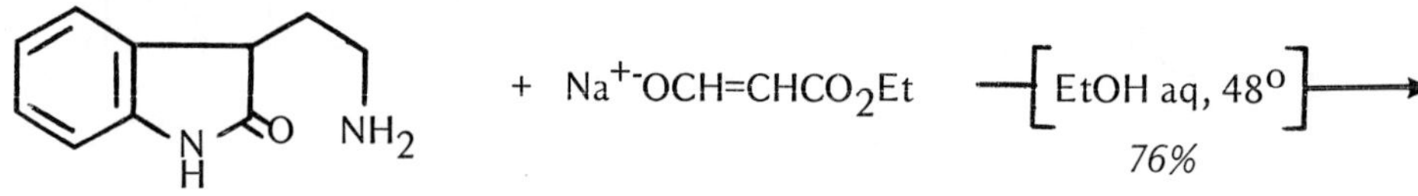
N
H
O
NH2
+ Na+-OCH=CHCO2Et
EtOH aq, 48°
76%

1. For a recent review of oxindole alkaloids, see J. S. Bindra, The Alkaloids, **14**, 84 (1973).

2. The preparation of oxindoles from yohimbines by rearrangement of chloroindolenine derivatives (i ⟶ ii) affords an unusually efficient route to the oxindole alkaloids from their corresponding indole counterparts, N. Finch and W. I. Taylor, J. Amer. Chem. Soc., **84**, 1318 (1962); **84**, 3871 (1962); J. Shavel and H. Zinnes, J. Amer. Chem. Soc., **84**, 1320 (1962).

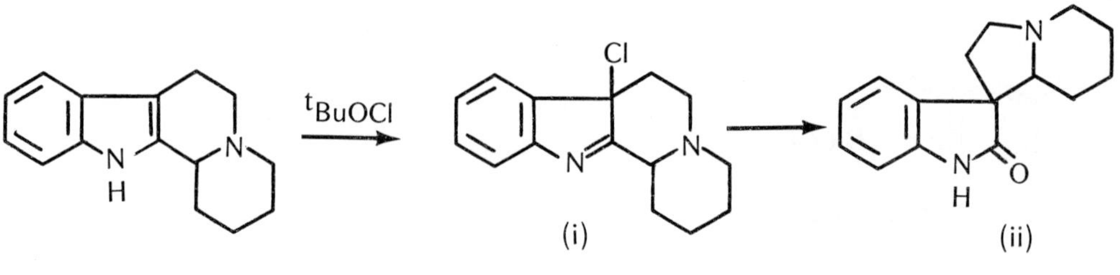
N
H
N
tBuOCl
Cl
N
N
(i)
N
N
H
O
(ii)

3. For earlier attempts towards obtention of rhychophylline, see Y. Ban and T. Oishi, Tetrahedron Lett., 791 (1961); Chem. Pharm. Bull. Japan, **11**, 441, 446, 451 (1963); and A. H. Warfield, Diss. Abstr., **26**, 1357 (1965).

4. Y. Ban, M. Seto and T. Oishi, Tetrahedron Lett., 2113 (1972).

[CHO / EtCHCO$_2$Et]

[H$_2$, PtO$_2$, AcOH]

[NaH, Tol, Δ]

CO_2Et

[dil HCl, Δ]

B-Series[5]

[(EtO)$_2$P(O)CHCO$_2$Me / NaH, Glyme, 50°] 60%

A-Series[5]

[H$_2$, Pd-C / EtOH]

---

5. Typically, the spiro oxindoles exist in pairs of interconvertible *normal* A and B series isomers, and any one series affords a mixture of stereoisomers upon acid or base catalyzed equilibration. The B series is predominant in the resulting mixture (A:B=7:3) on acid catalyzed isomerization, and the A series predominates on base treatment. Equilibration occurs at the β-aminolactam group by cleavage and reformation of the C-3, C-7 bond.

B-Series $\underset{B:}{\overset{H^+}{\rightleftharpoons}}$ *A-Series*

1. HCO_2Et, $(Me_3Si)_2NNa$
2. CH_2N_2, MeOH, Et_2O

dil AcOH

A-Series
Isorhynchophylline

B-Series
Rhynchophylline

The synthesis of rhynchophyllol by van Tamelen *et al.*[6] illustrates a biogenetic-type approach[7] to oxindole alkaloids. A key feature of this synthesis is generation of the desired tetracyclic structure by an intramolecular Mannich cyclization of dialdehyde (B) produced by oxidative cleavage of the substituted cyclopentanediol (A).

Δ

6. E. E. van Tamelen, J. P. Yardley, M. Miyano and W. B. Hinshaw, J. Amer. Chem. Soc., **91**, 7333 (1969).

[1. LAH, THF, Δ; 2. $BzCO_2Cl$, K_2CO_3] →

N—CO_2Bz

[OsO_4] →

N—CO_2Bz

HO

HO

[1. NBS, AcOH; 2. H_2, Pd-C] →

NH

HO

HO

(A)

[aq $NaIO_4$, 0°; → dil HCl, Δ] →

NH

CHO

OHC

(B)

Footnote 8 →

N

CHO

[$NaBH_4$; MeOH, 0°] →

N

HO

Rhynchophyllol

---

7. For a treatise on biogenetic-type synthesis, see E. E. van Tamelen, Fortschr. Chim. Org. Naturs., **19**, 242 (1961).

8. Cyclization of the piperidinoid D ring involves fixation of the two side chain units in the more stable *trans* relationship.

SCOPINE

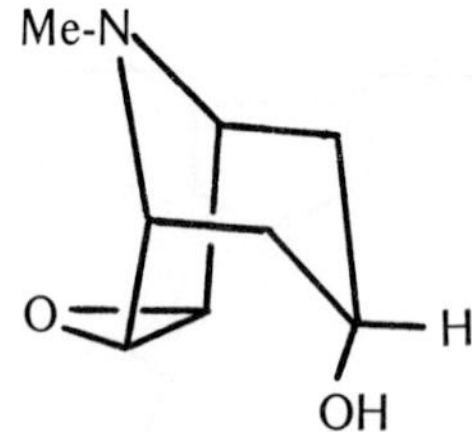

The synthesis of scopine[1] illustrates a novel general synthesis of tropane alkaloids[2] based on the iron carbonyl promoted cyclocoupling reaction between α,α'-dibromo ketones and 1,3-dienes leading to 4-cycloheptenones.[3] Tetrabromoacetone provides the starting three-carbon unit, serving as a synthetic equivalent of the unsubstituted oxyallyl species (i) which is trapped using N-carbethoxypyrrole.[4]

CO_2Me-N (pyrrole) + $Br_2CHCOCHBr_2$ —[$Fe_2(CO)_9$, C_6H_6, 50°]→ (*52%*) MeO_2C-N (H, H, Br, Br, O) + MeO_2C-N (Br, H, H, Br, O)

---

1. R. Noyori, Y. Baba and Y. Hayakawa, J. Amer. Chem. Soc., **96**, 3336 (1974).

2. Review: G. Fodor, The Alkaloids, **13**, 351 (1971).

3. R. Noyori, S. Makino and H. Takaya, J. Amer. Chem. Soc., **93**, 1272 (1971).

4. The direct reaction of dibromoacetone, unlike other ordinary dibromoketones, could not serve as a precursor of the oxyallyl species (i) [R. Noyori *et al.*, J. Amer. Chem. Soc., **94**, 7202 (1972)]. Furthermore, since trapping of the reactive 3-C species (i) with N-methylpyrrole results in electrophilic substitution rather than formation of the desired [$\pi^4 + \pi^2$] cycloadduct (ii) [R. Noyori, Y. Baba, S. Makino and H. Takaya, Tetrahedron Lett., 1741 (1973)], the reaction was carried out using N-carbomethoxypyrrole, and the carbomethoxy function later transformed into a methyl group.

$BrCH_2COCH_2Br$ + $Fe_2(CO)_9$ ⟶ [O^-, +] Fe^{II} —(N-Methylpyrrole)→ O, Me-N

(i) (ii)

MeO_2C-N

[Zn-Cu, MeOH; NH_4Cl, 25°]

100%

[$(^iBu)_2AlH$, THF; -78° → 25°]

Me-N

H

OH

[1. Acetylate; 2. CF_3CO_3H; 3. B^-]

Me-N

O

H

OH

Scopine

SEYCHELLENE

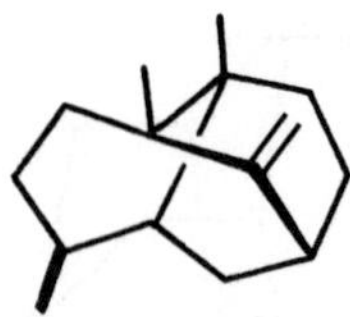

In principle, several approaches to the complex tricyclo[5·3·1·0^{3,8}]undecane skeleton of seychellene can be devised by strategic bond disconnection.[1,2] The synthesis by Piers *et al.*[2] illustrates one of these approaches involving the base-catalyzed intramolecular cyclization of a suitably functionalized *cis*-decalone (B) to norseychellanone.

O
A B
O
Wieland-Miescher ketone[3]

1. $NaBH_4$, EtOH, 0°
2. DHP
80%

OTHP
O

1. Me_2CuLi, Et_2O [4]
2. AcCl
79%

OTHP
AcO

CO_3H
Cl

---

1. E. J. Corey, XXIIIrd Congress Pure Appl. Chem., Vol. 2, Butterworths, London (1971) p. 45; Quart. Rev., **25**, 455 (1971).

2. E. Piers, W. de Waal and R. W. Britton, Chem. Commun., 1069 (1969); J. Amer. Chem. Soc., **93**, 5113 (1971).

3. The Wieland-Miescher ketone, chosen as starting material for the synthesis, contains not only the desired keto group in the B ring, but also possesses the requisite functionality in the A ring for introduction of the methyl groups and the $-CH_2OTs$ moiety (*Cf.*, **B**).

4. Conjugate addition generated a specific enolate anion which could be trapped with electrophilic reagents such as acetyl chloride, cf. J. A. Marshall and A. R. Hochstetler, J. Amer. Chem. Soc., **91**, 648 (1969).

OTHP

AcO

O

[160°]
73%

OTHP

O

OAc

[Ph_3P=CH_2, DMSO]
73%

OTHP

OAc

[$(Ph_3P)_3RhCl$, C_6H_6, H_2][5]
90%

OTHP

OAc

[1. KOH, aq EtOH, Δ
2. CrO_3, Py, 0°]

OTHP

5
2
9

O

(A)[6]

[MeLi, Et_2O]

OTHP

OH

[$SOCl_2$, Py, C_6H_6, 0°]

5. The homogenous catalyst tris(triphenylphosphine)chlororhodium was chosen for hydrogenation to avoid the possibility of hydrogenolysis of the allylic acetate.

6. The presence of an α-oriented substituent at C-5 ensures that the alternative ketone (chair-chair conformation i) bearing a 2α-methyl group, is destabilized (1,3-diaxial interaction between C-5, C-9 substituents) with respect to (A) (conformation ii) in which the C-2 methyl group is in the desired β-configuration.

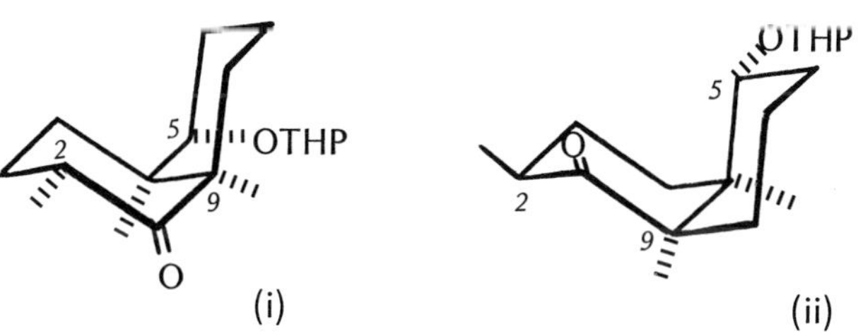

OTHP

[1. B_2H_6, THF→H_2O_2, NaOH
2. TsCl, Py, rt]

73%

OTHP

TsO (B)

[1. MeOH, TsOH,Δ
2. CrO_3, Py
3. NaH, DMSO, 75°]

O

TsO

O

TsO

90%

O

[MeLi, Et_2O,Δ]

Norseychellanone

OH

[$SOCl_2$, Py, 0°]

Seychellene[7]

7. For a different synthesis of seychellene, see R. N. Mirrington and K. J. Schmalzl, Tetrahedron Lett., 3219 (1970); J. Org. Chem., **37**, 2877 (1972).

An interesting one-step construction of the tricyclic seychellene skeleton developed at Tohoku University[8] involves as the key step an intramolecular Diels-Alder addition of monocyclic cyclohexadienone derivative (B). The desired dienophile in (B) is generated by pyrolysis of the dimethylamino N-oxide (A).

I, $NaNH_2$, NH_3

35%

:M

Br

NBS, CCl_4

64%

aq $CrCl_2$, Me_2CO

TsOH, EtOH, Δ

Me_2NH

63% for three steps

NMe_2

H_2O_2, MeOH

A

N → O

150°

≡

8. N. Fukamiya, M. Kato and A. Yoshikoshi, Chem. Commun., 1120 (1971); J. Chem. Soc. Perkin I, 1843 (1973).

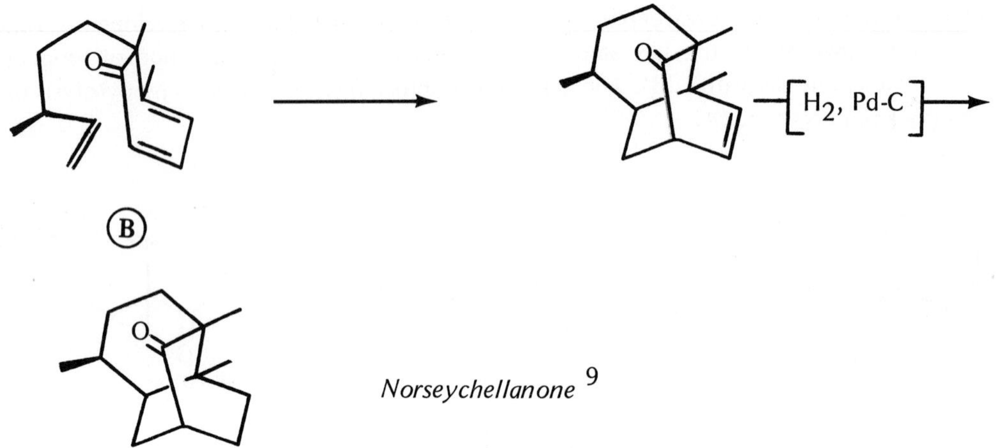

9. For another synthesis of seychellene based on intramolecular Diels-Alder addition (i→ii), see G. Frater, Chimia, **26**, 657 (1972); Helv., **57**, 172 (1974); H. Greuter, G. Frater and H. Schmid, Helv., **55**, 526 (1972).

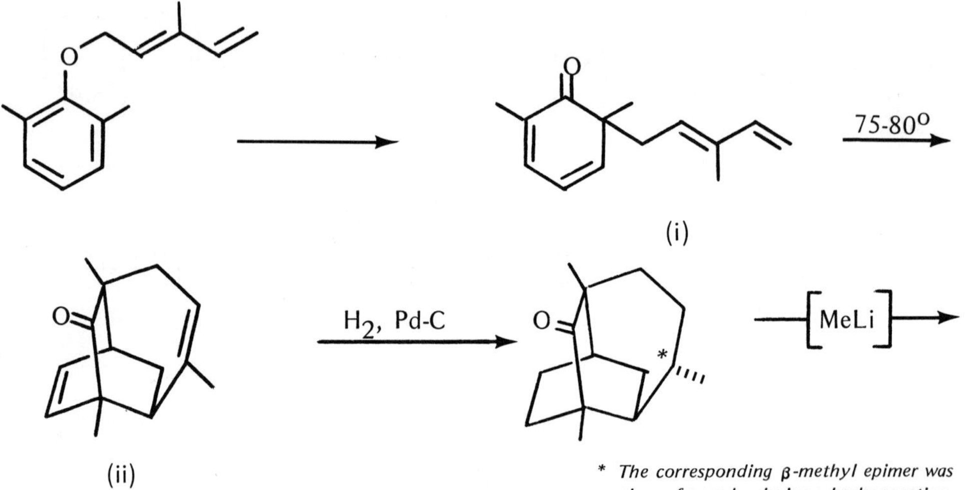

* *The corresponding β-methyl epimer was also formed during hydrogenation. However, methyl lithium reacted preferentially with the α-methyl ketone (shown above) allowing the selective preparation of (iii).*

(iii)

Seychellene

SIRENIN-SESQUICARENE

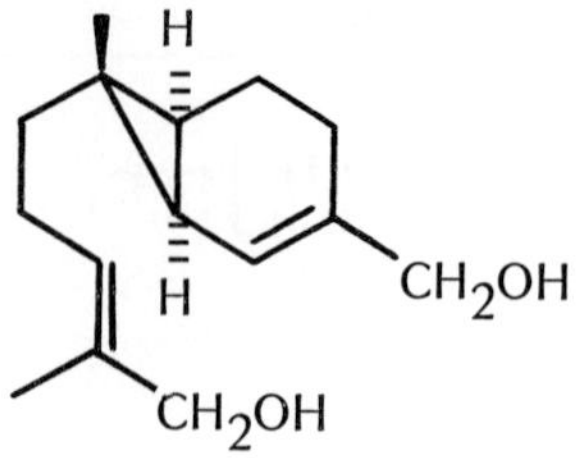

Sirenin is a powerful sperm-attractant produced by the female gametes of the water mold *Allomyces*, and is unique in that it was the first plant sex hormone to be fully characterized.[1] Systematic bond disconnection of the molecule suggests that the intramolecular α-ketocarbene-olefin reaction[2] is especially suited for stereospecific construction of the bicyclo[4·1·0]heptane skeleton (i $\rightarrow$ ii). Indeed, several syntheses[3-8] of sirenin have been accomplished along these lines, and one such approach reported by Rapoport *et al.*[3] is outlined below.

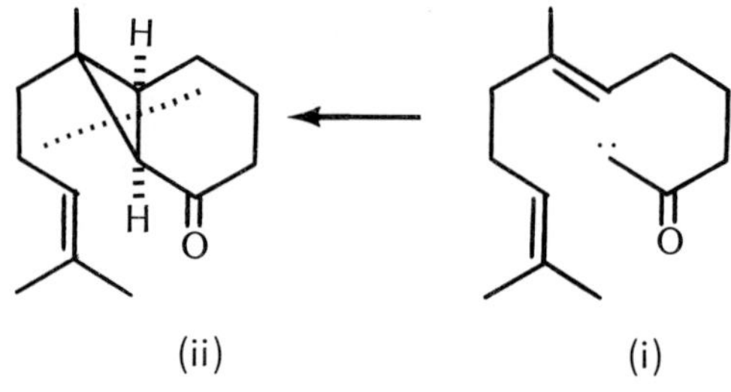

1. L. Machlis, W. H. Nutting and H. Rapoport, J. Amer. Chem. Soc., **90**, 1674, 6434 (1968).

2. G. Stork and J. Ficini, J. Amer. Chem. Soc., **83**, 4678 (1961); M. M. Fawzi and C. D. Gutsche, J. Org. Chem., **31**, 1390 (1966).

3. U. T. Bhalerao, J. J. Plattner and H. Rapoport, J. Amer. Chem. Soc., **91**, 4933 (1969); **92**, 3429 (1970).

4. E. J. Corey, K. Achiwa and J. A. Katzenellenbogen, J. Amer. Chem. Soc.,**91**,4318 (1969).

5. P. A. Grieco, J. Amer. Chem. Soc., **91**, 5660 (1969).

6. K. Mori and M. Matsui, Tetrahedron Lett., 4435 (1969).

7. J. J. Plattner and H. Rapoport, J. Amer. Chem. Soc., **93**, 1758 (1971).

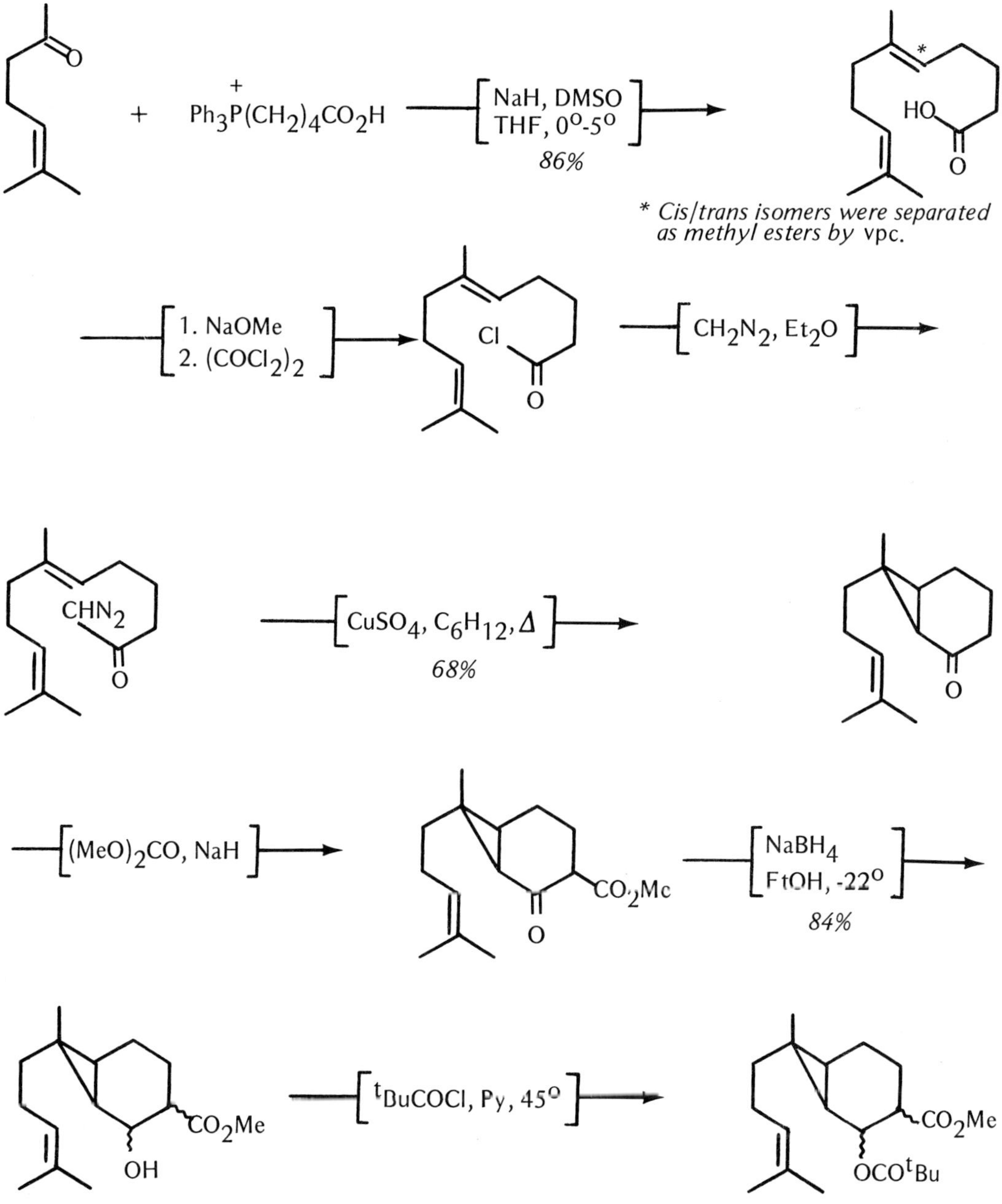

O

+ $Ph_3\overset{+}{P}(CH_2)_4CO_2H$ — [NaH, DMSO; THF, 0°-5°] *86%* →

*

HO

O

* *Cis/trans isomers were separated as methyl esters by* vpc.

— [1. NaOMe; 2. $(COCl_2)_2$] →

Cl

O

— [CH_2N_2, Et_2O] →

CHN_2

O

— [$CuSO_4$, C_6H_{12}, Δ] *68%* →

O

— [$(MeO)_2CO$, NaH] →

CO_2Me

O

— [$NaBH_4$; EtOH, -22°] *84%* →

CO_2Me

OH

— [^tBuCOCl, Py, 45°] →

CO_2Me

OCO^tBu

[KO^tBu, Tol] 87% for two steps → (A) [SeO_2, EtOH, Δ → MnO_2, Hex] 63% →

CO_2Me

CO_2Me, CHO [AlH_3, Et_2O, 0°] → CH_2OH, CH_2OH

Sirenin

An efficient route to the bicyclic ester (A) has been devised by Corey and Achiwa.[8]

Br [$[CH_2C{\equiv}CSiMe_3]Li$][9] 50% → $SiMe_3$

Geranyl bromide

8. E. J. Corey and K. Achiwa, Tetrahedron Lett., 2245 (1970).

9. For the development of Lithio-l-trimethylsilypropyne as a reagent for synthesis of l-en-5-yne derivatives, see E. J. Corey and H. A. Kirst, Tetrahedron Lett., 5041 (1968).

[$AgNO_3$, EtOH → NaCN]

[^nBuLi, THF → $(HCHO)_n$]
82%

CH_2OH

[$Ni(CO)_4$, AcOH aq EtOH, 70°]
30%

OH
CO_2H

[CH_2N_2, Et_2O]

OH
CO_2Me

[MnO_2, Hex]

CHO
CO_2Me

[NH_2NH_2, TEA, EtOH]

NNH_2
CO_2Me

[MnO_2, CH_2Cl_2, 0°]

N_2
CO_2Me

[CuI, THF, 35°]
50%

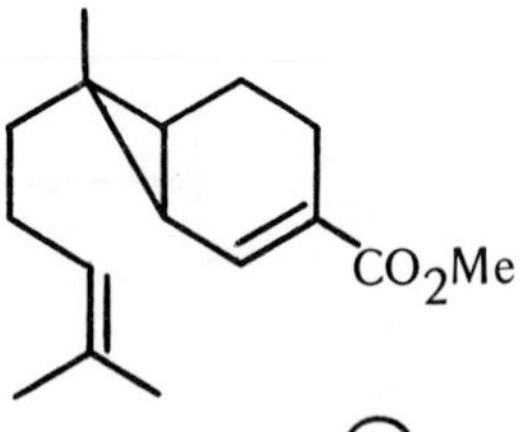

CO_2Me

(A)

Several syntheses involving intramolecular carbenoid addition as a key step for construction of the structurally related hydrocarbon sesquicarene[10-14] have been described. One efficient synthesis of sesquicarene is outlined below,[10] and commences from *cis-trans* farnesol which is readily available by fractional distillation of commercial farnesol.

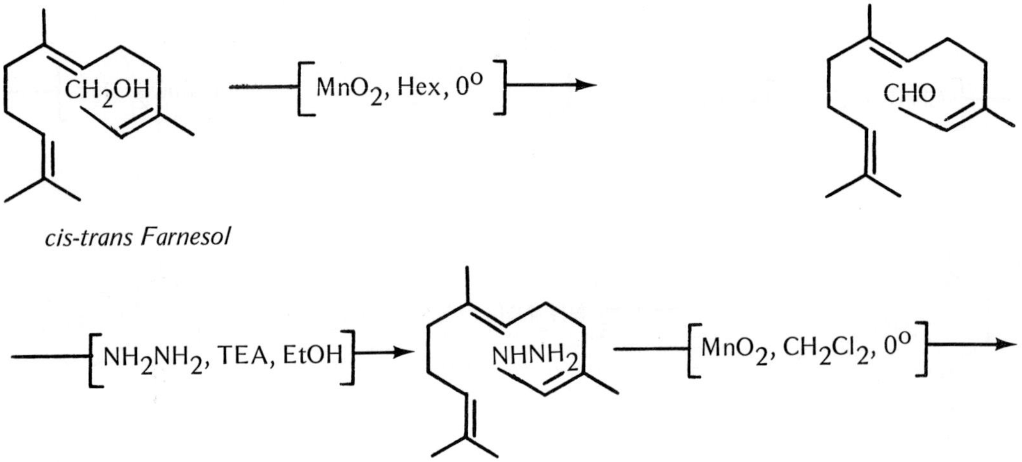

CH_2OH —[MnO_2, Hex, 0°]→ CHO

cis-trans Farnesol

—[NH_2NH_2, TEA, EtOH]→ $NHNH_2$ —[MnO_2, CH_2Cl_2, 0°]→

10. E. J. Corey and K. Achiwa, Tetrahedron Lett., 3257 (1969).

11. E. J. Corey and K. Achiwa, Tetrahedron Lett., 1837 (1969).

12. R. M. Coates and R. M. Freidirger, Chem. Commun., 871 (1969); Tetrahedron, **26**, 3487 (1970).

13. K. Mori and M. Matsui, Tetrahedron Lett., 2729 (1969); Tetrahedron, **26**, 2801 (1970).

14. Y. Nakatani and T. Yamanishi, Agr. Biol. Chem. Tokyo, **33**, 1805 (1969).

N_2

[CuI, THF, 35°]

H

H

Sesquicarene

(*25% overall yield based on farnesol*)

SPORIDESMIN-A

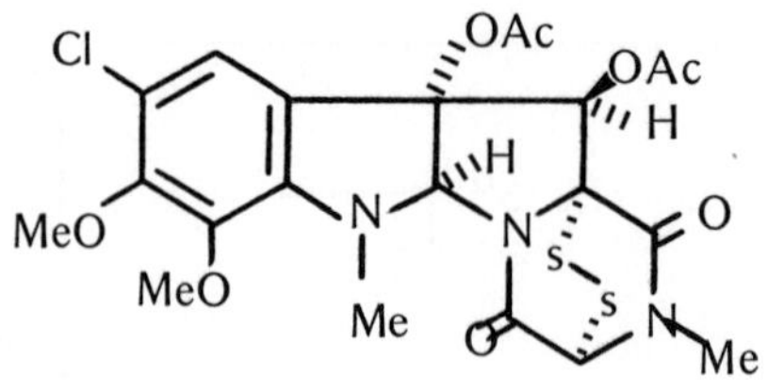

The synthesis of sporidesmin-A[1] is based on a new general method for the synthesis of epidithioketopiperazines developed by Kishi *et al*[2], in which a thioacetal grouping serves as a latent episulfide bridge. Introduction of the indole part of the molecule has been carried out by acylation of the bridgehead monocarbanion derived from (A) followed by oxidative ring closure of ring C (C→D).[3]

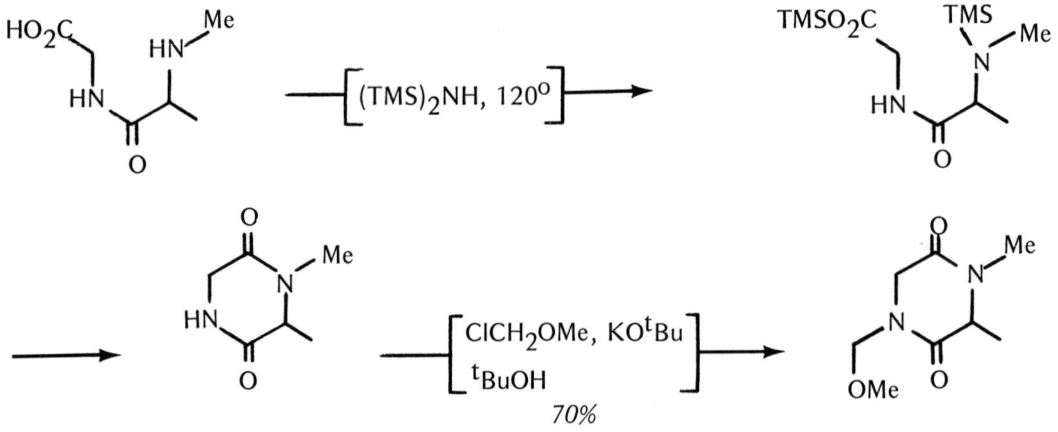

[$(TMS)_2NH$, 120°]

[$ClCH_2OMe$, KO^tBu, ^tBuOH] 70%

1. Sporidesmins are toxic metabolites of *Pithomyces chartarum*, and the cause of serious hepatoxic disease in sheep, J. W. Ronaldson, A. Taylor, E. P. White and R. J. Abraham, J. Chem. Soc., 3172 (1963); S. Safe and A. Taylor, J. Chem. Soc. Perkin I, 472 (1972) and references cited therein.

2. Y. Kishi, T. Fukuyama and S. Nakatsuka, J. Amer. Chem. Soc., **95**, 6491 (1973).

3. Y. Kishi, S. Nakatsuka, T. Fukuyama, M. Havel, J. Amer. Chem. Soc., **95**, 6493 (1973).

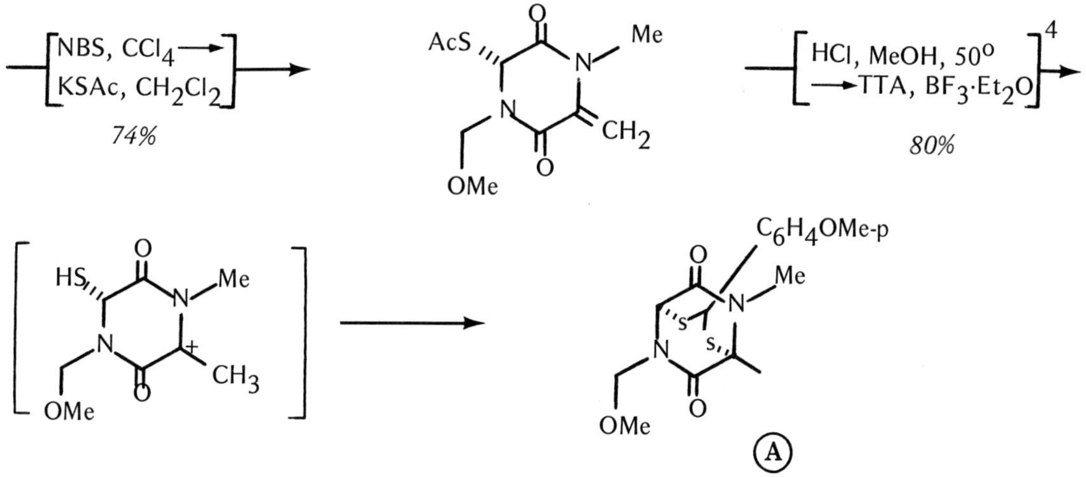

[NBS, CCl_4 → KSAc, CH_2Cl_2] 74%

[HCl, MeOH, 50° → TTA, $BF_3 \cdot Et_2O$][4] 80%

(A)

Preparation of indole part:

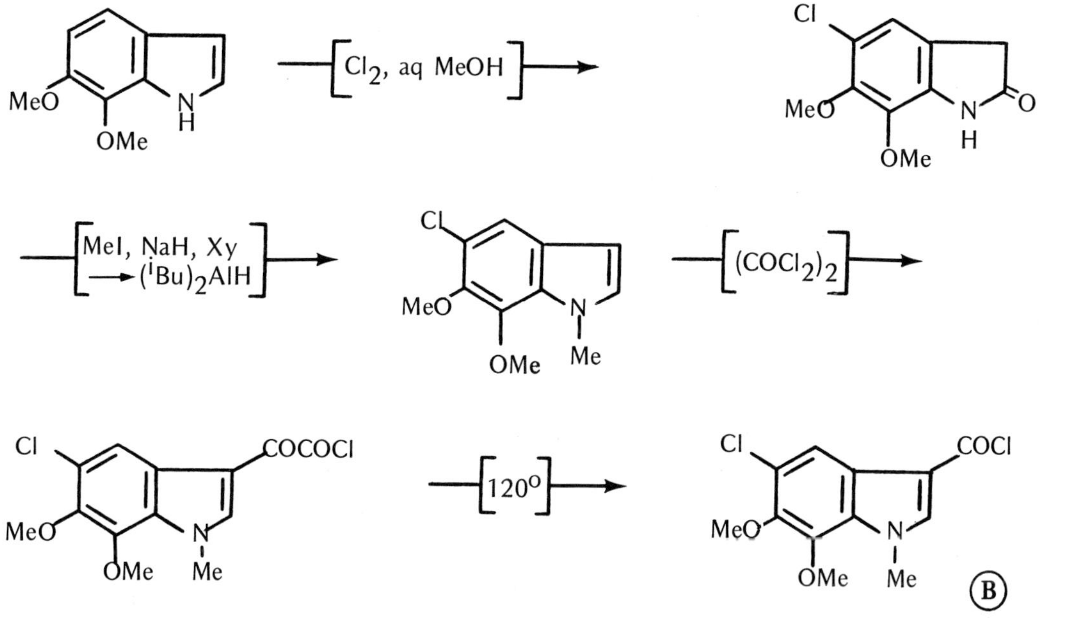

[Cl_2, aq MeOH]

[MeI, NaH, Xy → $(^{i}Bu)_2AlH$]

[$(COCl)_2$]

[120°]

(B)

4. TTA= Trithiane derivative of anisaldehyde.

Preparation of sporidesmine-A:

(A) —[BuLi, THF, -78°]→ [anion intermediate: C_6H_4OMe-p, Me, O, S, S, N, OMe] —[**B**, -110°]→ *61%*

[Cl, MeO, OMe, N, Me, O, O, C_6H_4OMe-p, Me, S, S, N, OMe, O] —[1. HCl, TFA, 70° * 2. NaOH]→ *50%*

*$(-CH_2OMe \rightarrow -CH_2OH)$

[Cl, MeO, OMe, N, Me, O, O, C_6H_4OMe-p, Me, S, S, HN, O] —[$(^iBu)_2AlH$, THF, -78°][5]→ *86%*

[Cl, MeO, OMe, N, Me, OH, H, O, C_6H_4OMe-p, Me, S, S, HN, O] —[Ac_2O, Py]→ *95%*

---

5. This reduction is rationalized as involving complex formation between the amide N-H group and the reducing agent. The crucial intramolecular hydride transfer then occurs from the α-side of the diketopiperazine which stays as far away as possible from the bulky indole unit (i).

R=*Indolyl*

[R, O, *β-side*, O, Me, N, N, H, O, S, S, *α-side*, H, C_6H_4OMe-p]

(i)

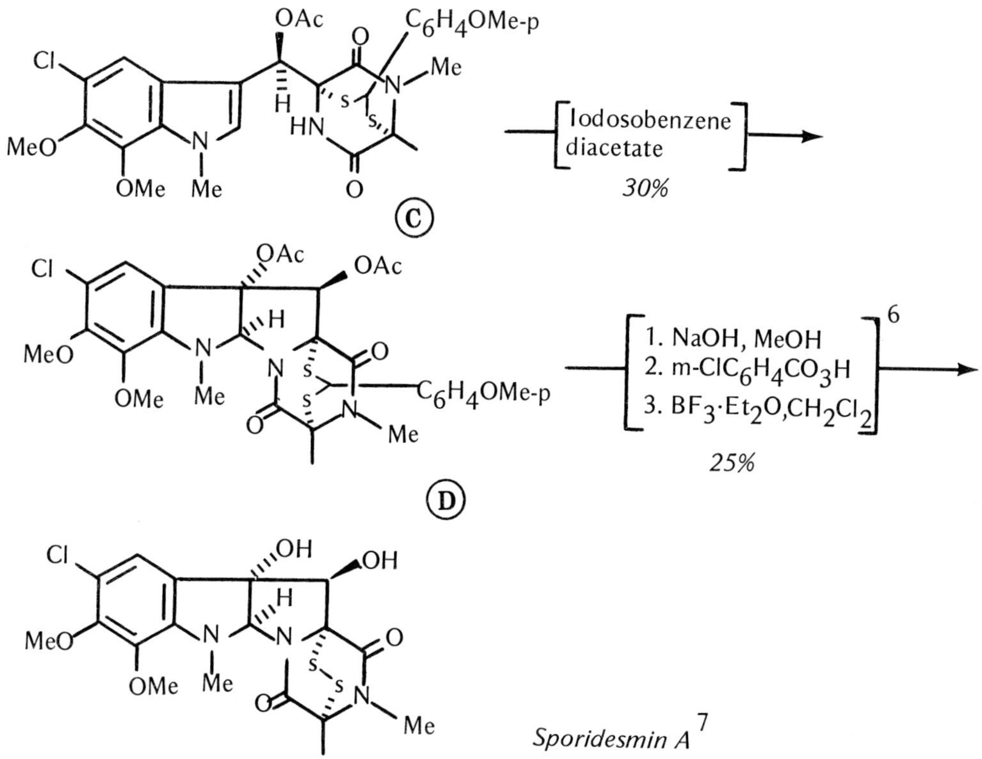

(C)

Iodosobenzene diacetate

30%

(D)

1. NaOH, MeOH
2. m-$ClC_6H_4CO_3H$
3. $BF_3 \cdot Et_2O, CH_2Cl_2$[6]

25%

Sporidesmin A[7]

---

6. The dithioacetal group is readily converted to the disulfide bridge by oxidation to sulfoxide, followed by treatment with acid to effect the carbon-sulfur bond fission. The p-methoxybenzene group is essential for this cleavage and facilitates resonance stabilization of the incipient carbonium ion.

7. For the synthesis of dehydrogliotoxin, a related epidithioketopiperazine, see Y. Kishi, T. Fukuyama and S. Nakatsuka, J. Amer. Chem. Soc., **95**, 6492 (1973).

TABERSONINE

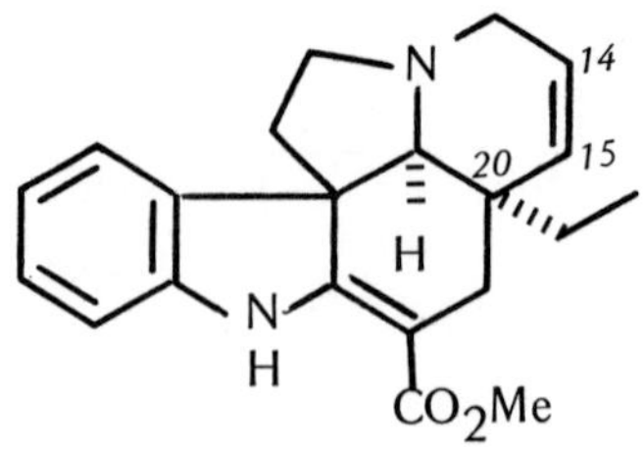

Tabersonine plays a critical role in the biosynthesis of *Iboga* and *Aspidosperma* alkaloids.[1] The primary challenge in devising a synthesis of tabersonine resides in developing an effective means for establishing the unsaturation at the C-14,15 position and the adjoining quaternary center at C-20, a problem efficiently solved by Ziegler and Bennett[2] utilizing the Claisen rearrangement for delivery of the requisite functionality.[3]

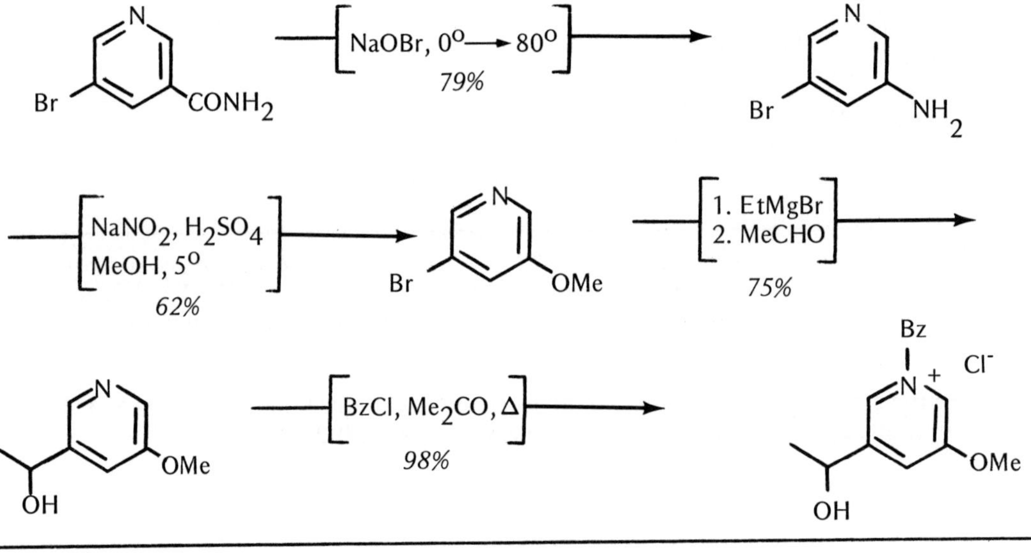

$NaOBr$, 0° → 80°; 79%

$NaNO_2$, H_2SO_4, MeOH, 5°; 62%

1. EtMgBr 2. MeCHO; 75%

BzCl, Me_2CO, Δ; 98%

1. A. I. Scott, Accounts Chem. Res., **3**, 151 (1970).
2. F. E. Ziegler and G. B. Bennett, J. Amer. Chem. Soc., **93**, 5930 (1971); **95**, 7458 (1973).
3. F. E. Ziegler and G. B. Bennett, Tetrahedron Lett., 2545 (1970).

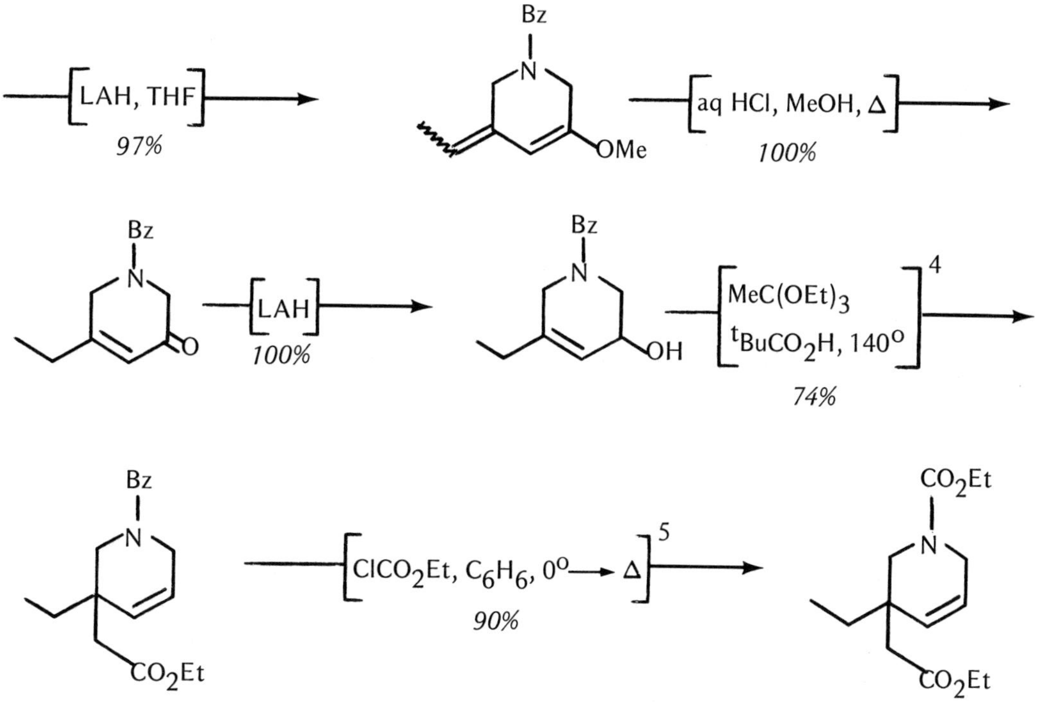

[LAH, THF] 97%

Bz N OMe

[aq HCl, MeOH, Δ] 100%

Bz N O

[LAH] 100%

Bz N OH

[$MeC(OEt)_3$ $^tBuCO_2H$, 140°][4] 74%

Bz N CO_2Et

[$ClCO_2Et$, C_6H_6, 0° → Δ][5] 90%

CO_2Et N CO_2Et

4. This stereoselective version of the Claisen rearrangement leading to *trans*-trisubstituted olefinic bonds has been developed by W. S. Johnson, L. Werthmann, W. R. Bartlett, T. J. Brocksom, T.-t. Li, D. J. Faulkner and M. R. Petersen, J. Amer. Chem. Soc., **92**, 741 (1970). The reaction evidently involves formation of a mixed orthoester (i), and loss of ethanol to form the ketene acetal (ii) which rearranges to the olefinic ester (iii).

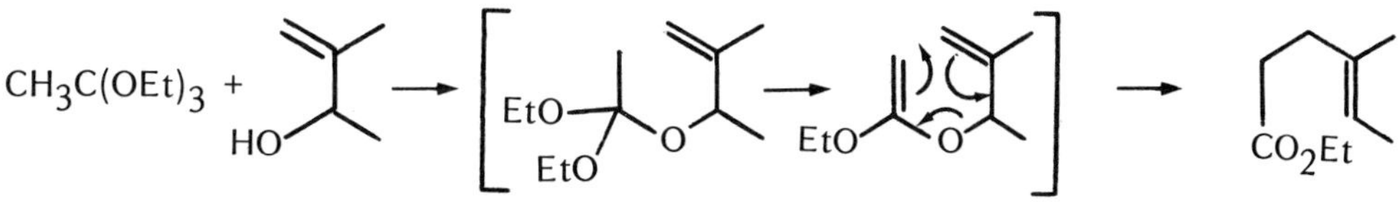

$CH_3C(OEt)_3$ + HO → [EtO EtO O (i) → EtO O (ii)] → CO_2Et (iii)

5. Since the presence of an olefin in the molecule rendered catalytic debenzylation impractical, removal of the benzyl group was effected using ethyl chloroformate, J. O. Hobson and J. G. McClusky, J. Chem. Soc. (C), 2015 (1967).

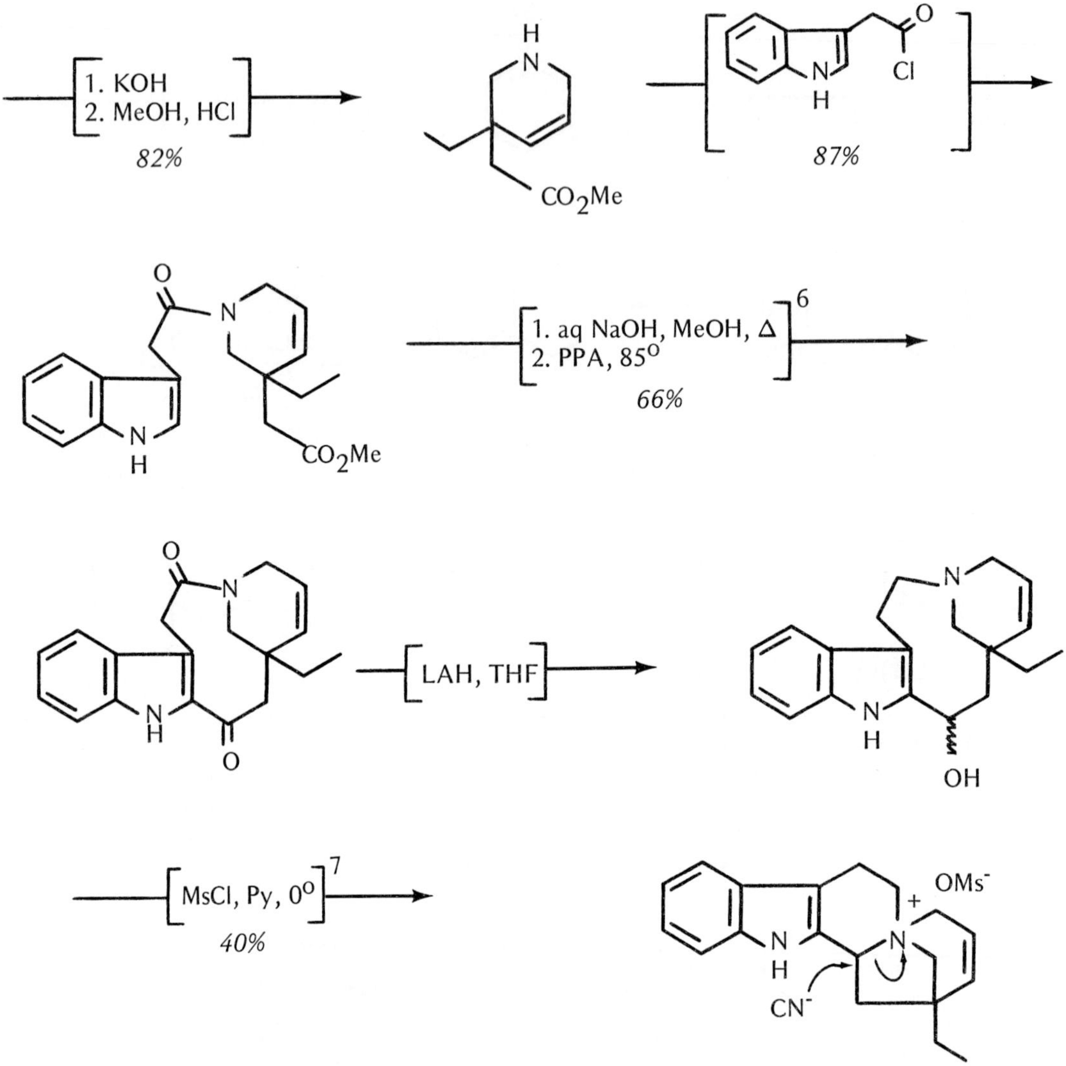

1. KOH
2. MeOH, HCl

82%

87%

1. aq NaOH, MeOH, Δ
2. PPA, 85°[6]

66%

LAH, THF

MsCl, Py, 0°[7]

40%

6. This technique for medium ring cyclization to 2-acylindoles has been developed earlier by F. W. Ziegler, J. A. Kloek and P. A. Zoretic [J. Amer. Chem. Soc., **91**, 2342 (1969)] in the synthesis of quebrachamine.

7. This method for introduction of the one-carbon unit was originally developed by Harley-Mason and later modified by Kutney: G. H. Foster and J. Harley-Mason, Chem. Commun., 1440 (1968); J. P. Kutney and F. Bylsma, J. Amer. Chem. Soc., **92**, 6090 (1970).

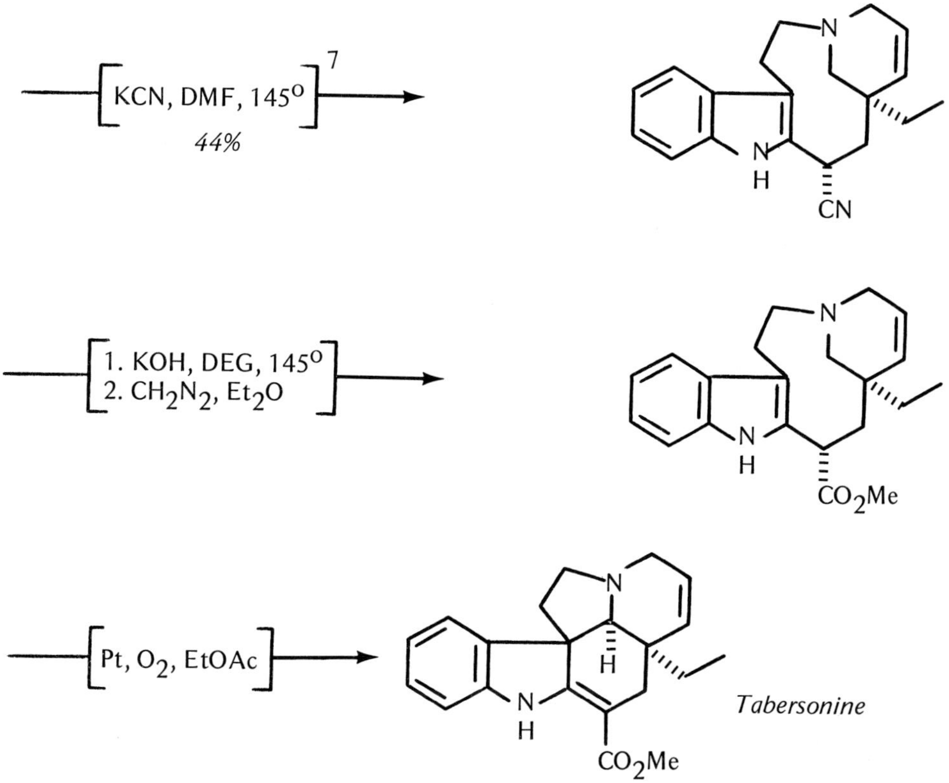

[KCN, DMF, 145°][7] →
44%

[1. KOH, DEG, 145°
2. CH_2N_2, Et_2O] →

[Pt, O_2, EtOAc] →

CN

CO_2Me

H

CO_2Me

Tabersonine

TESTOSTERONE

OH

17

H

H H

O

Biogenetic-like olefinic cyclizations in nitroalkane solvents result in nucleophilic trapping of an intermediary polycyclic vinyl action by the nitroalkane to afford oxime ethers.[1] The ingenious application of this finding to secure an entry into 17-oxygenated steroids is portrayed in the facile synthesis of testosterone benzoate by Johnson *et al.*[2]

HO

(A)[3]

$\left[Cl_3CCO_2H, {}^iPrNO_2, 0^o\right]^4$

45%

---

1. D. R. Morton, M. B. Gravestock, R. J. Parry and W. S. Johnson, J. Amer. Chem. Soc., **95**, 4417 (1973).

2. D. R. Morton and W. S. Johnson, J. Amer. Chem. Soc., **95**, 4419 (1973).

3. The key trienyol (A), which has also served as an intermediate for the synthesis of progesterone, has been obtained in optically active form and shown to undergo cyclization without racemization to give optically active tetracyclic products, R. L. Markezich, W. E. Willy, B. E. McCarry and W. S. Johnson, J. Amer. Chem. Soc., **95**, 4414 (1973); B. E. McCarry, R. L. Markezich and W. S. Johnson, J. Amer. Chem. Soc., **95**, 4416 (1973).

4.

+

$EtNO_2$

H

O

N+

O

O

O-N=C(CH_3)$_2$

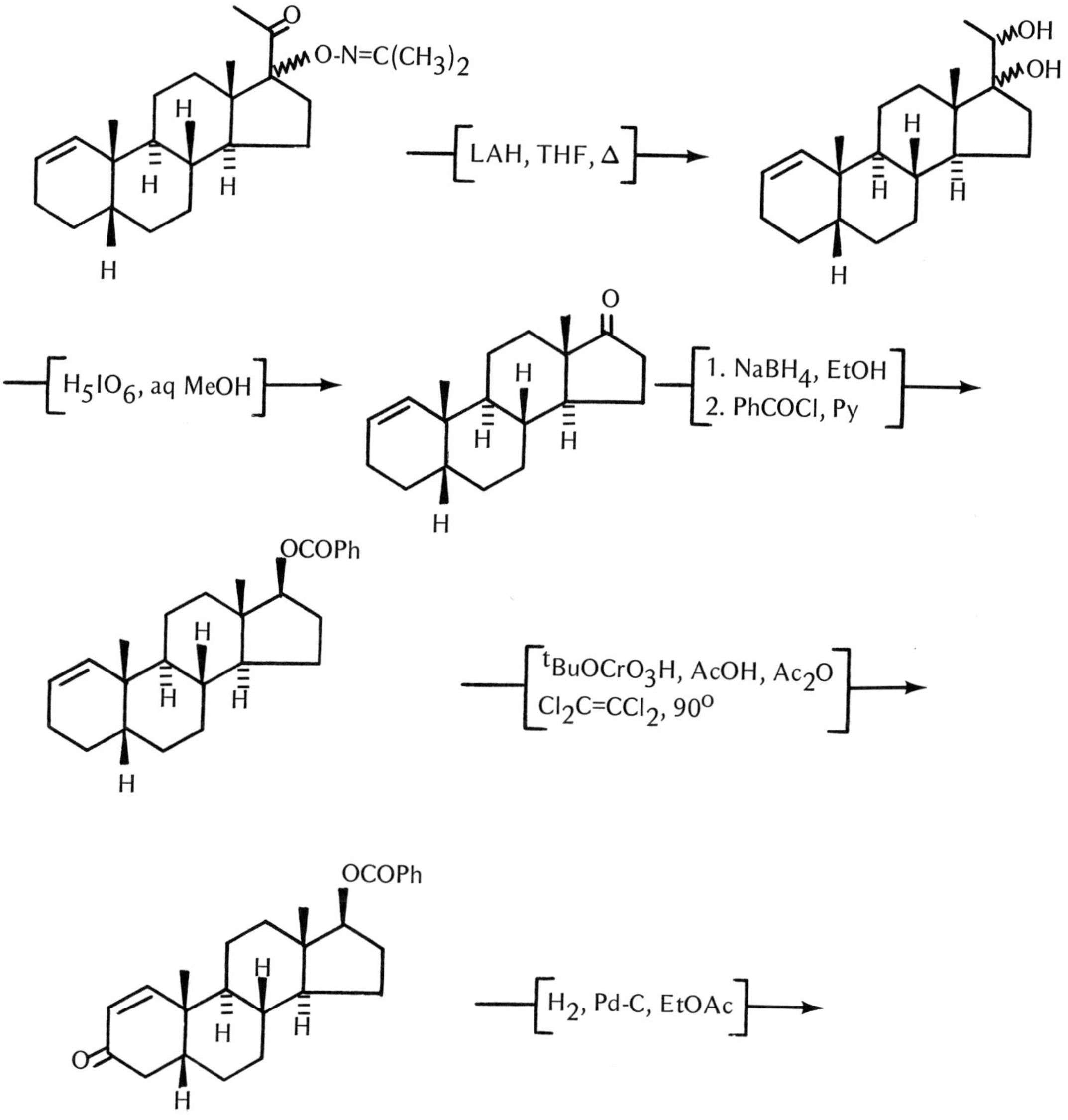

O-N=C(CH$_3$)$_2$

[LAH, THF, Δ]

OH
OH

[H_5IO_6, aq MeOH]

[1. $NaBH_4$, EtOH
2. PhCOCl, Py]

OCOPh

[^{t}BuOCrO$_3$H, AcOH, Ac$_2$O
Cl$_2$C=CCl$_2$, 90°]

OCOPh

[H_2, Pd-C, EtOAc]

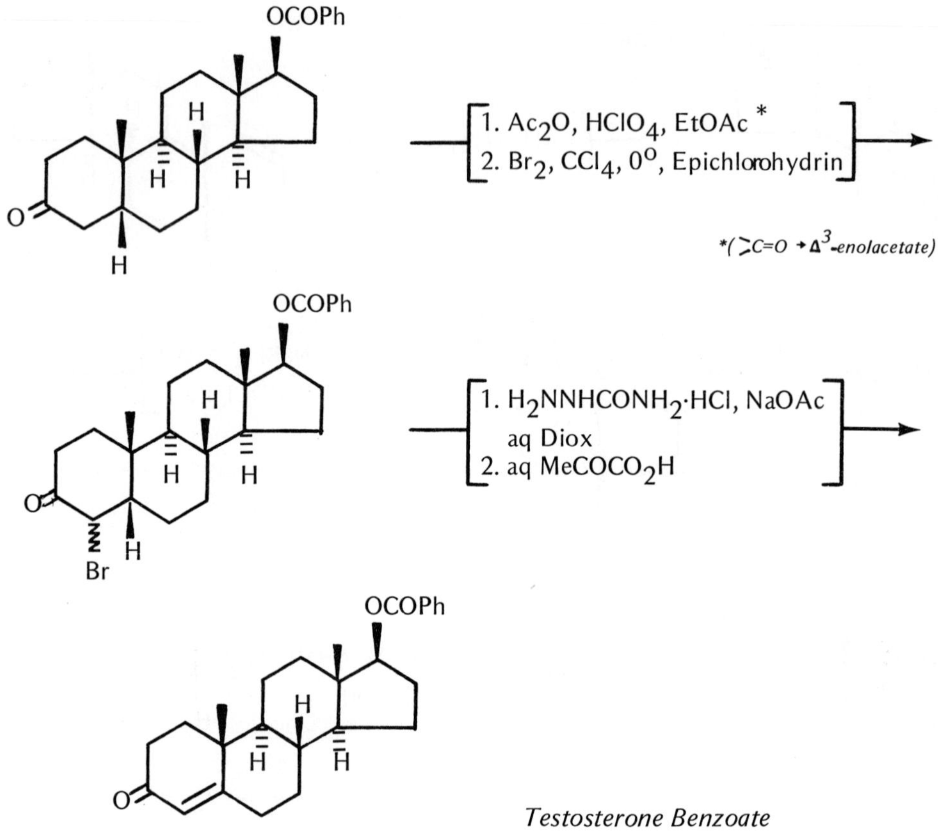

OCOPh

H

H H

O

H

1. Ac_2O, $HClO_4$, EtOAc *
2. Br_2, CCl_4, 0°, Epichlorohydrin

(>C=O → Δ^3-enolacetate)

OCOPh

H

H H

O

Br H

1. $H_2NNHCONH_2 \cdot HCl$, NaOAc
aq Diox
2. aq $MeCOCO_2H$

OCOPh

H

H H

O

Testosterone Benzoate

TETRODOTOXIN

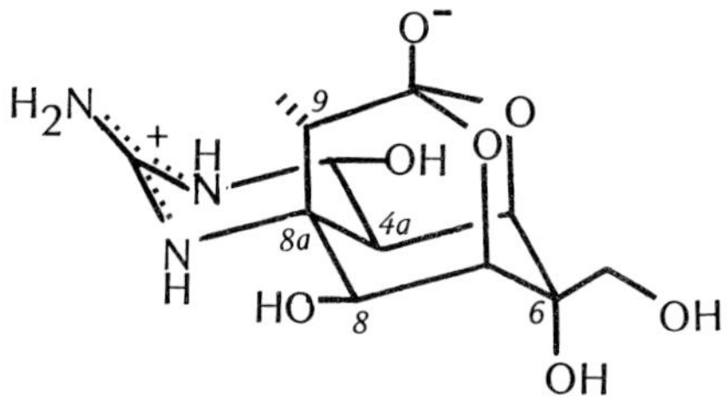

Tetrodotoxin is an incredibly powerful poison isolated from the ovaries and liver of the Japanese puffer fish the *tora fugu* (tiger puffer), and the closely related *ma fugu* (common puffer).[1] A noteworthy feature of the synthesis of tetrodotoxin by Kishi *et al.*[2-4] is the impressive aggrandization of the six chiral centers on the cyclohexane nucleus in the key tetrodamine intermediate (B), by effective use of proximity effects. Introduction of the elements of future functional groups at C-4a and C-8a has been accomplished by the Diels-Alder reaction of an α-oximinoethylbenzoquinone (A) with butadiene. The furan ring in (D) serves to mask reactive functionalities during conversion of the amino group into a guanidine. Finally, cleavage of the furan ring and hydrolysis of protecting groups furnishes the toxin.

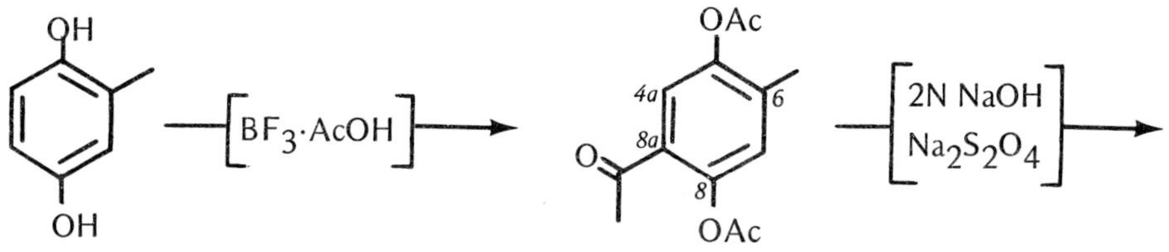

$BF_3 \cdot AcOH$

2N NaOH, $Na_2S_2O_4$

---

1. The fearsome array of functionalities and the complex cage-like structure of tetrodotoxin was deduced in three independent studies: T. Goto, Y. Kishi, S. Takahashi and Y. Hirata, Tetrahedron, **21**, 2059 (1965); K. Tsuda *et al.*, Chem. Pharm. Bull. (Tokyo), **12**, 634 (1964); R. B. Woodward, Pure Appl. Chem., **9**, 49 (1964).

2. Y. Kishi, F. Nakatsubo, M. Aratani, T. Goto, S. Inoue, H. Kakoi and S. Sugiura, Tetrahedron Lett., 5127 (1970).

3. Y. Kishi, F. Nakatsubo, M. Aratani, T. Goto, S. Inoue and H. Kakoi, Tetrahedron Lett., 5129 (1970).

4. Y. Kishi, M. Aratani, T. Fukuyama, F. Nakatsubo, T. Goto, S. Inoue, H. Tanino, S. Sugiura and H. Kakoi, J. Amer. Chem. Soc., **94**, 9217, 9219 (1972).

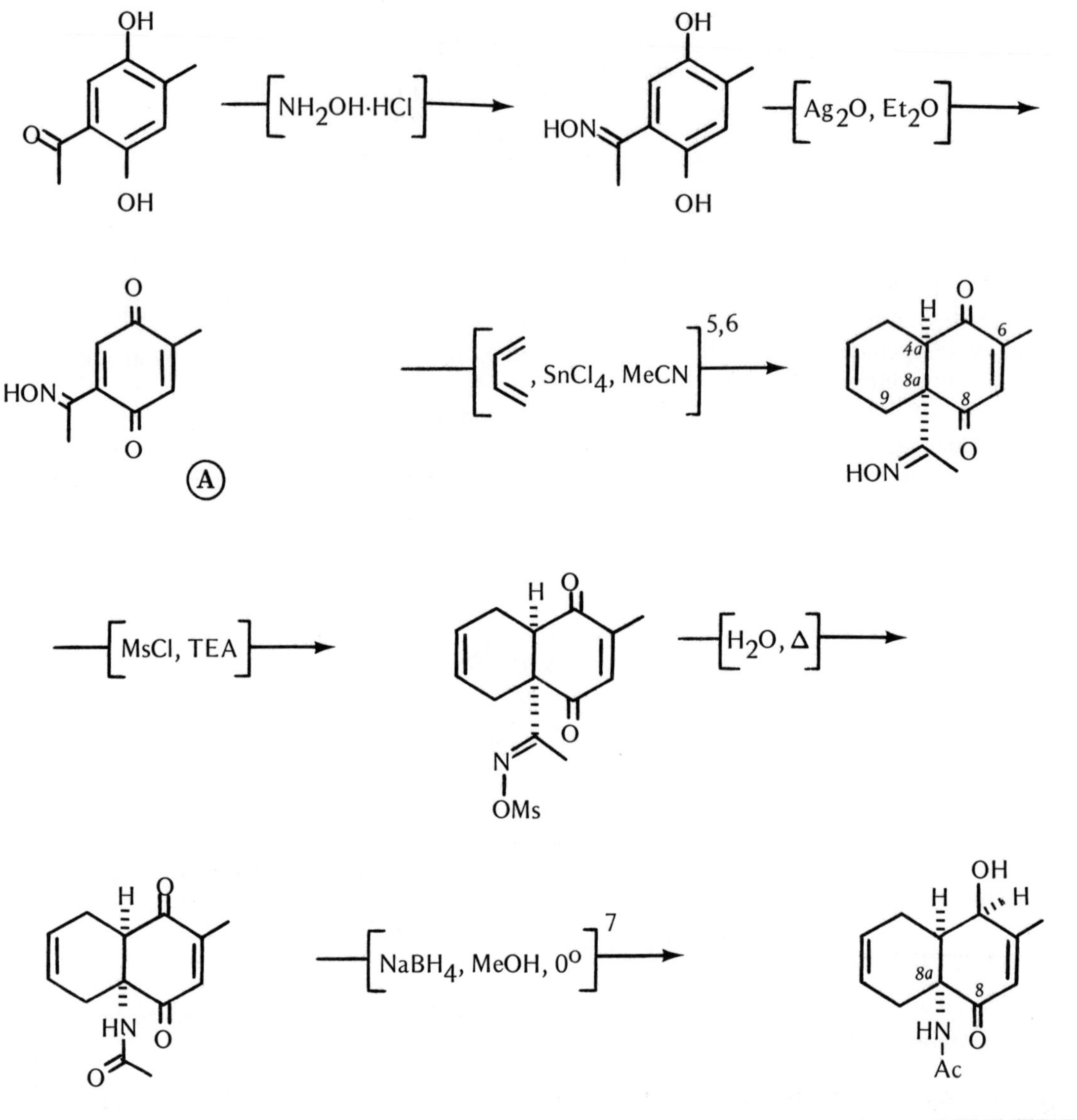

[$NH_2OH{\cdot}HCl$] → [Ag_2O, Et_2O] → A

[diene, $SnCl_4$, MeCN][5,6] → [MsCl, TEA] → [H_2O, Δ] → [$NaBH_4$, MeOH, 0°][7]

5. Attack by the diene could be directed to ethylene linkage with the α-oximinoalkyl group by addition of a Lewis acid which enhanced the electron withdrawing ability of the substituent.

6. It was planned to introduce the amino function at C-8a by the use of a Beckmann rearrangement.

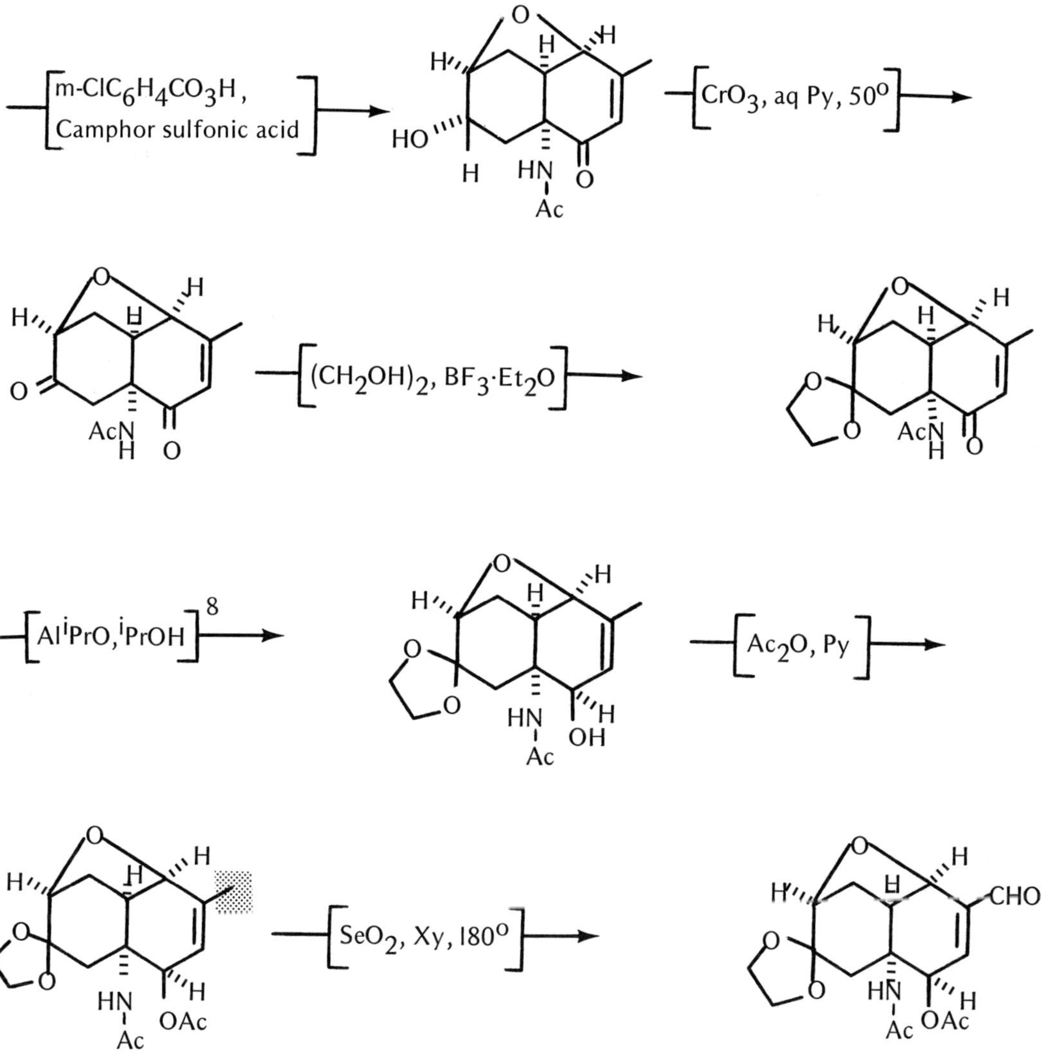

[m-$ClC_6H_4CO_3H$, Camphor sulfonic acid] → [CrO_3, aq Py, 50°] →

[$(CH_2OH)_2$, $BF_3·Et_2O$] →

[Al^{i}PrO, ^{i}PrOH][8] → [Ac_2O, Py] →

[SeO_2, Xy, 180°] →

7. The reduction was highly stereoselective since the reagent could approach the carbonyl group only from the outer side of the cage-like molecule, and regiospecific since the carbonyl at C-8 was protected by presence of the vicinal axial acetamido group at C-8a.

8. The reducing agent approaches from outer side of the cage-like molecule.

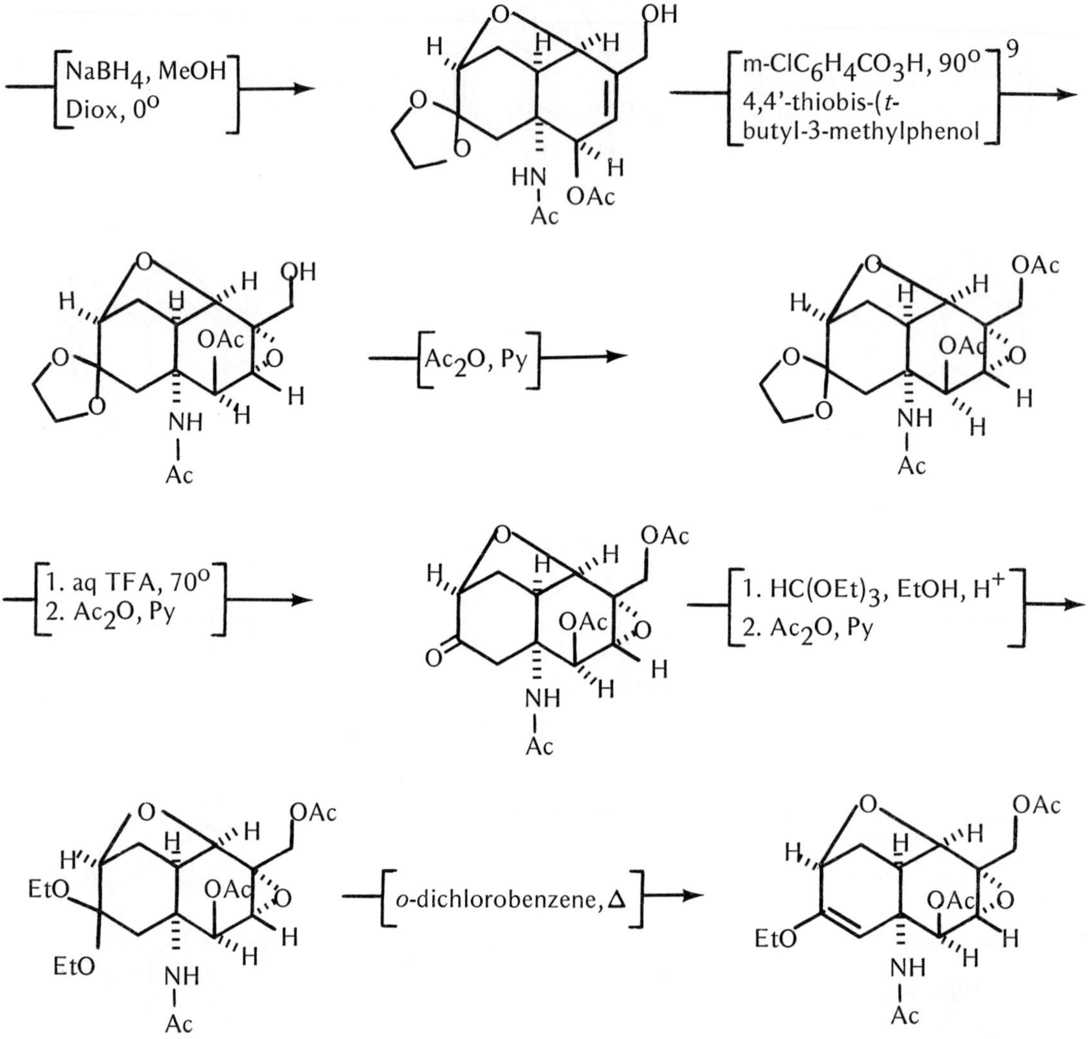

[$NaBH_4$, MeOH; Diox, 0°]

[m-$ClC_6H_4CO_3H$, 90°; 4,4'-thiobis-(*t*-butyl-3-methylphenol][9]

[Ac_2O, Py]

[1. aq TFA, 70°; 2. Ac_2O, Py]

[1. $HC(OEt)_3$, EtOH, H^+; 2. Ac_2O, Py]

[*o*-dichlorobenzene, Δ]

9. Addition of a radical inhibitor prevented thermal decomposition of the peracid, thus allowing epoxidation of the otherwise poorly reactive olefin to take place at elevated temperatures, Y. Kishi, M. Aratani, H. Tanino, T. Fukuyama, T. Goto, S. Inoue, S. Sugiura and H. Kakoi, Chem. Commun., 64 (1972).

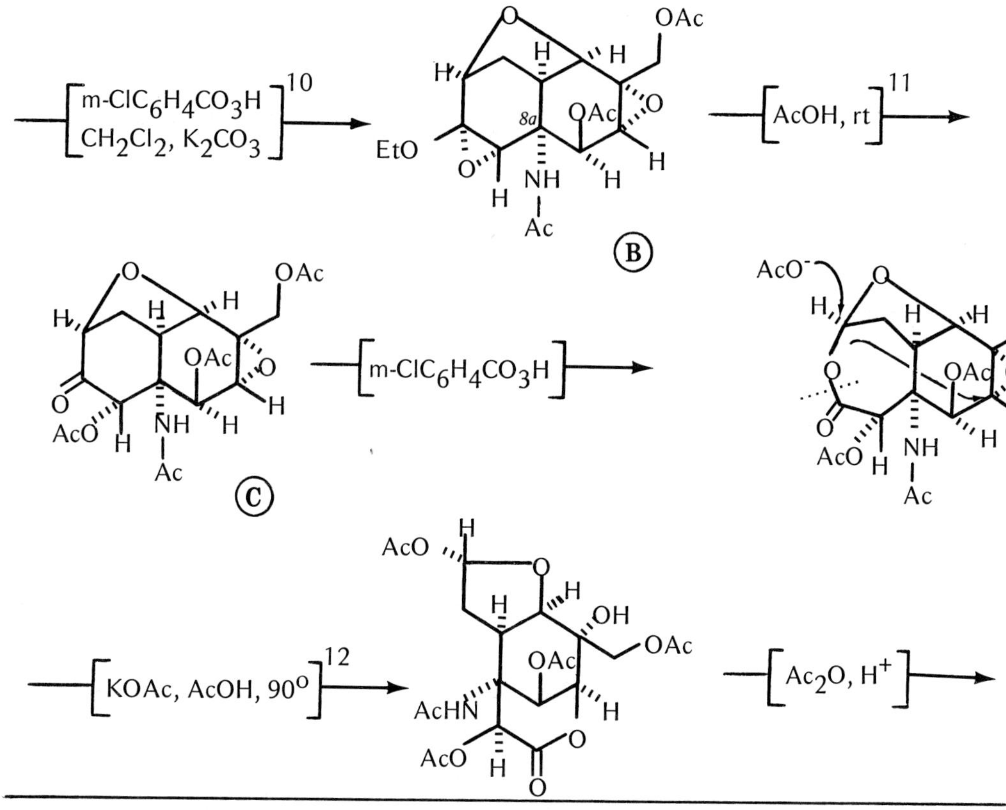

[m-$ClC_6H_4CO_3H$, CH_2Cl_2, K_2CO_3][10] → B → [AcOH, rt][11] → C → [m-$ClC_6H_4CO_3H$] → [KOAc, AcOH, 90°][12] → [Ac_2O, H^+] →

10. Addition of peracid to the double bond occurs from the α-side due to steric reasons.

11. Stereospecific opening of the epoxy ether by acetic acid may be rationalized as depicted below [Cf. C. L. Stevens and S. J. Dykstra, J. Amer. Chem. Soc., **75**, 5975 (1953)]:

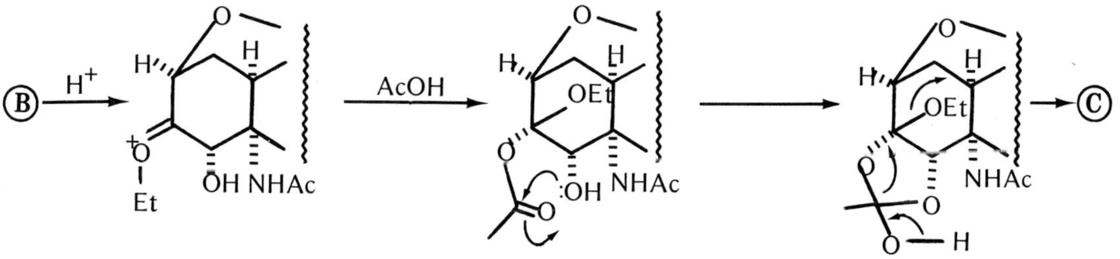

B —H^+→ —AcOH→ ——→ C

12. This remarkable transformation involves opening of the seven-membered lactone group, to generate at one terminus a carboxylate group which attacks the epoxide ring by an intramolecular S_N2 reaction, while at the other terminus the oxonium ion is attacked by acetate ion from the less hindered side.

290-300° (under vacuum)

D[13]

$Et_3O^+BF_4^-$, CH_2Cl_2, Na_2CO_3 → aq AcOH

AcN=C(SEt)$_2$, 120°

MeCONH$_2$, 150°, 1 hr

13. Because of the rather facile elimination (i→ii) generation of the aldehyde group at C-4a had to be deferred until after the guanidino group had been introduced at the C-8a position.

(i) (ii)

AcHN, AcHN, N, OAc, OAc, OAc, H, H, H, O, O, AcO, O

[NH_3, CH_2Cl_2, MeOH, rt]

H_2N, AcHN, N, OAc, OAc, OAc, H, H, H, O, O, AcO, O

[OsO_4, THF, -20°]

HO, HO, AcHN, H_2N, N, OAc, OAc, OAc, H, H, H, O, O, AcO, O

1. $NaIO_4$, aq THF, 0° → $(CH_2OH)_2$
2. NH_4OH, MeOH

OH, H, OH, OH, HN, OH, H_2N, +, N, H, H, HO, H, O, O, O^-

Tetrodotoxin[14]

14. In an alternative route to tetrodotoxin, hydroxylation of the dihydrofuran in (D) and protection of the resulting diol as an acetonide preceded introduction of the guanidino group. However, this approach was complicated by the unwanted formation of a cyclic monoacetylguanidino group (cf. i) involving the C_9-acetoxy group. This has been circumvented in later work, by proceeding with the C_9-*p*-anisoate derivative of (D), H. Tanino, S. Inoue, M. Aratani and Y. Kishi, Tetrahedron Lett., 335 (1974).

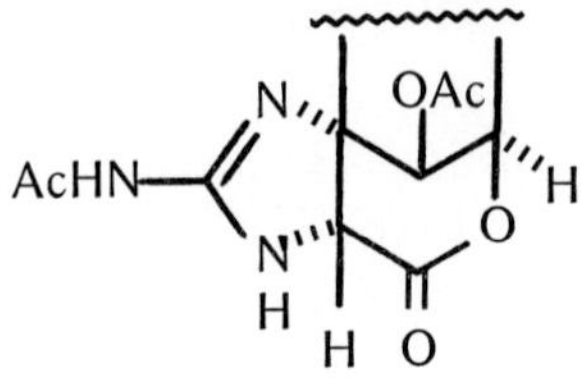

(i)

THUJAPLICIN

The synthetic method for tropolone involving 1,2-cycloaddition of halogenated ketenes to cyclopentadiene followed by solvolytic fragmentation of the resulting adduct provides a very convenient synthesis of tropolone derivatives from readily available starting materials.[1,2] The synthesis of β-thujaplicin (hinokitiol) illustrates this approach.[3]

[EtMgBr, THF → ^{i}PrTs, 0°] →

[$Cl_2CHCOCl$ / TEA, Pent, 0°] 79% →

Cl Cl H O H

[KOAc, aq AcOH, Δ][4] 44% →

O OH

β-Thujaplicin

---

1. This versatile tropolone synthesis was developed at the Chemical Division of Pittsburgh Plate Glass Company by H. C. Stevens, D. A. Reich, D. R. Brandt, K. R. Fountain and E. J. Gaughan, J. Amer. Chem. Soc., **87**, 5257 (1965).

2. For a review of the application of halogenated ketenes in synthesis, see W. T. Brady, Synthesis, **3**, 415 (1971).

3. K. Tanaka and A. Yoshikoshi, Tetrahedron, **27**, 4889 (1971).

4. Two mechanisms have been proposed for this solvolysis: T. Asao, T. Machiguchi, T. Kitamura and Y. Kitahara, Chem. Commun., 89 (1970); Bull. Chem. Soc. Japan, **43**, 2662 (1970); and P. D. Bartlett and T. Ando, J. Amer. Chem. Soc., **92**, 7518 (1970).

TRICHODERMIN

H, H, O, O, OAc

The trichothecanes are a group of modified sesquiterpene mold metabolites which exhibit pronounced phytotoxic activity against certain pathogenic fungi. The synthesis of a member of this group, trichodermin, by Raphael *et al.*,[1] represents the first synthetic venture towards construction of the novel structural framework characterizing this class of compounds. The synthesis is based on construction of the key *cis*-fused α-lactone (A),[2] followed by expansion of the lactone ring. An intramolecular conjugate cyclization (B⟶C) serves to establish a pyran ring in the bicyclic hydroxyaldehyde, already bearing the necessary functionalities for elaboration of the 5-membered fused ring C *via* reductive rearrangement of enol lactone (D).

Preparation of A ring component:

MeO — [1. Na, NH_3; 2. MeOH, TsOH] ⟶ MeO, MeO — [1. CHN_2CO_2Et, CuBr; 2. TsOH, Me_2CO] ⟶

O, CO_2Et — [NaOAc, EtOH] ⟶ O, CO_2Et — [MeMgCl] ⟶

OH, CO_2Et — [1. NaOH; 2. H_2SO_4] ⟶ H, O, O (A)

1. E. W. Colvin, R. A. Raphael and J. S. Roberts, Chem. Commun., 858 (1971).

2. An alternative synthesis of the *cis*-fused α-lactone has been devised by S. C. Welch and R. Y. Wong, Tetrahedron Lett., 1853 (1972).

[$(^iPr)_2NLi$, MeI] → [$Li^{+-}C{\equiv}CCH(OEt)_2$] →

$CH(OEt)_2$ H O OH → [$NaBH_4$] → $CH(OEt)_2$ OH OH

[Na, NH_3, EtOH] → H OH $CH(OEt)_2$ OH (B) → [AcOH, NaOAc, H_2O] →

H O CHO OH (C) [1. CrO_3, Py 2. CrO_3, H_2SO_4, Me_2CO] → H O CO_2H O

[NaOAc, Ac_2O] → H O H O O (D) [1. $Li(O^tBu)_3AlH$ 2. Ac_2O, Py] 3,4 →

3. Cf. J. Martin, W. Parker and R. A. Raphael, J. Chem. Soc., 289 (1964).

4. For a detailed study of the mechanism and scope of this interesting route to bridged ring ketols, see J. Martin, B. Shroot and T. Stewart, J. Chem. Soc. (C), 101 (1967).

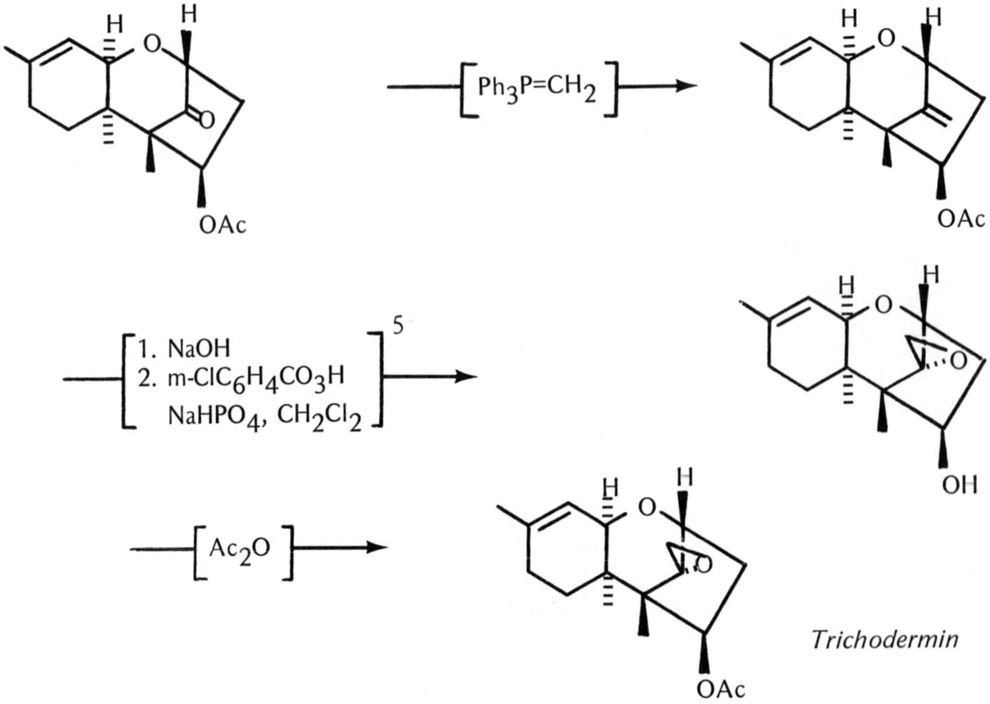

[$Ph_3P{=}CH_2$]

[1. NaOH
2. m-$ClC_6H_4CO_3H$
$NaHPO_4$, CH_2Cl_2] [5]

[Ac_2O]

Trichodermin

---

5. Introduction of the α-epoxide is undoubtedly a consequence of intramolecularly assisted delivery of peracid by the strategically placed hydroxyl group in the molecule, cf. H. B. Henbest and R. A. L. Wilson, J. Chem. Soc., 1958 (1957).

ULEINE

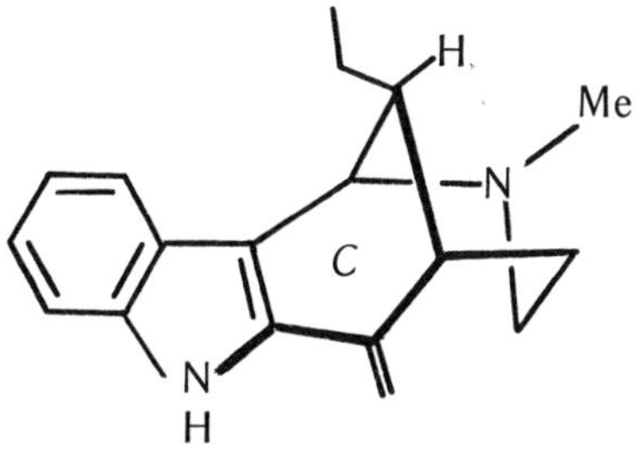

Uleine belongs to a relatively small group of indole alkaloids lacking the usual two-carbon chain biogenetically derived from tryptophan.[1] The synthesis of uleine outlined below[2] illustrates one approach to this class of compounds involving isomerization of the key indol-2-yl-Δ^3-piperideine ketone (A) to a Δ^2-piperideine, followed by protonation of the cyclic enamine to an immonium species (D) and Mannich-type closure on the indole β-position for formation of ring C.

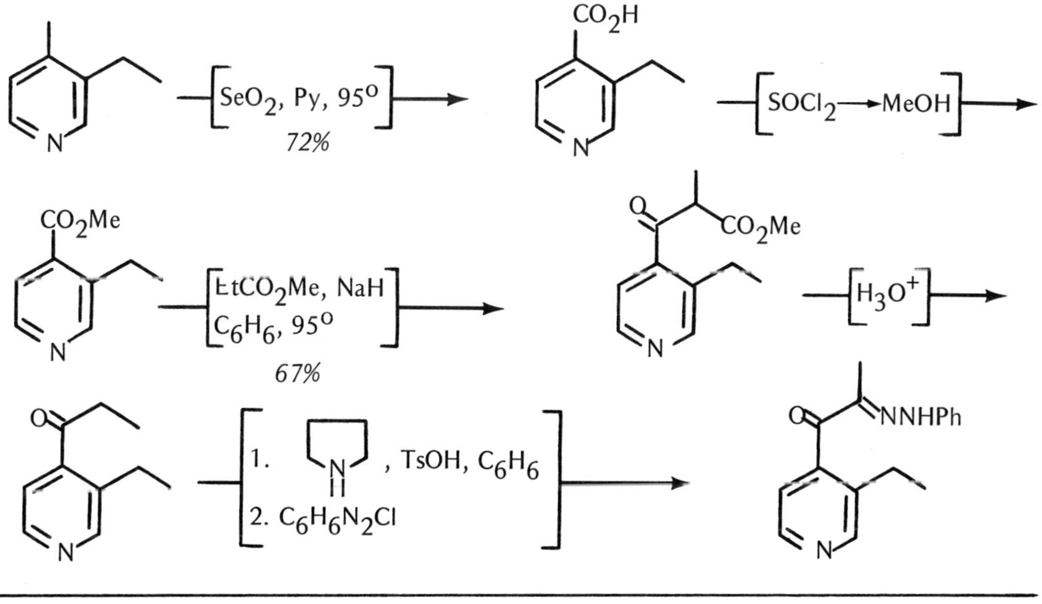

SeO_2, Py, 95°; 72%

$SOCl_2 \rightarrow$ MeOH

$EtCO_2Me$, NaH, C_6H_6, 95°; 67%

H_3O^+

1. pyrrolidine, TsOH, C_6H_6
2. $C_6H_6N_2Cl$

---

1. Review: B. Gilbert, The Alkaloids, **8**, 469 (1965).

2. N. D. V. Wilson, A. Jackson, A. J. Gaskell and J. A. Joule, Chem. Commun., 584 (1968); J. Chem. Soc. (C), 2738 (1969).

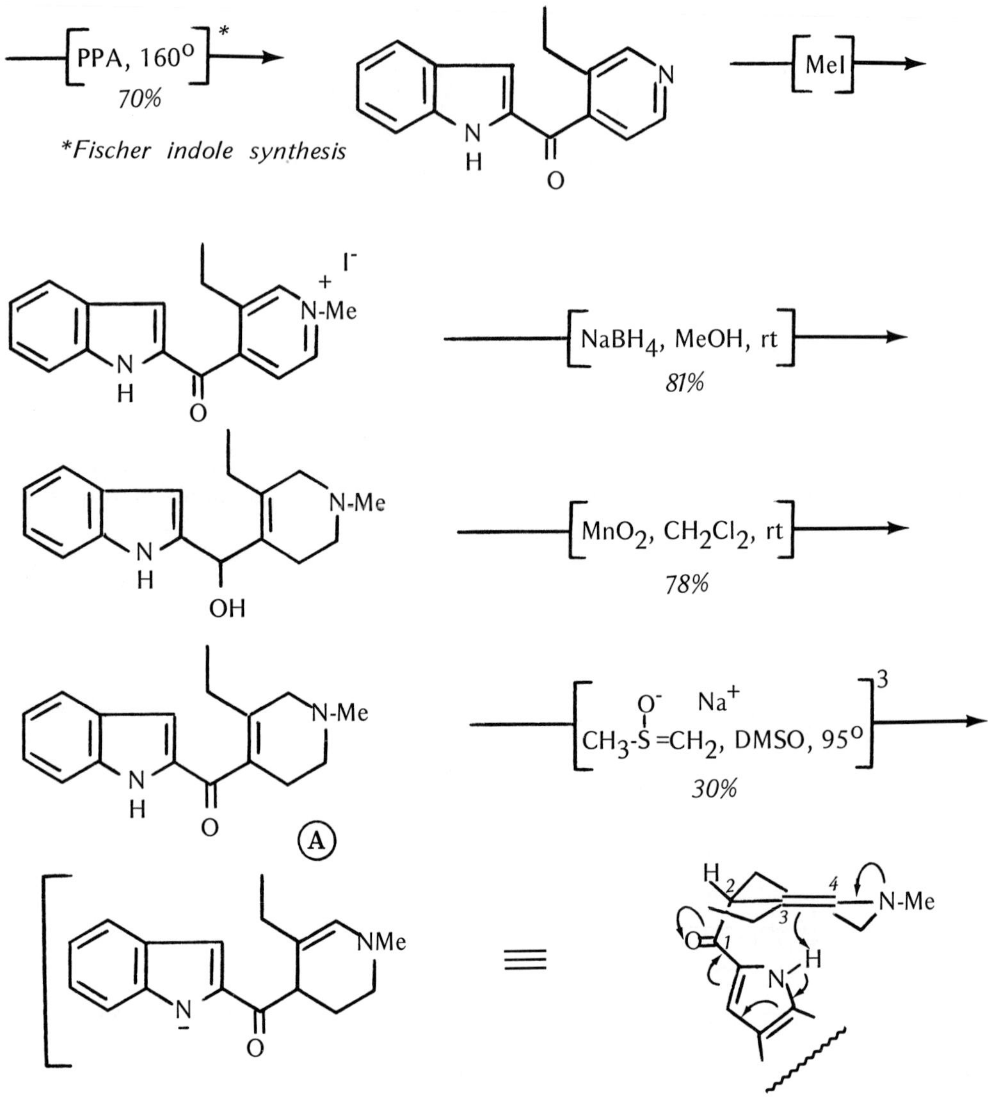

[PPA, 160°]* 70%

*Fischer indole synthesis

[MeI]

[$NaBH_4$, MeOH, rt] 81%

[MnO_2, CH_2Cl_2, rt] 78%

[CH_3-S(O⁻)=CH_2 Na⁺, DMSO, 95°][3] 30%

(A)

3. The planned utilization of (A) involved deconjugation, by kinetically controlled protonation of the enolate of (A), to give the desired Δ^2-piperideine. Instead, a product (C) derived, formally at least, by cyclization of the immonium species (B) on the indole nitrogen was produced. However, this product could be converted to the dasycarpidone nucleus by treatment with aqueous acetic acid.

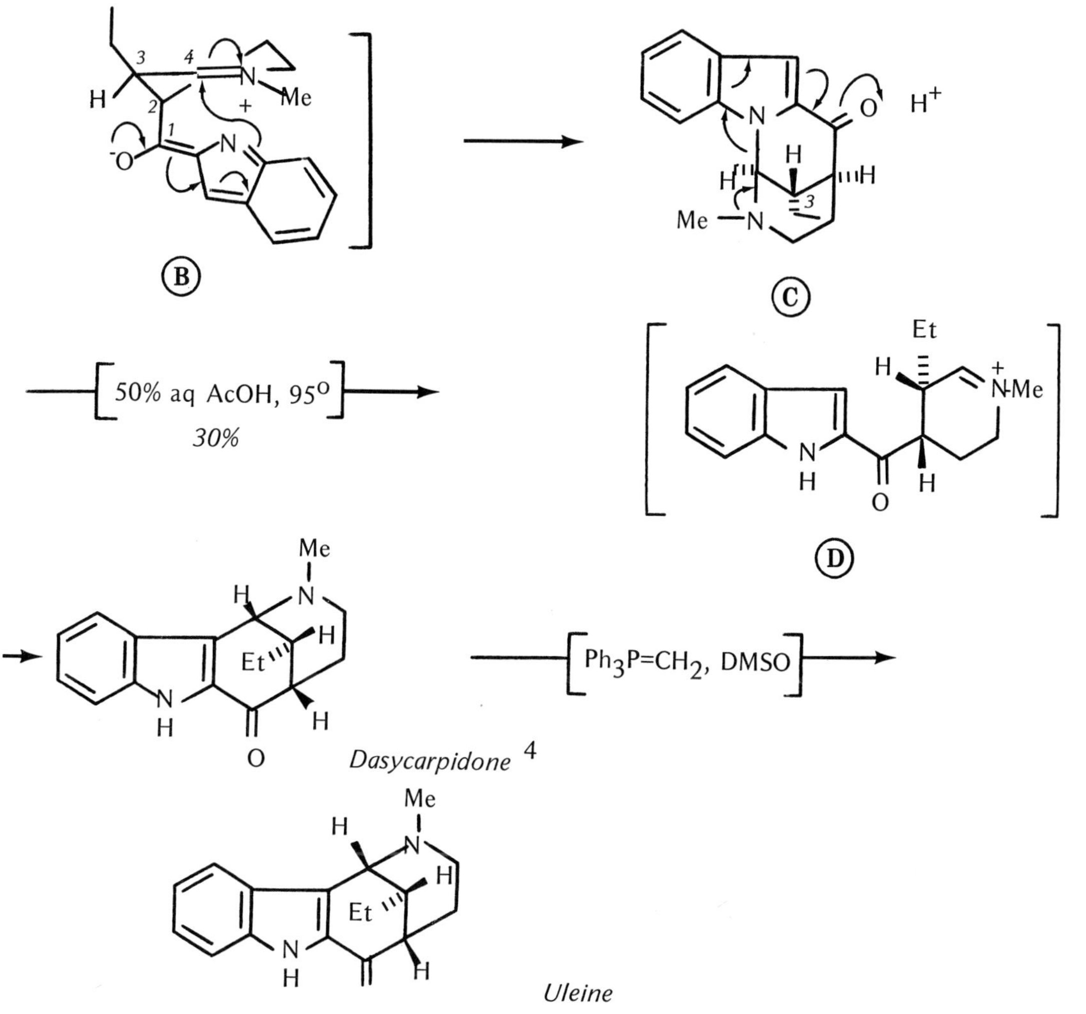

B

C

50% aq AcOH, 95°
30%

D

$Ph_3P{=}CH_2$, DMSO

Dasycarpidone 4

Uleine

---

4. A mixture of dasycarpidone and 3-*epi*-dasycarpidone (1:5) was obtained and separated chromatographically.

A second approach to uleine is based on cyclization of indolyl-3-piperidyl intermediates (e.g., F) at the indole α-position to furnish the desired tetracyclic nucleus.[5-7] The stereoselective synthesis of uleine by Büchi *et al.*, outlined below, illustrates this approach.[7]

Preparation of amino ketone:

[DMF, 100°] 87%

[aq HCl, Δ] 98%

E

Preparation of uleine:

E [1N NaOH, EtOH] 66%

[8]

---

5. L. J. Dolby and H. Biere, J. Amer. Chem. Soc., **90**, 2699 (1968).

6. T. Kametani and T. Suzuki, J. Org. Chem., **36**, 1291 (1971); J. Chem. Soc. (C), 1053 (1971).

7. G. Büchi, S. J. Gould and F. Naf, J. Amer. Chem. Soc., **93**, 2492 (1971).

8. Formation of a piperidone with two equatorial substituents has been rationalized in terms of the cyclization proceeding within a chair-like intermediate in which the bulky substituents on the two double bonds are *trans* oriented (i).

(i)

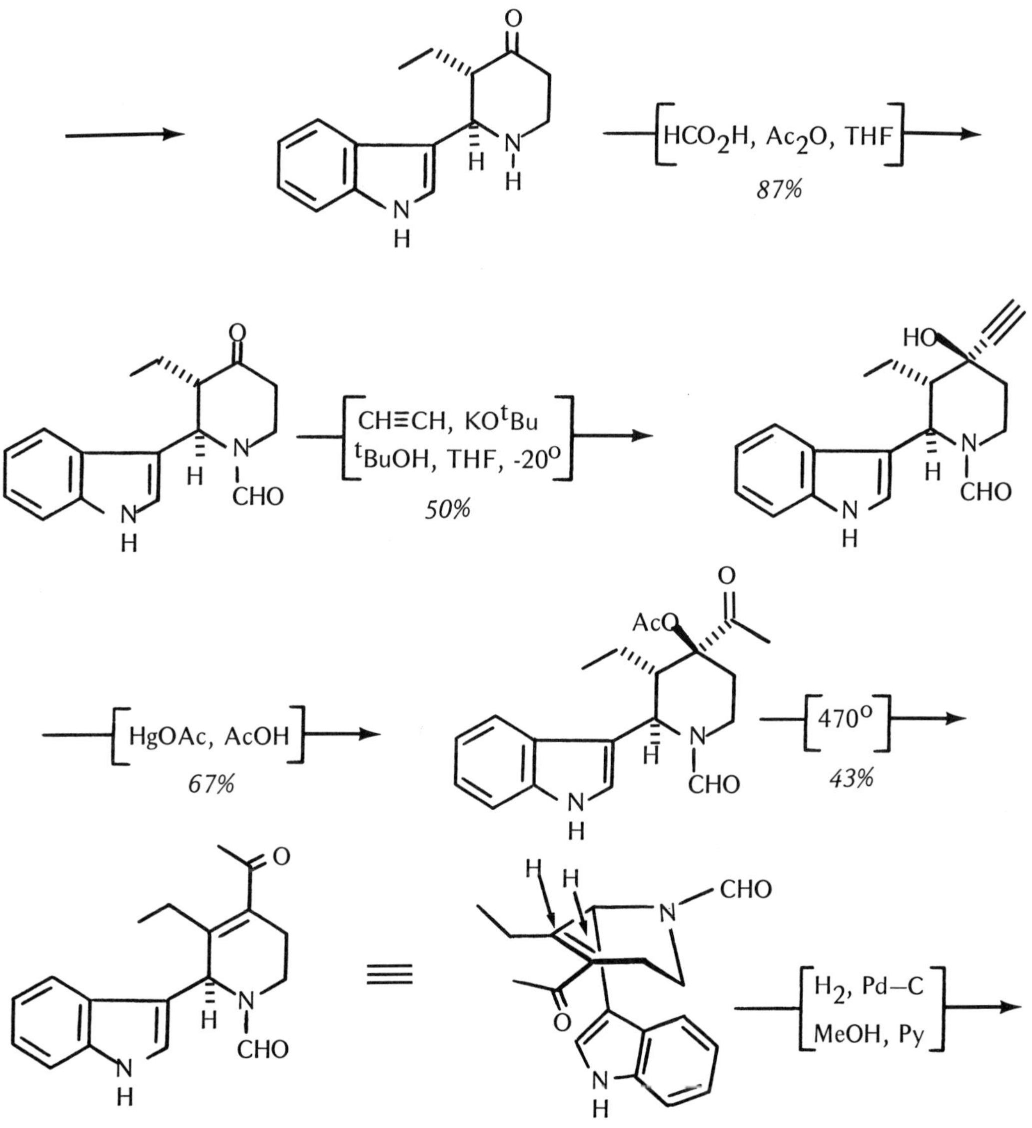

HCO_2H, Ac_2O, THF

87%

$CH{\equiv}CH$, KO^tBu

^tBuOH, THF, -20°

50%

HgOAc, AcOH

67%

470°

43%

H_2, Pd–C

MeOH, Py

$BF_3 \cdot Et_2O$, CH_2Cl_2, rt

72%

LAH, Glyme

73%

Uleine

β-VETIVONE

β-Vetivone is a constituent of the essential oil of vetiver, and for a long time was thought to possess a hydroazulenic structure. However, during work on stereoselective routes to substituted hydroazulenes, Marshall and his group discovered that the vetivane skeleton must be reformulated in terms of a spiro[4·5]decane system,[1,2] and confirmation of the new structure was obtained through an unambiguous total synthesis of β-vetivone based on the spiro tricyclic enone (A).[3]

[hν, Diox][4]

[AcOH, Ac_2O, H_2SO_4] 66%

(A)

[HCO_2Et, NaH, C_6H_6, MeOH]

CHOH

[^nBuSH, Et_2O, H_2SO_4] 81%

CHS^nBu

1. J. A. Marshall, N. H. Andersen and P. C. Johnson, J. Amer. Chem. Soc., **89**, 2748 (1967).

2. J. A. Marshall and P. C. Johnson, J. Amer. Chem. Soc., **89**, 2750 (1967).

3. J. A. Marshall and P. C. Johnson, J. Org. Chem., **35**, 192 (1970).

4. P. J. Kropp, J. Amer. Chem. Soc., **87**, 3914 (1965).

[$NaBH_4$, MeOH] →

HO — CHS^nBu

[$HgCl_2$, H_3O^+, Me_2CO] →

83% for two steps

CHO

[MeLi, EtO, -35°] →

94%

OH

[MnO_2, Hex] →

90%

O

[1. Li-NH_3, Et_2O; 2. CrO_3] →

92%

O

[MeLi, Et_2O] →

98%

HO

Hinesol and epi-hinesol

[Ac_2O, NaOAc, Δ] →

65%

AcO

[Na_2CrO_4, AcOH, Ac_2O, 40°] →

52%

O

AcO

[BF_3·Et_2O, rt][5] →

90%

O

β-Vetivone

5. A small amount of the isopropenyl isomer is also formed.

Recently Yamada *et al*[6] described a facile method of preparing the properly functionalized spiro[4·5] decane system by acid catalyzed cyclization of cyclohexenone derivatives, and applied it to the synthesis of β-vetivone. The clever use of proximity effects is a noteworthy feature of this ingenious synthesis.

CHO

CO_2H

1. $Ph_3P{=}CHOMe$
2. $(CH_2OH)_2$, H^+

73%

CO_2H

Li-NH_3, ^tBuOH
THF, -33° ⟶
$(CO_2H)_2$

80%

CO_2H

6N HCl, DME, Δ

90%

H O

1. $(CH_2OH)_2$, H^+
2. LAH, THF, 25°
3. 2N HCl, DME, 25°

77%

OH

H

1. Ac_2O, Py
2. $(CH_2OH)_2$, H^+
3. KOH, MeOH

83%

OH

H

CrO_3, Py

---

6. K. Yamada, H. Nagase, Y. Hayakawa, K. Aoki and Y. Hirata, Tetrahedron Lett., 4963 (1973).

[(MeO)$_2$CO, NaH, 50°]

CO$_2$Me

[1. NaBH$_4$, MeOH
2. MsCl, Py, 5°
3. NaOMe, MeOH]

82%

CO$_2$Me

[H$_2$, PtO$_2$, MeOH]

98%

CO$_2$Me

[MeLi, DME, 40°]

OH

[POCl$_3$, Py]

[2N HCl, DME]

[Ph$_3$P, Br$_2$, MeCN]*

88%

Br

**Regeneration of the double bond in the six-membered ring was accomplished without migration of the labile double bond of the isopropylidene group.*

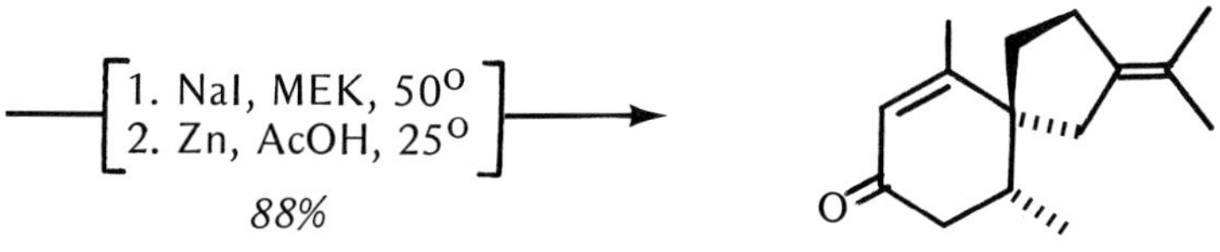

[1. NaI, MEK, 50°; 2. Zn, AcOH, 25°] 88%

β-Vetivone

Another approach to the spiro[4·5]decane skeleton is illustrated by the exceptionally simple stereospecific synthesis of β-vetivone[7] based on a novel spiroannelation reaction of cyclic 1,3-diketone enol ethers developed at Columbia.[8]

Preparation of Dihalide:

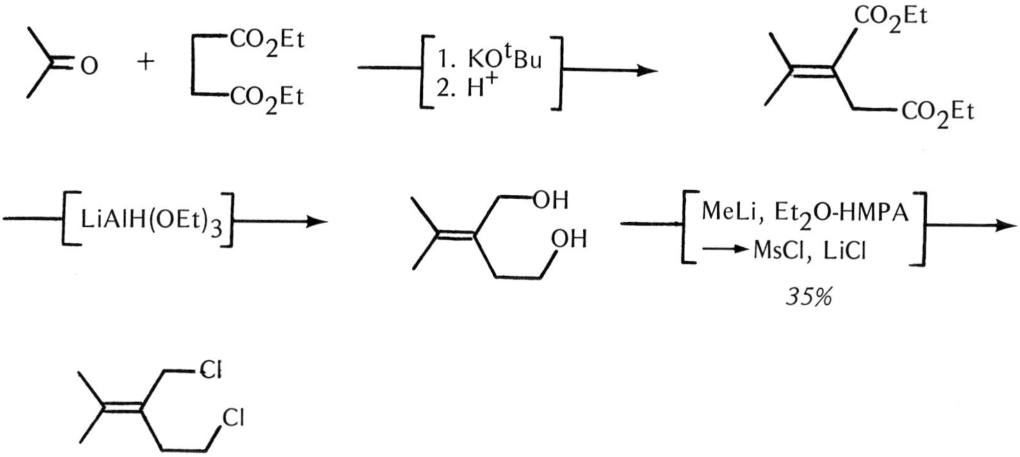

[1. KO^tBu; 2. H^+]

[$LiAlH(OEt)_3$]

[MeLi, Et_2O-HMPA → MsCl, LiCl] 35%

Preparation of β-Vetivone:

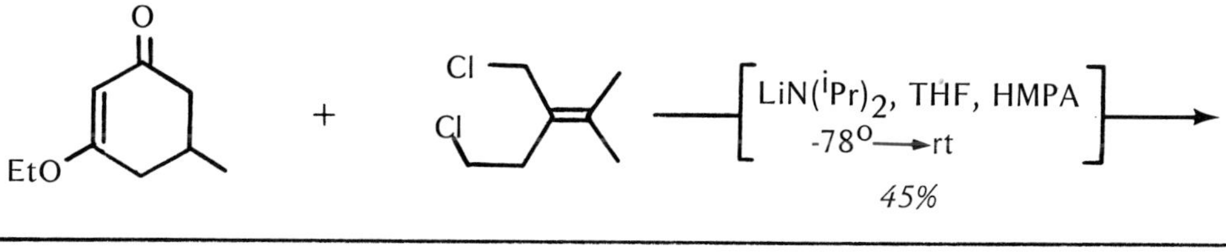

EtO

[$LiN(^iPr)_2$, THF, HMPA; -78° → rt] 45%

---

7. G. Stork, R. L. Danheiser and B. Ganem, J. Amer. Chem. Soc., **95**, 3414 (1973).

8. For regiospecific alkylation of kinetic enolates of cyclic β-diketone enol ethers, see G. Stork and R. L. Danheiser, J. Org. Chem., **38**, 1775 (1973).

OEt

OEt

Footnote 9

Cl

Cl

OEt

[MeLi, Et_2O, 0°
→ 1N HCl, rt]

60%

β-Vetivone[10,11]

9. Stereochemistry of the spiroannelation was determined by geometry of the enolate ion which forces the ring methyl into an axial conformation as completion of the ring occurs trans to that methyl.

10. For a route to spiro[4·5]decanes based on intramolecular keto-carbene insertion in substituted cyclohexenes, see M. Mongrain, J. LaFontaine, A. Belanger and P. Deslongchamps, Canad. J. Chem., **48**, 3273 (1970); and P. M. McCurry, Tetrahedron Lett., 1845 (1971). An interesting synthesis of spiro systems by α,α'annelation of enamines has been developed by D. J. Dunham and R. G. Lawton, J. Amer. Chem. Soc., **93**, 2074 (1971).

11. For alternative syntheses of β-vetivone and the related spirovetivane hinesol, see P. M. McCurry and R. K. Singh, Tetrahedron Lett., 3325 (1973); J. A. Marshall and S. F. Brady, J. Org. Chem., **35**, 4068 (1970); K. Yamada, K. Aoki, H. Nagase, Y. Hayakawa and Y. Hirata, Tetrahedron Lett., 4967 (1973).

VINDOROSINE

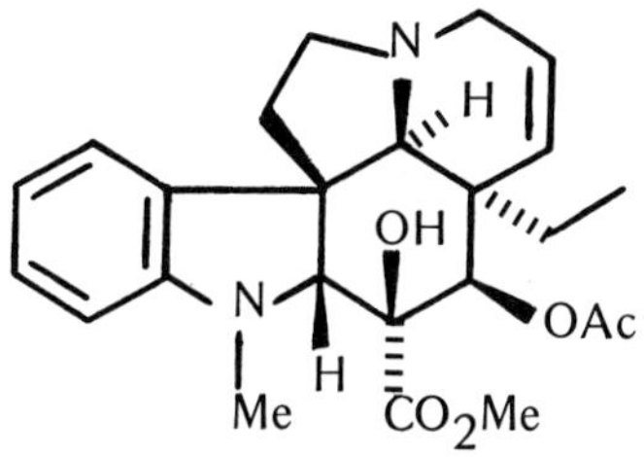

The synthesis of vindorosine,[1] a highly functionalized pentacyclic *Aspidosperma* alkaloid is based on intramolecular indole β-cyclization[2] of the enamino ketone (A) followed by trapping of the resulting 3,3-spiroindolenine by nucleophilic enol addition at C-2.

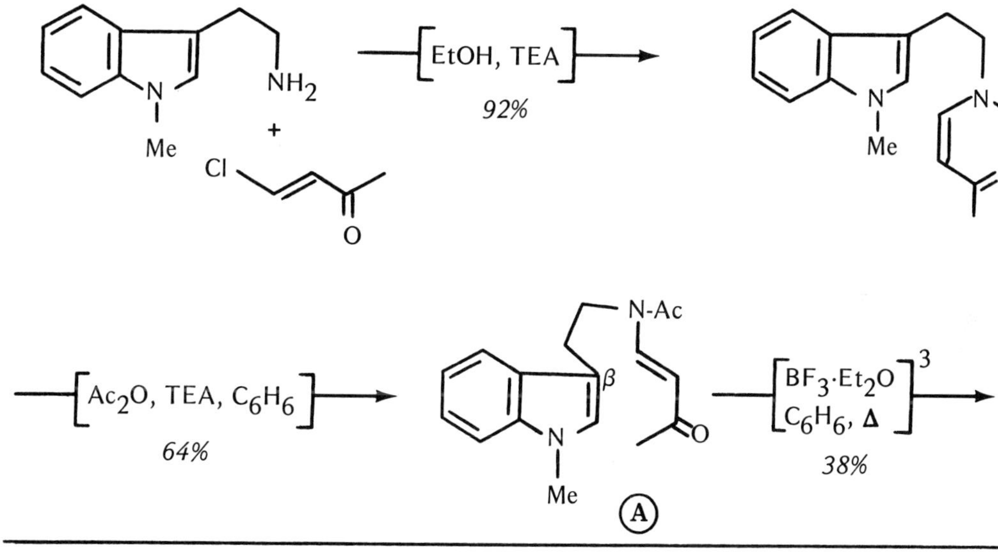

1. G. Büchi, K. E. Matsumoto and H. Nishimura, J. Am. Chem. Soc., **93**, 3299 (1971).

2. A. H. Jackson and A. E. Smith, Tetrahedron, **24**, 403 (1968); J. R. Williams and L. R. Unger, Chem. Commun., 1605 (1970).

3. Cf. J. E. D. Barton and J. Harley-Mason, Chem. Commun., 298 (1965).

N-Ac

Me

N-Ac

N

Me H

10% aq HCl

92%

NH

N

Me H

1. CHO, MeOH, NaOMe
2. BF_3, AcOH, rt

40%

N

H

N

Me H

EtI, KO^tBu, ^tBuOH, DME

44%

N

H

N

Me H

$(MeO)_2CO$, NaH

100%

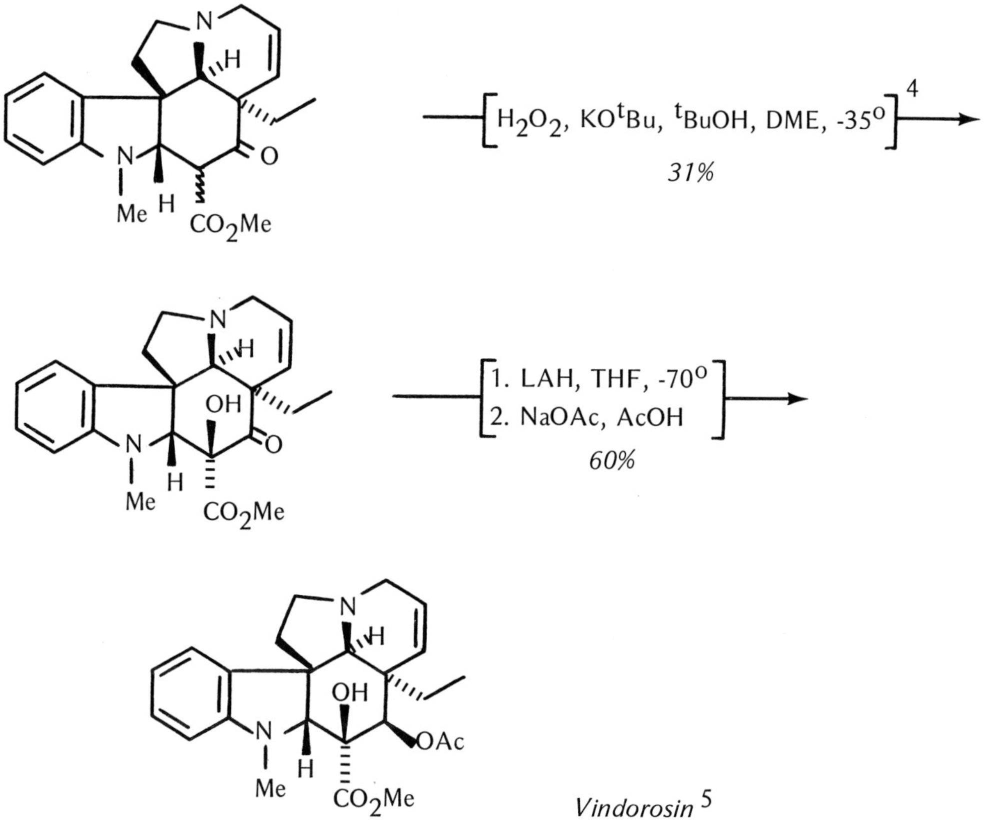

$\left[H_2O_2, KO^tBu, {}^tBuOH, DME, -35^o\right]^4$

31%

1. LAH, THF, -70°
2. NaOAc, AcOH

60%

Vindorosin [5]

4. The overall course and stereochemical outcome of this oxidation has been rationalized as below:

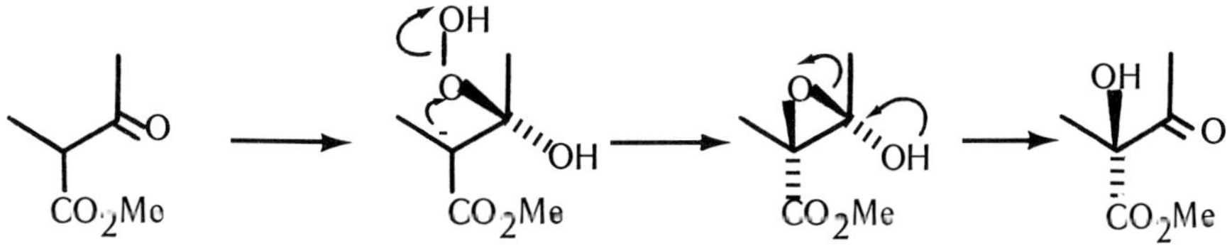

5. For a related, facile two-step entry into the pentacyclic *Aspidosperma* alkaloid skeleton (i→ii), see E. Wenkert, J. S. Bindra and B. Chauncy, Synth. Commun., **2**, 285 (1972).

$BF_3 \cdot Et_2O$

100°

(i)

(ii)

METHYL VINHATICOATE

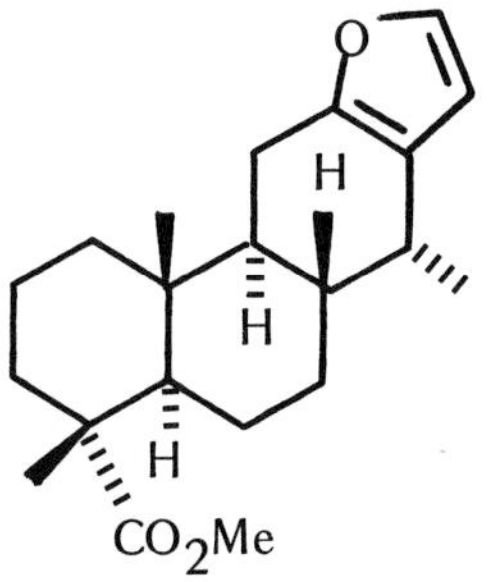
O
H
H
H
CO_2Me

The synthesis of methyl vinhaticoate, a furanoid diterpene constituent of South American hardwood, illustrates a novel furan synthesis based on the 1,4-addition of carbethoxycarbene to α-methoxymethylene ketones.[1,4]

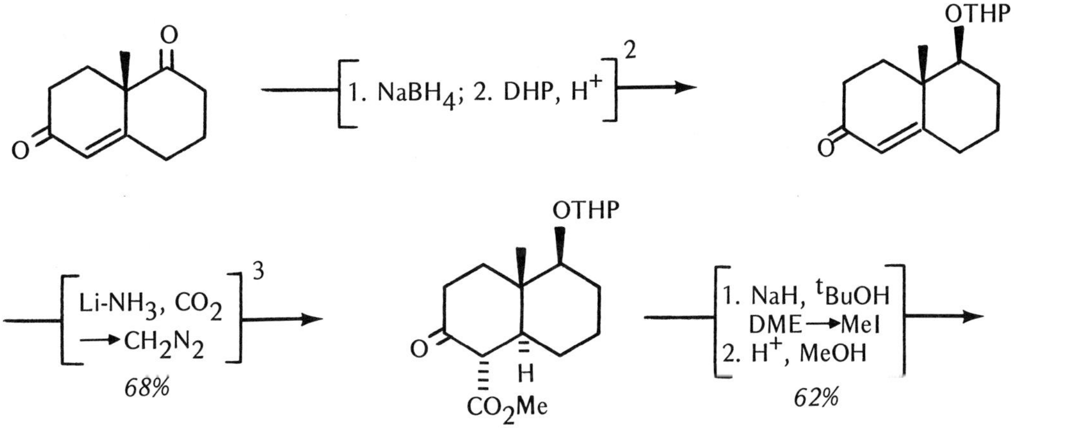
O
O
1. $NaBH_4$; 2. DHP, H^+
2
OTHP
O
$Li\text{-}NH_3$, CO_2
→ CH_2N_2
3
68%
OTHP
O
H
CO_2Me
1. NaH, ^tBuOH
DME→MeI
2. H^+, MeOH
62%

1. T. A. Spencer, R. M. Villarica, D. L. Storm, T. D. Weaver, R. J. Friary, J. Posler and P. R. Shafer, J. Amer. Chem. Soc., **89**, 5497 (1967); T. A. Spencer, R. A. J. Smith, D. L. Storm and R. M. Vallarica, J. Amer. Chem. Soc., **93**, 4856 (1971).

2. Elaboration of the tricyclic enone (A) *via* the AB→ABC approach has previously been described, T. A. Spencer, T. D. Weaver, R. M. Villarica, R. J. Friary, J. Posler and M. A. Schwartz, J. Org. Chem., **33**, 712 (1968).

3. The "reductive carbomethoxylation" procedure, resulting in formation of the desired *trans* ring fusion and introduction of carbomethoxy function at C_4 is based on enolate anion trapping procedure of G. Stork, P. Rosen, N. Goldman, R. V. Coombs and J. Tsuji, J. Amer. Chem. Soc., **87**, 275 (1965).

OH

O

H

CO_2Me

$(CH_2SH)_2$, $BF_3 \cdot Et_2O$

OH

S

S

H

CO_2Me

Raney Ni, EtOH

90% for two steps

OH

H

CO_2Me

1. CrO_3
2. NaH, HCO_2Et

90%

O

CHOH

H

CO_2Me

MVK, NaOMe, MeOH, △

54%

O

H

H

CO_2Me

Ⓐ [2]

Li-NH_3, EtOH

42%

OH

H

H

H

CH_2OH

1. CrO_3
2. CH_2N_2

O

H

H

H

CO_2Me

1. Br_2, AcOH
2. $CaCO_3$, DMA, △
→ Zn, EtOH, △ [4]

34%

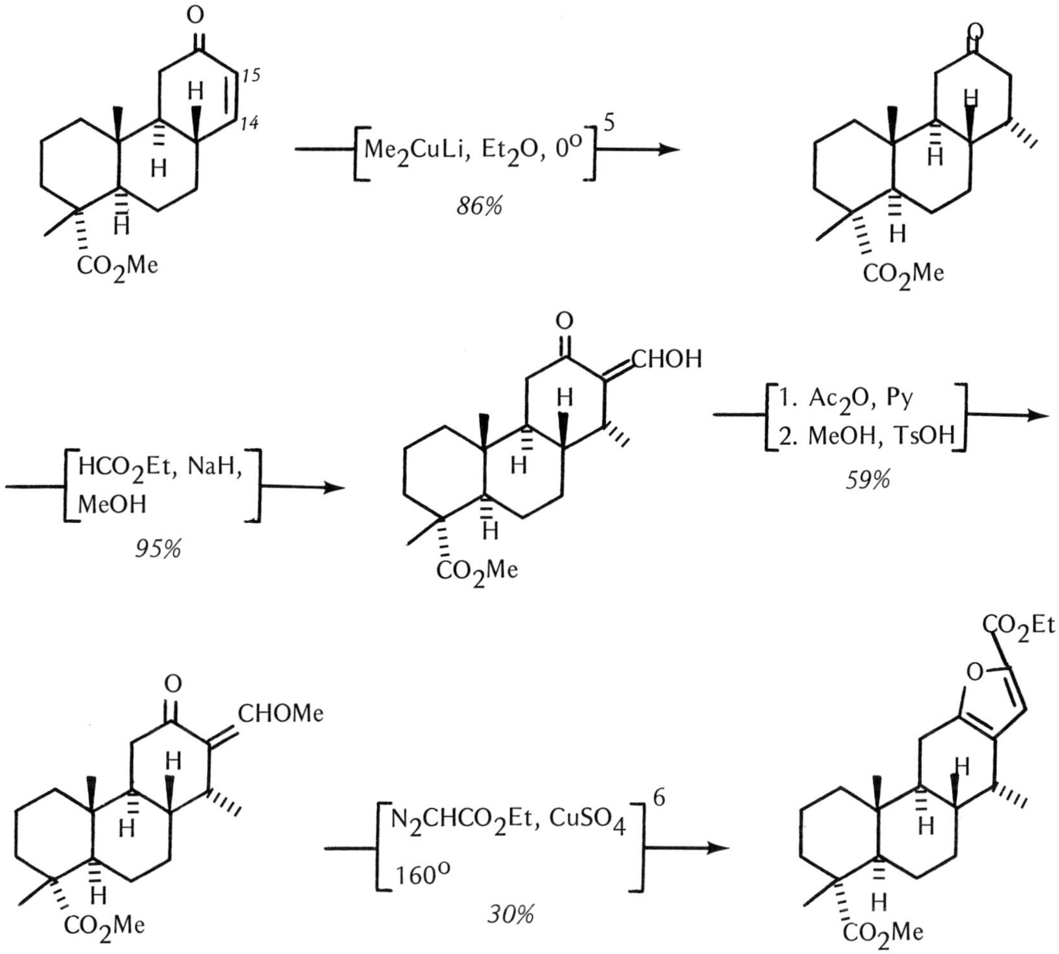

$\left[Me_2CuLi, Et_2O, 0^o\right]^5$
86%

$\left[HCO_2Et, NaH, MeOH\right]$
95%

CHOH

$\left[1. Ac_2O, Py; 2. MeOH, TsOH\right]$
59%

CHOMe

$\left[N_2CHCO_2Et, CuSO_4, 160^o\right]^6$
30%

CO_2Et

CO_2Me

4. The problem of introducing the methyl group at C-14 was solved by establishing a 14,15-double bond (bromination-dehydrobromination sequence) followed by introduction of the desired methyl group by conjugate addition of an organometallic reagent.

5. Conjugate methylation product was a 1:1 mixture of α and β-methyl isomers. The α-isomer crystallized spontaneously from the reaction mixture. Its configuration was assigned by NMR studies.

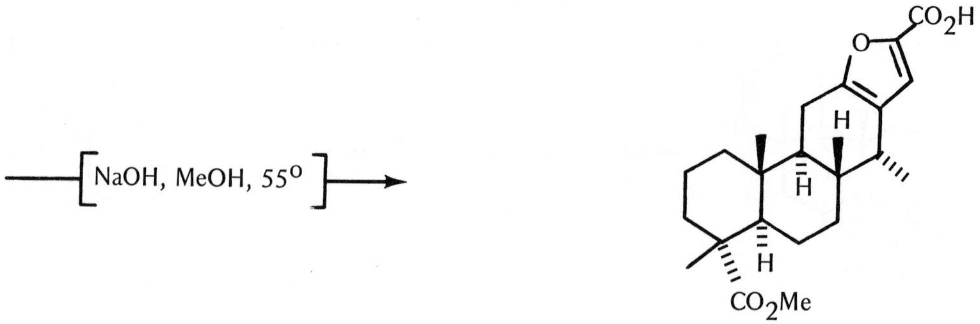

—[NaOH, MeOH, 55°]→

CO_2H O H H H CO_2Me

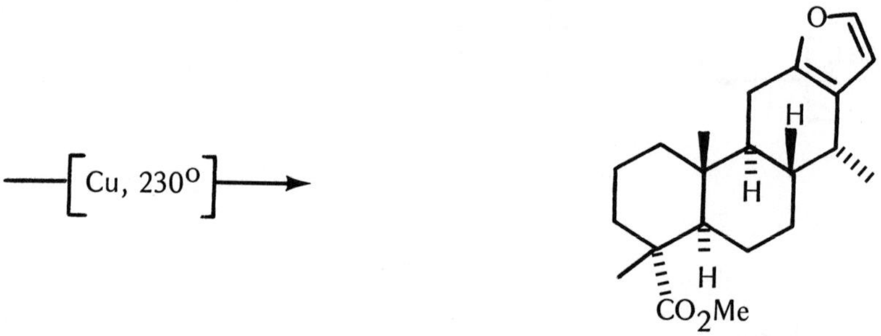

—[Cu, 230°]→

O H H H CO_2Me

Methyl Vinhaticoate

---

6. For the synthesis of furans by 1,4-addition of carbethoxycarbene to α-methoxymethylene ketones, see D. L. Storm and T. A. Spencer, Tetrahedron Lett., 1865 (1967).

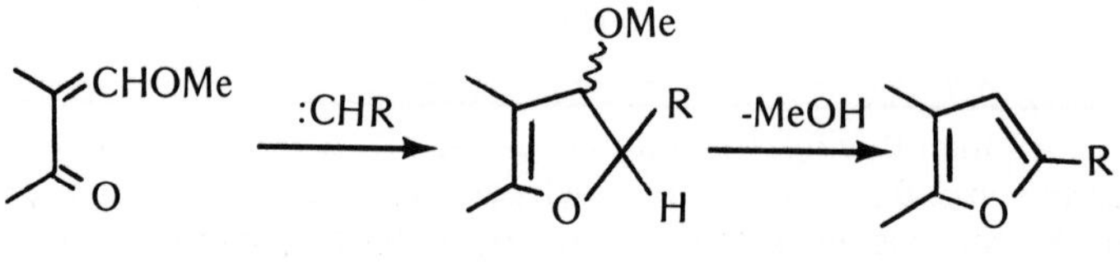

CHOMe O —:CHR→ OMe R O H —-MeOH→ O R

VITAMIN B_{12}

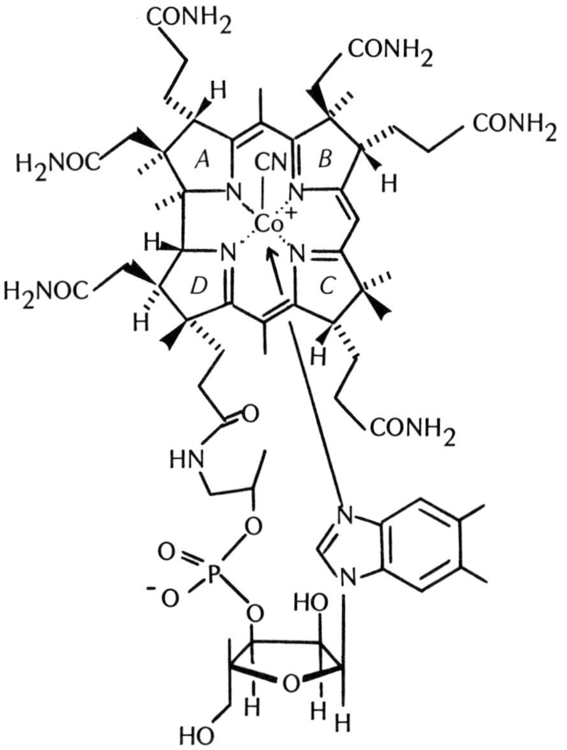

The synthesis of vitamin B_{12} represents perhaps the most significant achievement of natural products synthesis in recent years. The completion of this mammoth task, which spanned nearly a decade, was the outcome of a joint project pursued in the laboratories of R. B. Woodward at Cambridge and Albert Eschenmoser at Zürich.[1] The synthesis actually culminates in the preparation of cobyric acid, the simplest of vitamin B_{12} derivatives, which has previously been converted to the vitamin by Bernhauer[17], and thus constitutes a formal total synthesis.[2,3]

---

1. For an engrossing series of progress reports on the work towards synthesis of vitamin B_{12}, see R. B. Woodward, Pure Appl. Chem., **17** 519 (1968); **25**, 283 (1971); **33** 145 (1973); A. Eschenmoser, Proc. of the Robert A. Welch Foundation Conferences on Chemical Research, XII, 9 (1968); Quart. Revs., **24** 366 (1970); XXIIIrd Int. Congress Pure Appl. Chem., **2** 69 (1971). For a review of Eschenmoser's earlier work on the synthesis of corrins, see Art. Org. Synth., 123 (1970).

The overall plan of synthesis involves the preparation and bringing together of three optically active building blocks of the properly specified absolute configuration, one of them representing the A/D portion of the molecule, one representing the ring B, and the third representing ring C sector. The central problem in corrin synthesis, namely, the construction of the vinylogous amidine system has been solved using the method of sulfide contraction *via* oxidative and alkylative coupling, which is used for coupling the precursors of rings B and C, and later for coupling the two bicyclic optically active components:

```
A   B        A   B        A — B
| + |  ───►  |   |  ───►  |   |
D   C        D — C        D — C
```

A noteworthy feature of the synthesis of the A/D component, bearing six contiguous asymmetric centres, is the aggrandization of new centres of stereochemistry under directive influence of those already present in prior intermediates (stereospecific synthesis by induction). The aromatic ring in the ring A precursor serves not only as a stable platform for construction of the complex A/D unit, but also as a latent ring A propionate side chain. The ring D nitrogen atom has been introduced as an oxime function and is ingeniously inserted into the desired position by a Beckmann rearrangement. The remaining special feature of the cobyric acid nucleus, the two methyl groups borne on the corrin nucleus at the meso carbons between rings A/B and C/D have been introduced by alkylation during the final stages of the synthesis.

Another feature of the cobyric acid molecule is that it contains all the peripheral carboxy functions in the primary amide form, except for the propionate chain on ring D. Therefore, the problem of differentiating one chain terminus from all the others adorning the corrin nucleus is quite crucial, and has been solved by masking the ring D carboxy function as a nitrile until the penultimate step.

---

2. The synthesis of vitamin B_{12} and the related work on corrins provides a beautiful illustration of how a synthetic project directed towards the synthesis of a complex natural product is bound to have an impact even beyond the immediate structural boundaries of the specific synthetic problem. Innumerable synthetic methods of general applicability emerged from this momentous work, and the principle of orbital symmetry conservation (Woodward-Hoffmann rules), which is of immense consequence to chemistry as a whole, arose directly from the early studies on vitamin B_{12} synthesis. For an essentially direct transcript of a lecture by R. B. Woodward describing the events that led to the discovery of the Woodward-Hoffmann rules, see Chem. Soc. Publ. No. **21**, 217 (1967).

3. Cornforth's pioneering work based on the chemistry of the isoxazole ring system marks the beginning of the history of the synthetic confrontation with the vitamin B_{12} molecule [J. W. Cornforth, report by P. B. D. de la Mare, Nature, **195**, 441 (1962)]. This approach has been exploited by R. V. Stevens in a brilliant series of studies which also have as their objective the synthesis of vitamin B_{12} (ref. 18). Noteworthy contributions to corrin syntheses have been made by the Cambridge School [R. Bonnett, V. M. Clark, A. Giddey and A. R. Todd, J. Chem. Soc., 2087 (1959)], and by A. W. Johnson and his group [Chem. in Brit., 253 (1967)].

Preparation of ring C component: [4]

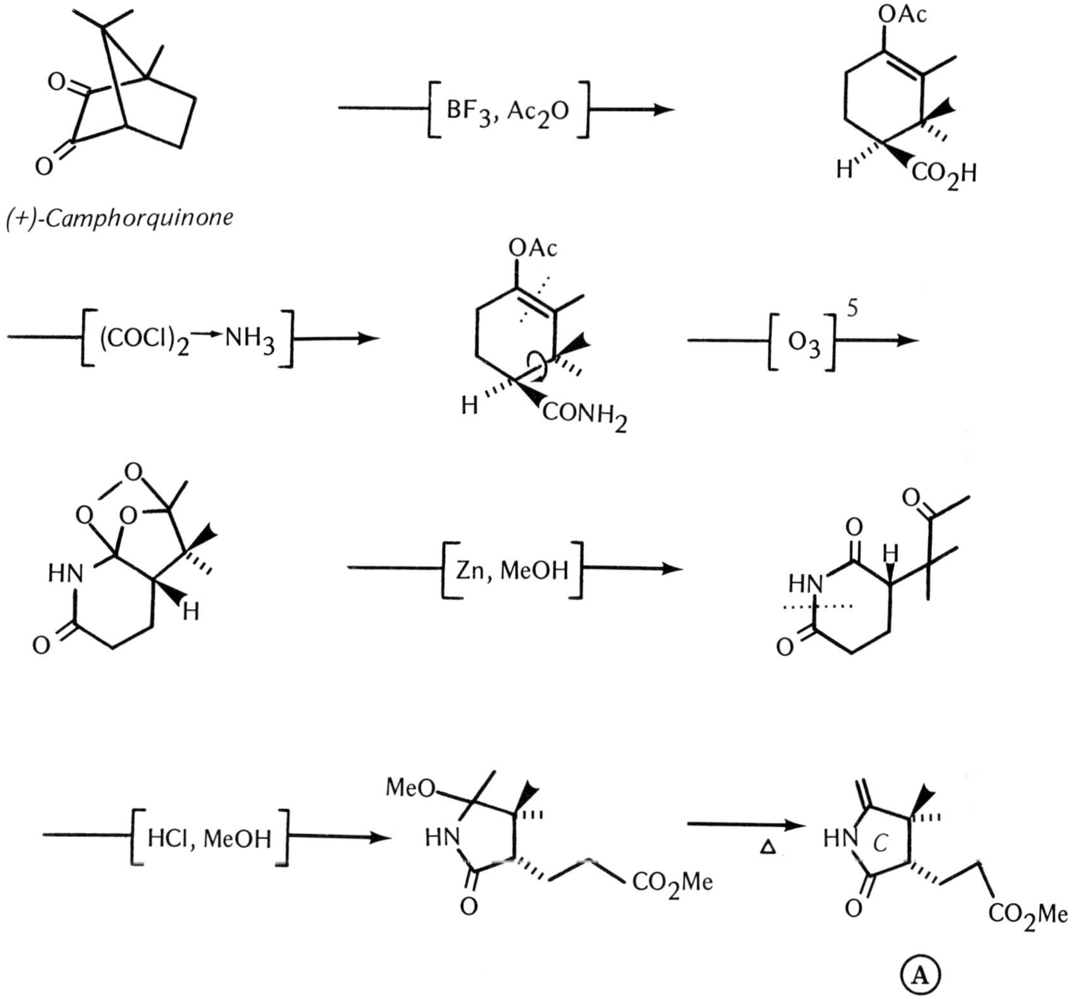

(+)-Camphorquinone

—[BF_3, Ac_2O]→

—[$(COCl)_2$ → NH_3]→

—[O_3]→ [5]

—[Zn, MeOH]→

—[HCl, MeOH]→

→ Δ

4. The synthesis of ring C component parallels earlier investigations by Cornforth *et al.* (ref. 3).

5. Ozone transforms the double bond to a mixed anhydride at one end and a ketone oxide at the other. Reaction of the amide group with the mixed anhydride results in a succinimide wherein the ketone oxide undergoes cycloaddition with one of the lactam carbonyls to yield the observed "false" ozonide.

Preparation of ring B component:

[butadiene], C_6H_6, $SnCl_4$ [6]

[CrO_3]

[1. $SOCl_2 \rightarrow CH_2N_2 \rightarrow MeOH$
2. NH_3–MeOH]

[P_2S_5]

(B)

---

6. The adduct is resolved *via* α-phenethylamine salt.

Combination of ring B/C components:

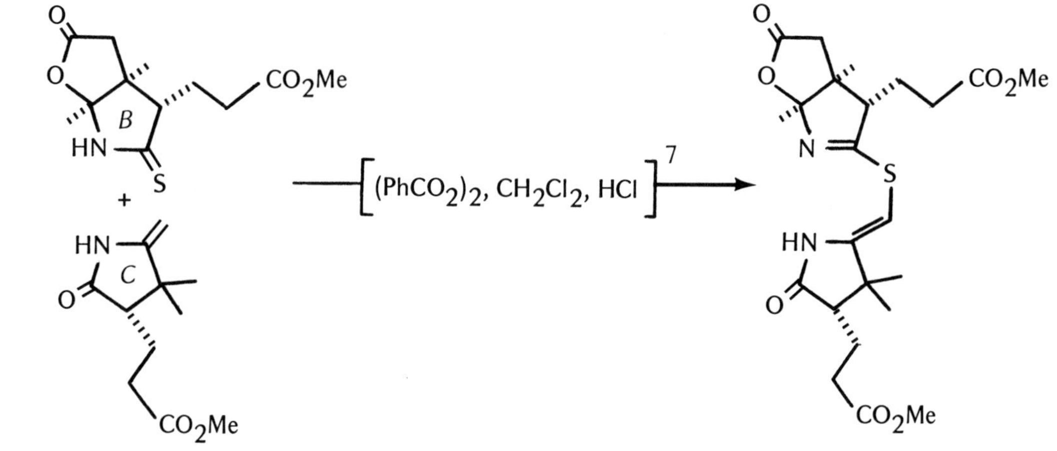

$[(PhCO_2)_2, CH_2Cl_2, HCl]^7$

---

7. The initial intermolecular coupling, to form a sulfur bridged intermediate (ii), consists in oxidizing the thiolactam with dibenzoylperoxide to a bisimidoyldisulfide (i) which suffers a nucleophilic attack by the methylidene carbon of the enamide. The critical C-C bond forming step can now proceed intramolecularly rather than intermolecularly in the thioiminoester-enamide (ii) leading to the episulfide (iii). The latter undergoes S-extrusion under influence of a suitable sulfur acceptor (e.g., phosphine or phosphite) resulting in the desired vinylogous amidine system (iv).

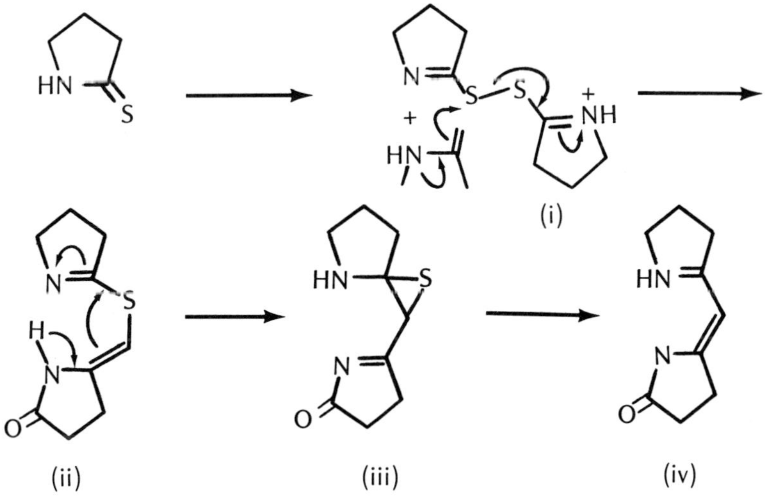

(i)

(ii) (iii) (iv)

$\left[P(OEt)_3, Xy \right]^{8}$ →

$\left[\text{Me-Hg-O}^i\text{Pr} \rightarrow; \; Me_3O^+BF_4^- \rightarrow; \; H_2S, (^iPr)_2NEt \right]^{9}$ →

Ⓒ

Thiodextrolin

8. Since C-8 is prone to epimerization the product obtained was a mixture of C-8 epimers. This, however, did not represent a problem since it was known that this very position is also configurationally labile in natural corrinoids, and would be expected to equilibrate quite readily even if separated at this stage. Moreover, since the natural configuration at this site was known to be the more stable one, restoration of correct stereochemistry at a later stage was not considered impractical.

9. Since P_2S_5 attacks both the lactone and the lactam function in the B/C component with equal facility, it was necessary to resort to intermediate conversion of the free lactam to a methyl-mercury complex which allowed specific activation of the lactam oxygen towards attack by the alkylating agent, without affecting the lactone carbonyl.

Preparation of rings A/D component:

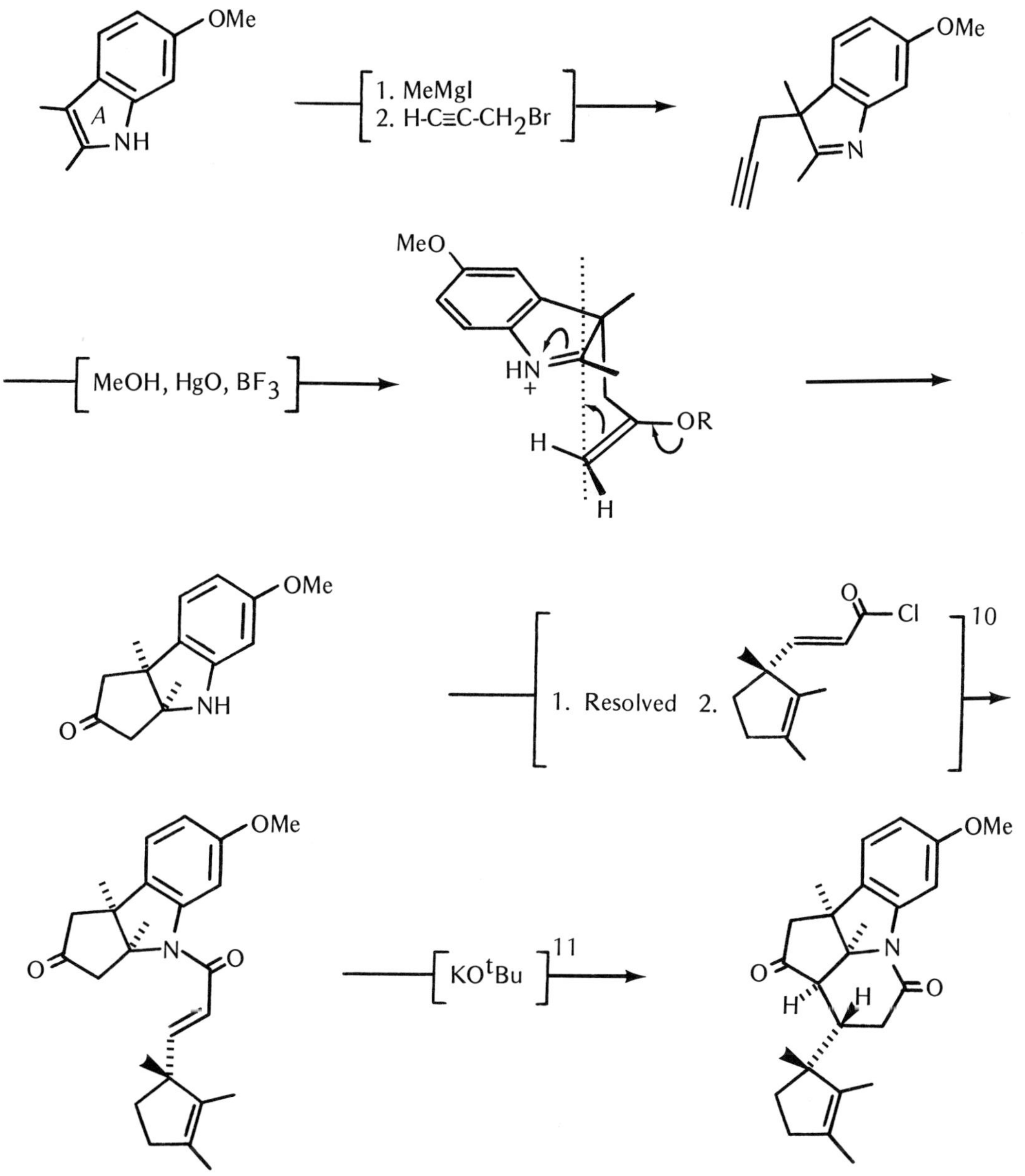
OMe
A
NH
1. MeMgI
2. H-C≡C-CH2Br
N
OMe
MeO
MeOH, HgO, BF3
HN
+
OR
H
H
OMe
NH
O
O
Cl
10
1. Resolved 2.
OMe
N
O
O
KOtBu
11
OMe
N
O
O
H
H

10. The resolution was carried out *via* the phenylethyl urea derivative. The diastereomers could easily be separated and then converted to the optically active amines by pyrolysis.

1. $(CH_2OH)_2$, H^+
2. $Et_3O^+BF_4^-$
3. NaOMe, MeOH

OMe
O
O
H
H
N
OMe
OMe

Tol, Δ

OMe
O
O
H
H
N
OMe

1. Li, NH_3, ^tBuOH, THF
2. H^+

O
H
N
O
H
H
O

1. NH_2OH
2. $NaNO_2$, AcOH*

HO
N
H
H
N
O
H
O

1. O_3, MeOH, -80°
2. HIO_4
3. CH_2N_2

The less hindered oxime is selectively cleaved by nitrous acid.

---

11. The alternative mode of cyclization, which would result in a β bridgehead hydrogen, is precluded on steric grounds.

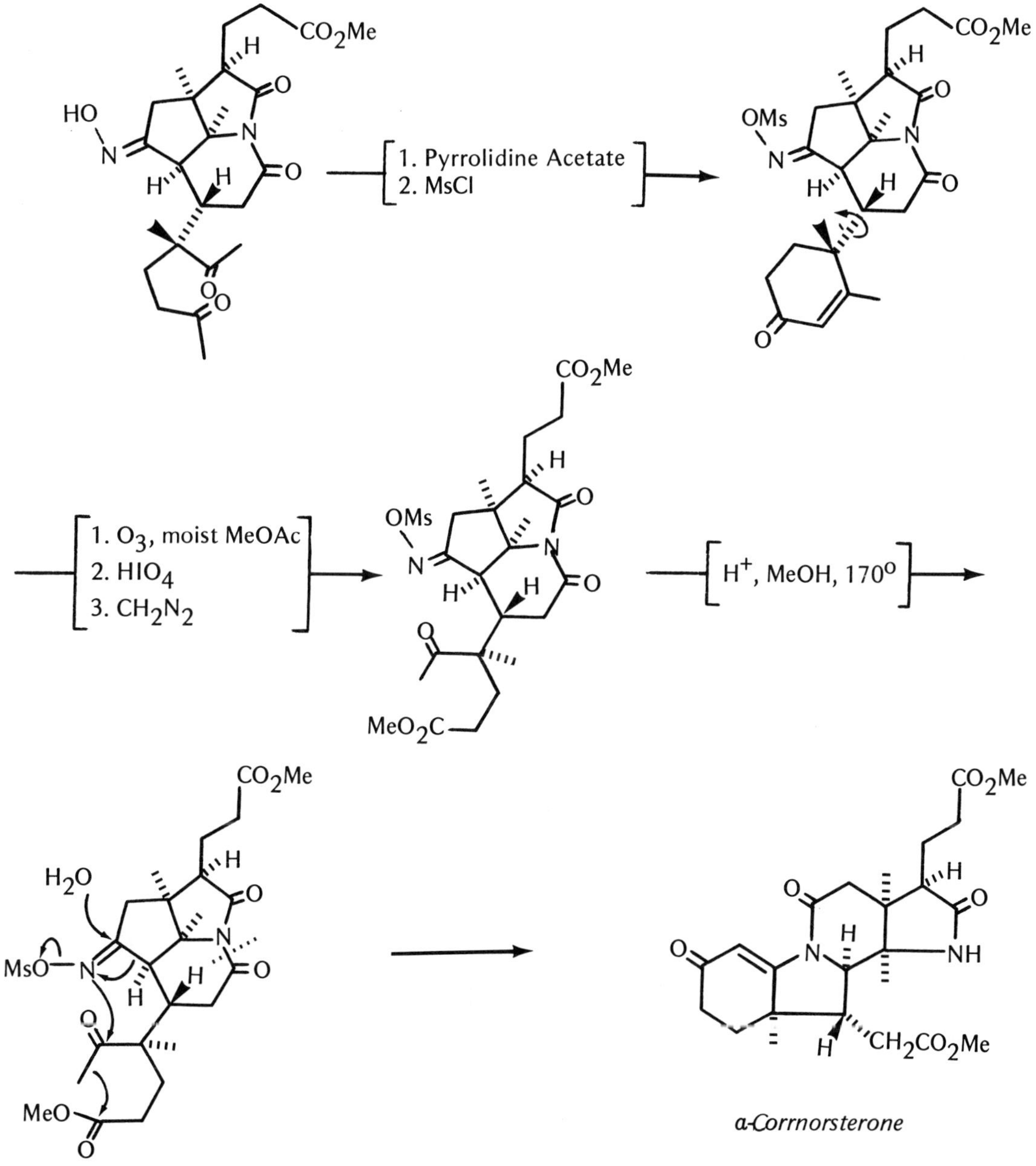
CO2Me
HO
N
H
O
1. Pyrrolidine Acetate
2. MsCl
OMs
1. O3, moist MeOAc
2. HIO4
3. CH2N2
H+, MeOH, 170°
MeO2C
H2O
MsO
MeO
NH
CH2CO2Me
α-Corrnorsterone

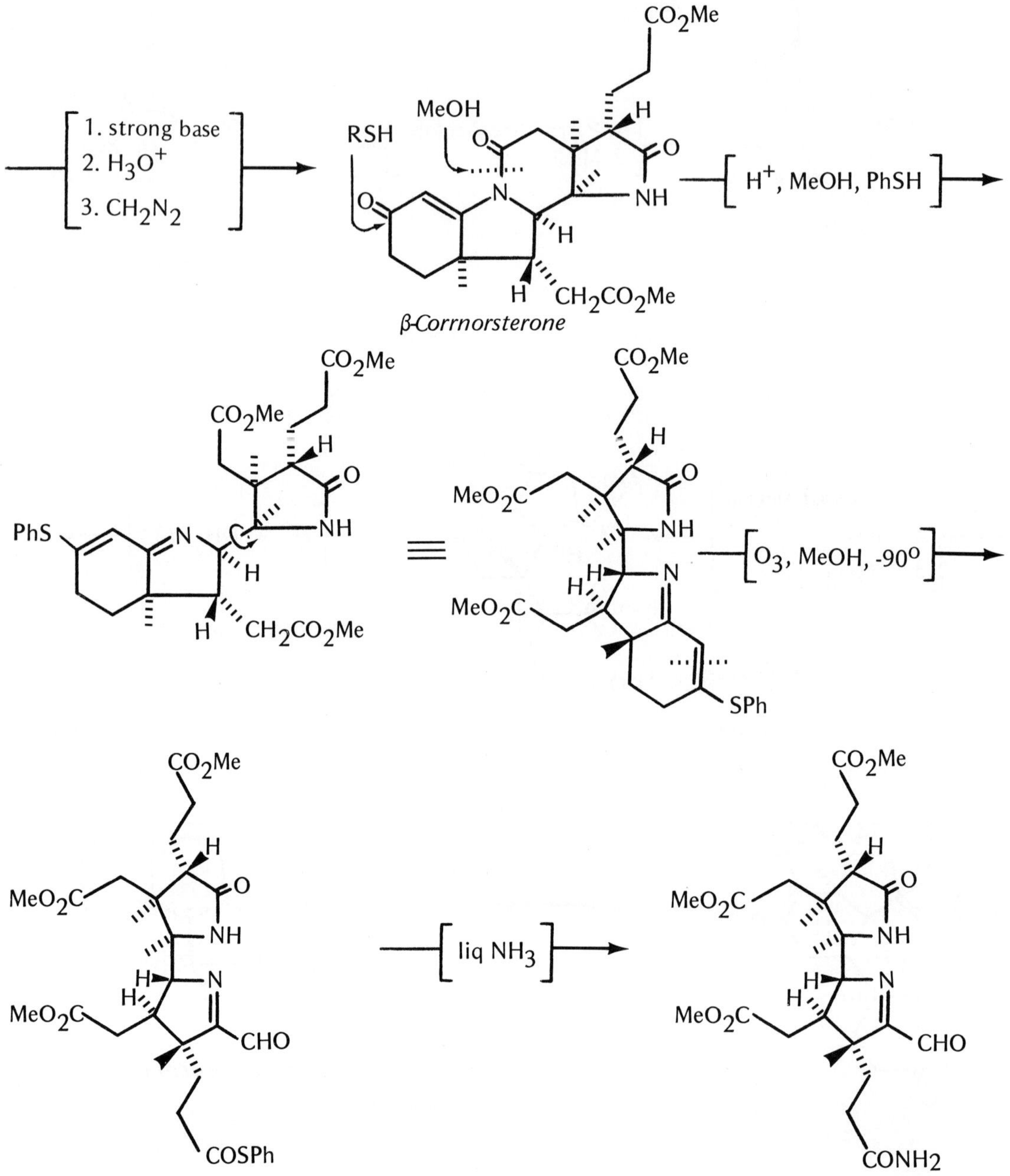
1. strong base
2. H_3O^+
3. CH_2N_2
MeOH
RSH
CO_2Me
CH_2CO_2Me
NH
β-Corrnorsterone
H^+, MeOH, PhSH
PhS
SPh
MeO_2C
O_3, MeOH, -90°
CHO
COSPh
liq NH_3
CONH2

[$NaBH_4$] →

CO_2Me, MeO_2C, H, O, NH, H, H, N, MeO_2C, CH_2OH, $CONH_2$

[$(Mes)_2O$, Py, 0°] →

CO_2Me, MeO_2C, H, O, NH, H, H, N, MeO_2C, CH_2OMs, CN

[LiBr, DMF] →

CO_2Me, MeO_2C, H, A, O, NH, H, H, N, D, MeO_2C, CH_2Br, CN

Cyanobromide

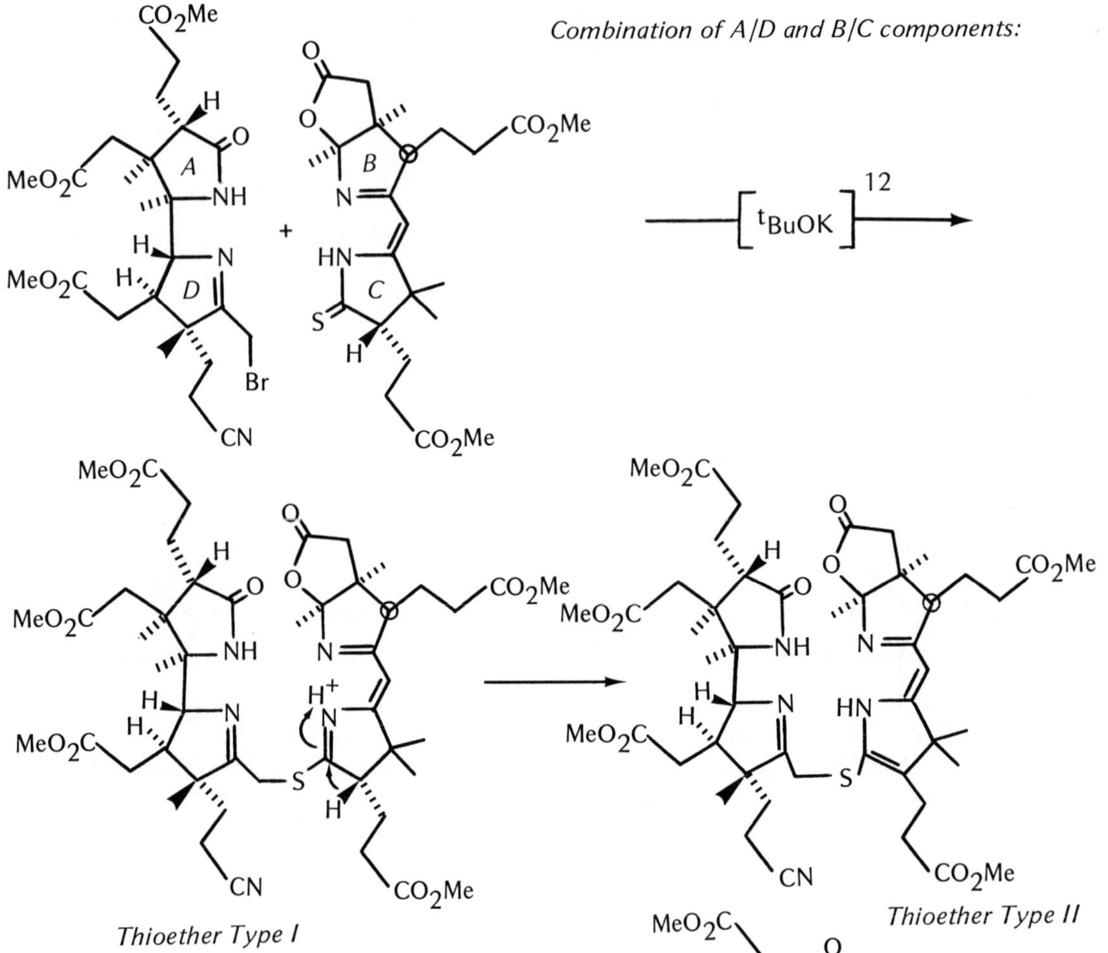

Combination of A/D and B/C components:

$\xrightarrow{[{}^{t}BuOK]^{12}}$

Thioether Type I

Thioether Type II

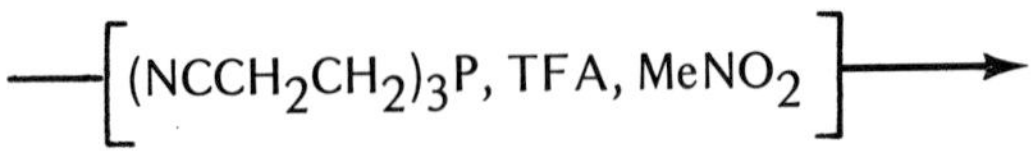

$\xrightarrow{[(NCCH_2CH_2)_3P,\ TFA,\ MeNO_2]}$

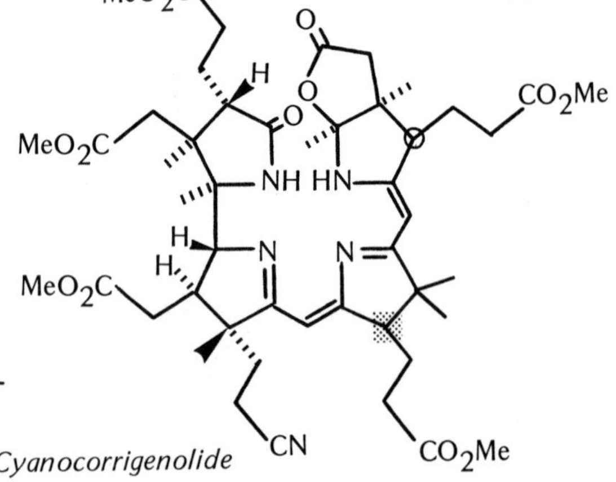

Cyanocorrigenolide

12. Sulfide contraction *via* alkylative coupling.

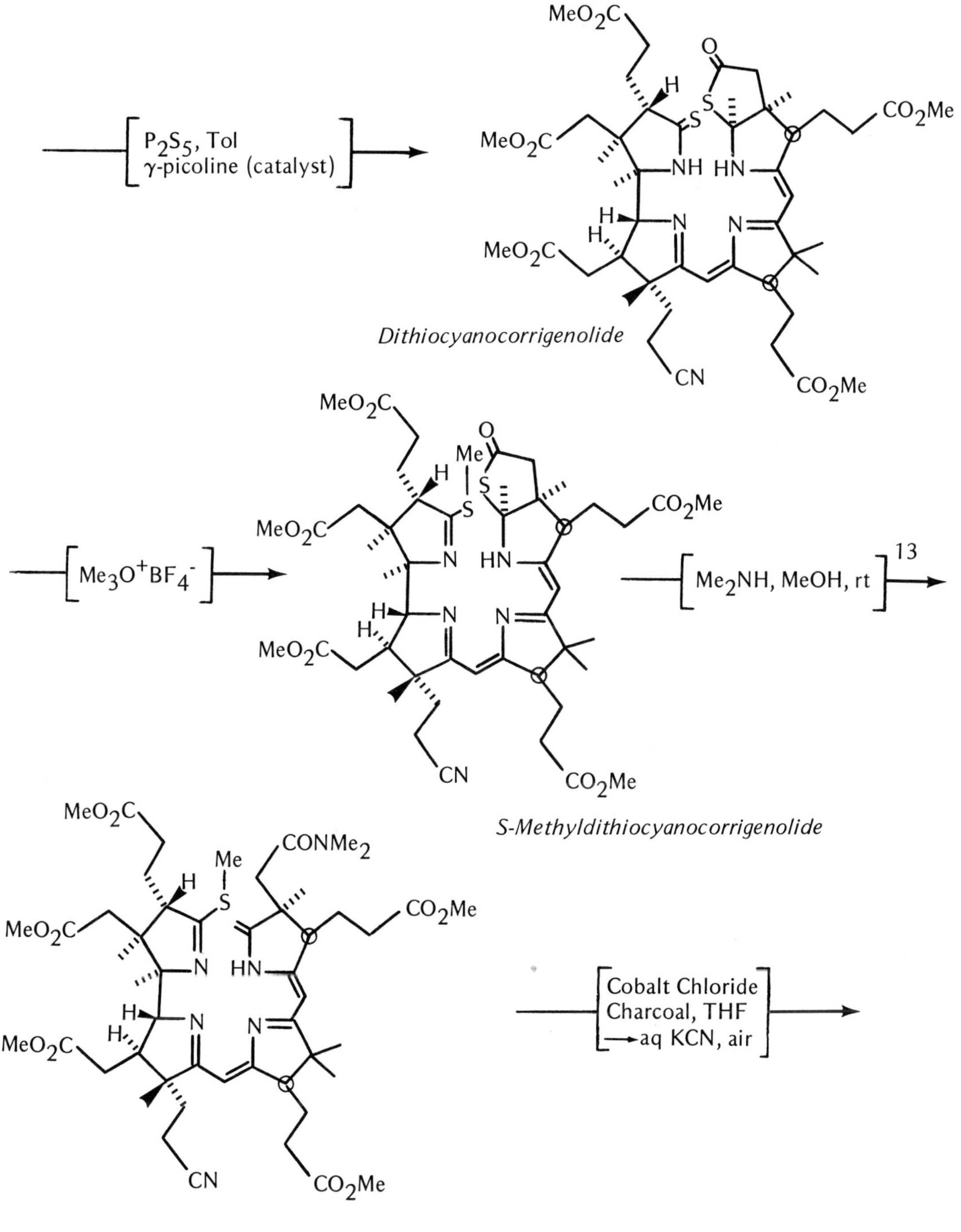

[P_2S_5, Tol
γ-picoline (catalyst)] →

Dithiocyanocorrigenolide

[$Me_3O^+BF_4^-$] →

[Me_2NH, MeOH, rt] →[13]

S-Methyldithiocyanocorrigenolide

[Cobalt Chloride
Charcoal, THF
→ aq KCN, air] →

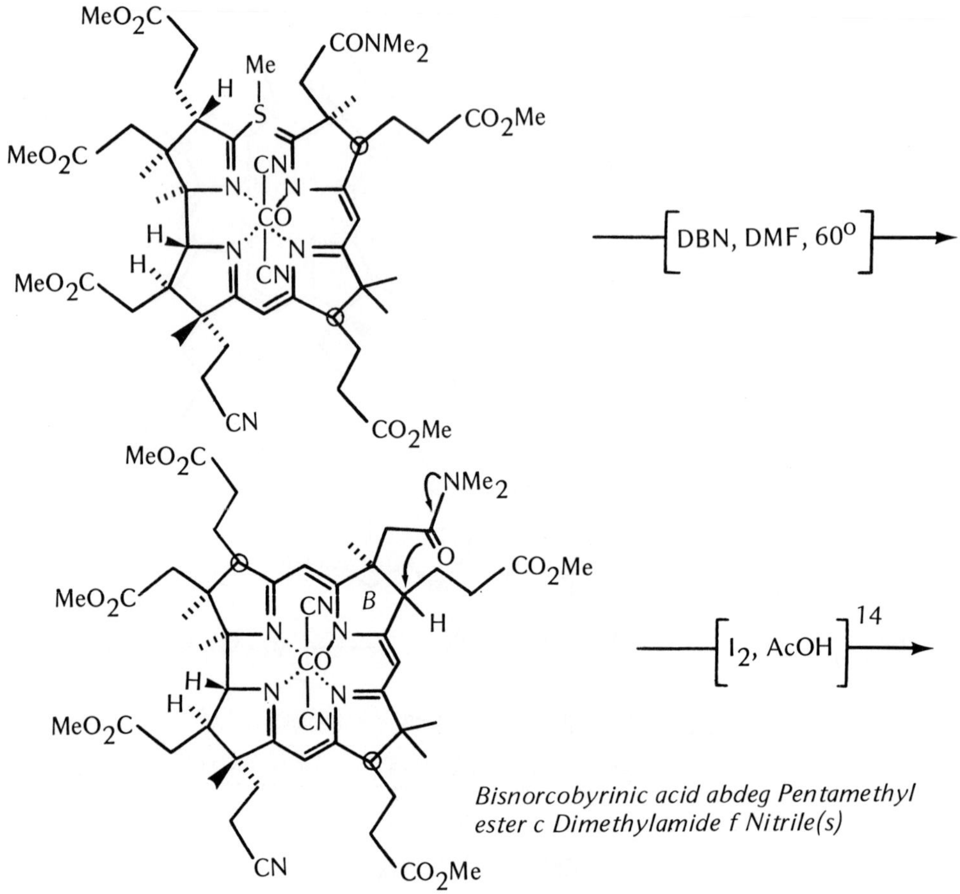

[DBN, DMF, 60°]

[I_2, AcOH][14]

Bisnorcobyrinic acid abdeg Pentamethyl ester c Dimethylamide f Nitrile(s)

---

13. The exocyclic ethylidene is not very stable, and under equilibrium conditions results in the undesired endocyclic compound.

14. Alkylation procedures that worked well at unhindered meso positions of model corrin complexes failed in the more complex case in hand, presumably due to severe steric crowding around all the three meso positions. However, using the oxidative lactone formation on ring B, a procedure which finds precedence in vitamin B_{12} chemistry, it was possible to introduce an additional degree of steric hindrance around C-10 and at the same time afford greater access by alkylating reagents to the other two meso positions.

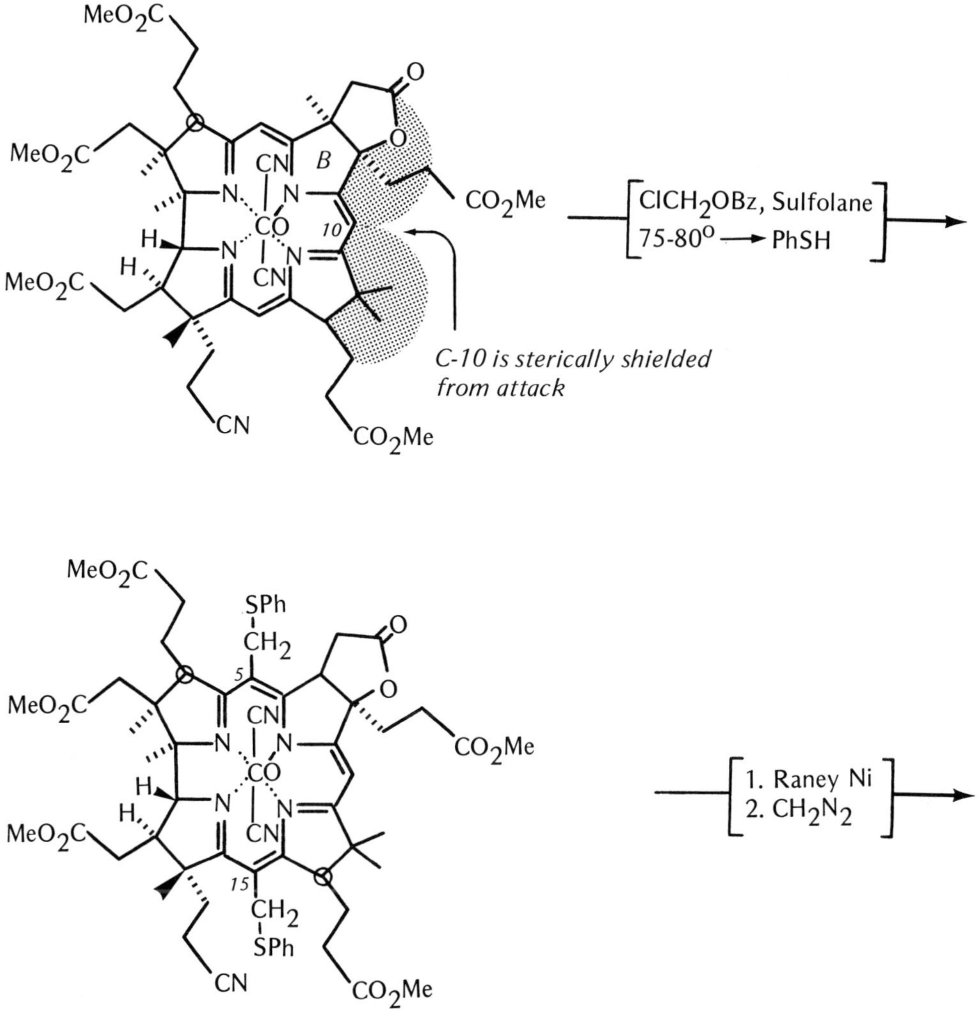

MeO$_2$C, CN, B, CO, 10, H, MeO$_2$C, CO$_2$Me, CN, CO$_2$Me

[ClCH$_2$OBz, Sulfolane 75-80° → PhSH]

C-10 is sterically shielded from attack

MeO$_2$C, SPh, CH$_2$, 5, CN, CO, H, 15, CH$_2$, SPh, CN, CO$_2$Me

[1. Raney Ni 2. CH$_2$N$_2$]

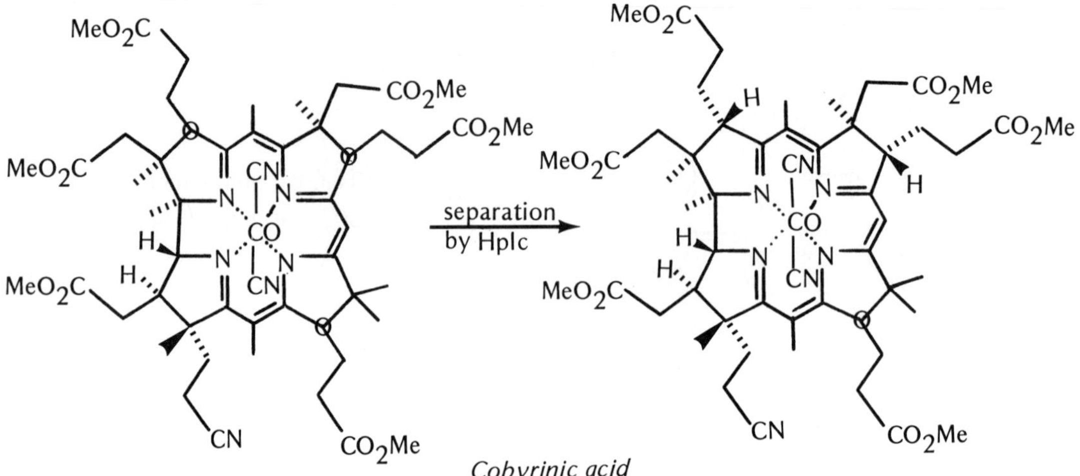

separation by Hplc

Cobyrinic acid
abcdeg Hexamethyl ester f nitrile(s)

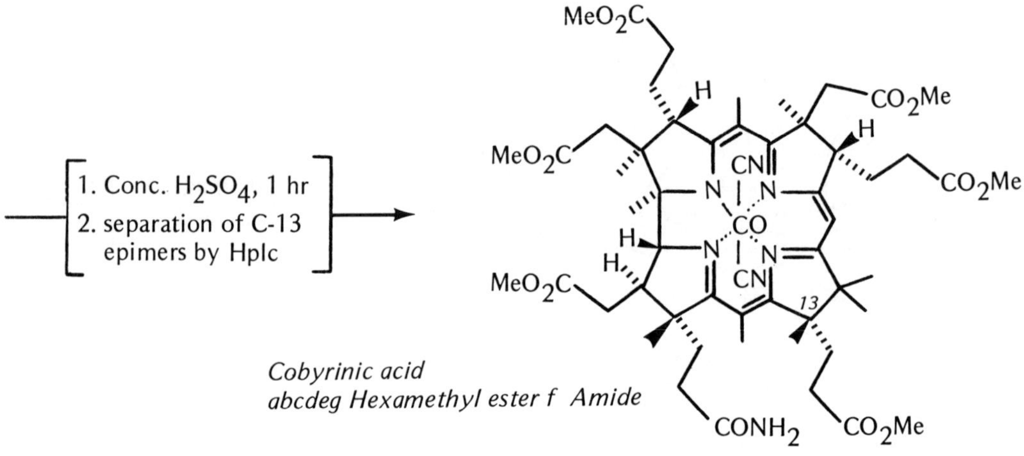

1. Conc. H_2SO_4, 1 hr
2. separation of C-13 epimers by Hplc

Cobyrinic acid
abcdeg Hexamethyl ester f Amide

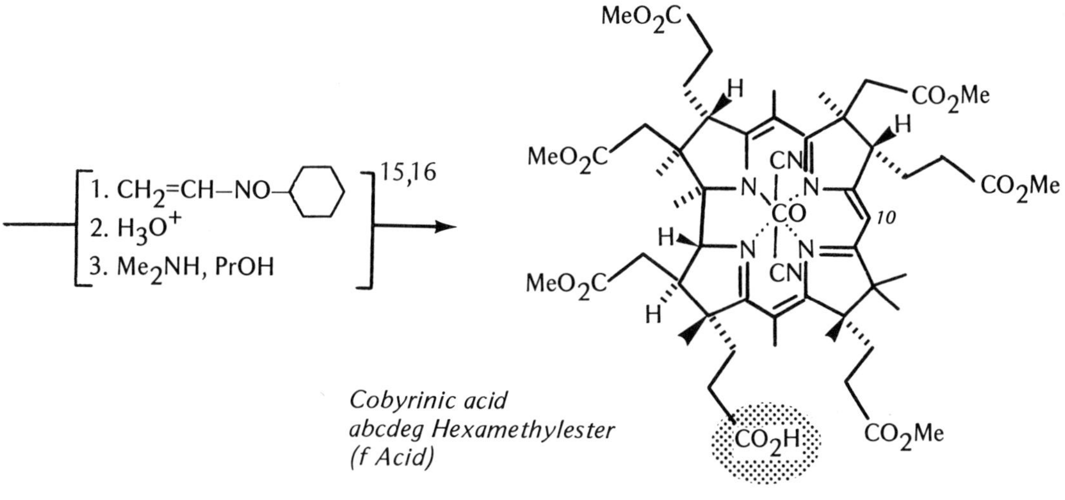

1. CH_2=CH–NO–cyclohexyl
2. H_3O^+
3. Me_2NH, PrOH

[15,16]

Cobyrinic acid abcdeg Hexamethylester (f Acid)

15. This elegant method for selective transformation of the f amide into the f acid presumably takes the following course:

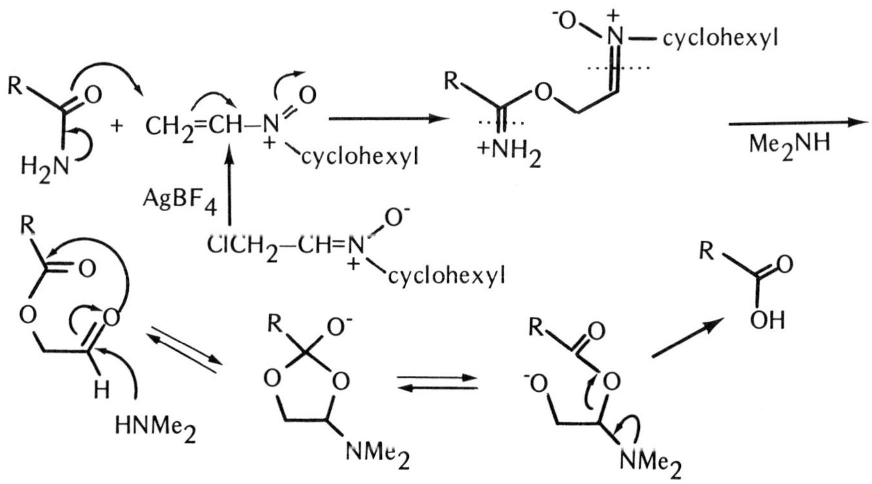

$AgBF_4$; $ClCH_2$–CH=N(O⁻)⁺–cyclohexyl → CH_2=CH–N(=O)⁺–cyclohexyl; Me_2NH; $HNMe_2$; NMe_2

16. Transformation of the f amide to the corresponding f acid was effected more conveniently by treatment with nitrogen tetroxide in carbon tetrachloride in the presence of sodium acetate. Conventional deamination of the amide using nitrous acid resulted in nitrosation at C-10.

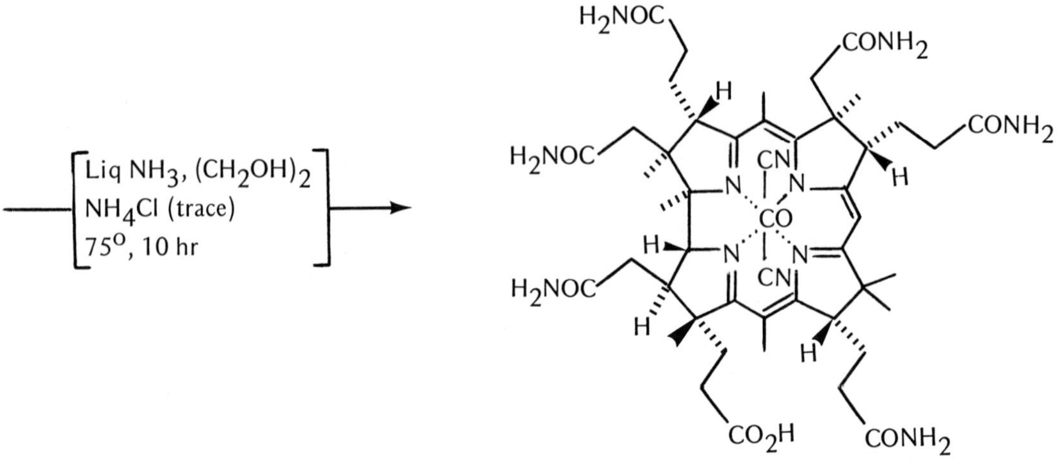

[Liq NH_3, $(CH_2OH)_2$
NH_4Cl (trace)
75°, 10 hr] →

Cobyric acid[17,18]

17. The final stages of the synthesis leading from cobyric acid to vitamin B_{12} have already been accomplished by W. Friedrich, G. Gross, K. Bernhauer and P. Zeller, Helv., **43** 704 (1960).

18. For a promising approach to natural corrins, see R. V. Stevens, C. G. Christensen, W. L. Edmonson, M. Kaplan, E. B. Reid and M. P. Wentland; J. Amer. Chem. Soc. **93** 6629 (1971); R. V. Stevens, L. E. DuPree Jr., W. L. Edmonson, L. L. Magid and M. P. Westland, J. Amer. Chem. Soc., **93** 6637 (1971).

YOHIMBINE

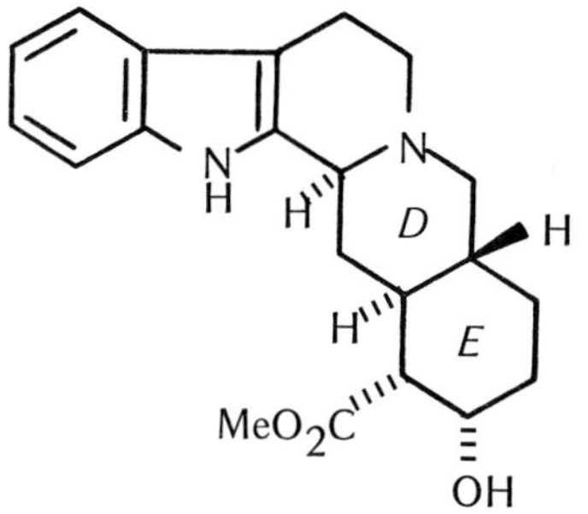

The synthetic challenge of yohimbine, historically perhaps the most venerable of the complex indole alkaloids, has attracted the attention of many workers.[1] Indeed, several approaches to the pentacyclic yohimbane system, including at least two total syntheses of yohimbine[2,3], have been reported during the last twenty years. Recently, an unusually simple and elegant synthesis of yohimbine was reported by Stork and Guthikonda.[4] This synthesis, based on the *trans*-hydroisoquinolone ester (A) representing rings D/E of the molecule, is outlined below.

---

1. H. J. Monteiro, The Alkaloids, **11**, 145 (1970); Art Org. Synth., 368 (1970).

2. The first total synthesis of yohimbine was reported by van Tamelen *et al.* in 1958. Details of this achievement have now appeared in full, E. E. van Tamelen, M. Shamma, A. W. Burgstahler, J. Wolinsky, R. Tamm and P. E. Aldrich, J. Amer. Chem. Soc., **91**, 7315 (1969).

3. Cz. Sźantay, L. Toke and K. Honty, Tetrahedron Lett., 1665 (1965); Chem. Ber., **102**, 3248 (1969). Subsequently these authors reported improvements in the Dieckmann ring closure of ring E in the late stages of this synthesis, Cz. Sźantay, K. Honty, L. Toke, A. Buzas and J. P. Jacquet, Tetrahedron Lett., 4871 (1971).

4. G. Stork and R. N. Guthikonda, J. Amer. Chem. Soc., **94**, 5109 (1972).

Preparation of C/D Component:

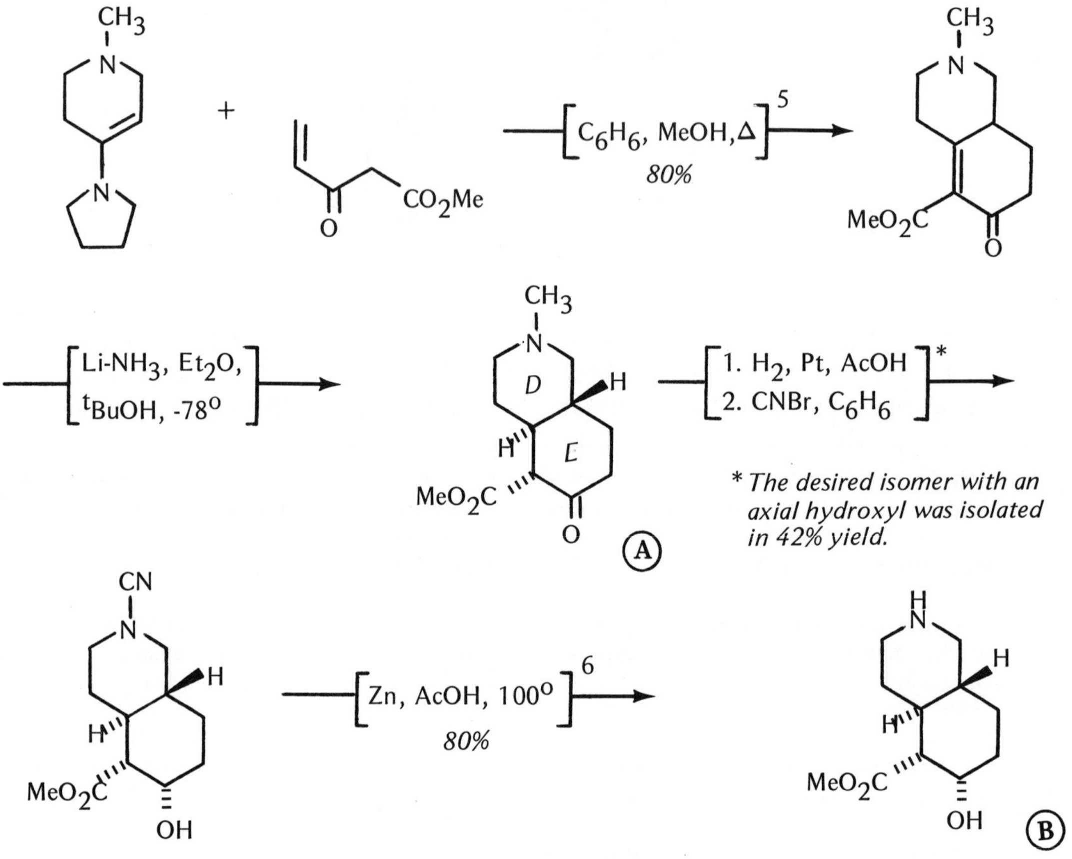
CH3
N
N
+
O
CO2Me
C6H6, MeOH, Δ
5
80%
CH3
N
MeO2C
O
Li-NH3, Et2O,
tBuOH, -78°
CH3
N
D
H
H
E
MeO2C
O
A
1. H2, Pt, AcOH
2. CNBr, C6H6
*
* The desired isomer with an axial hydroxyl was isolated in 42% yield.
CN
N
H
H
MeO2C
OH
Zn, AcOH, 100°
6
80%
H
N
H
H
MeO2C
OH
B

5. This useful annelation procedure for fusion of 2-carboxy-Δ^2-cyclohexenone systems to a preexisting ring was previously marred by the difficult accessibility of 3-oxo-4-pentenoic acid ester. An efficient synthesis of this compound has now been devised by G. Stork and R. N. Guthikonda, Tetrahedron Lett., 2755 (1972).

6. Reductive decyanation: T. Fehr, P. A. Stadler and A. Hofmann, Helv., **53**, 2197 (1970).

Preparation of Yohimbine:

Br

+

H N H

H

MeO_2C

OH Ⓑ

[K_2CO_3, MeOH, Δ] →

N N H

H

H

MeO_2C

OH Ⓒ

[1. HgOAc, EDTA, aq AcOH, 125°
2. $NaBH_4$, MeOH, 0°] [7,8] →
32%

N H N

H H

H

MeO_2C

OH

Yohimbine

---

7. Only two equivalents of mercuric acetate are required in the reaction since mercury, rather than mercurous acetate, is produced in the presence of EDTA; *cf.* J. Knabe and H. P. Herbort, Arch. Pharm., **300**, 774 (1967).

8. Oxidation of (C) by mercuric acetate results in kinetic formation of the ψ stereochemistry at C-3. However, at 120° in the presence of EDTA the kinetically produced ψ-yohimbine must be reoxidized to the iminium salt, which then gives the desired yohimbine stereochemistry at C-3 upon reduction with borohydride.

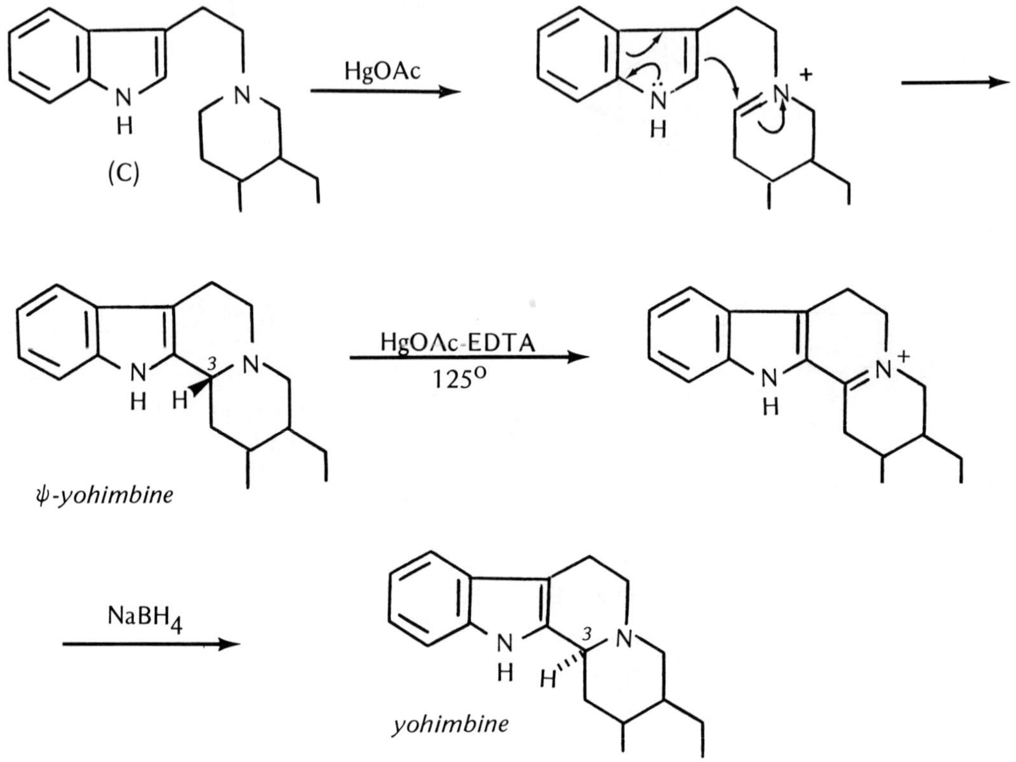

(C) —HgOAc→ → ψ-yohimbine —HgOAc-EDTA, 125°→ —$NaBH_4$→ yohimbine

SUBJECT INDEX

A 5
B 6
C 7
D 8
E 9
F 0
G 1
H 2
I 3
J 4